Informatik aktuell

Herausgeber: W. Brauer
im Auftrag der Gesellschaft für Informatik (GI)

Dieter Hogrefe (Hrsg.)

Formale Beschreibungstechniken für verteilte Systeme

Springer-Verlag
Berlin Heidelberg New York
London Paris Tokyo
Hong Kong Barcelona
Budapest

Herausgeber

Dieter Hogrefe
Universität Bern, Institut für Informatik
Länggassstraße 51, CH-3012 Bern

CR Subject Classification (1991): C.2.2, D.2, D.3, F.1, I.6

ISBN-13: 978-3-540-55568-1   e-ISBN-13: 978-3-642-77580-2
DOI: 10.1007/978-3-642-77580-2

Dieses Werk ist urheberrechtlich geschützt. Die dadurch begründeten Rechte, insbesondere die der Übersetzung, des Nachdrucks, des Vortrags, der Entnahme von Abbildungen und Tabellen, der Funksendung, der Mikroverfilmung oder der Vervielfältigung auf anderen Wegen und der Speicherung in Datenverarbeitungsanlagen, bleiben, auch bei nur auszugsweiser Verwertung, vorbehalten. Eine Vervielfältigung dieses Werkes oder von Teilen dieses Werkes ist auch im Einzelfall nur in den Grenzen der gesetzlichen Bestimmungen des Urheberrechtsgesetzes der Bundesrepublik Deutschland vom 9. September 1965 in der jeweils geltenden Fassung zulässig. Sie ist grundsätzlich vergütungspflichtig. Zuwiderhandlungen unterliegen den Strafbestimmungen des Urheberrechtsgesetzes.

© Springer-Verlag Berlin Heidelberg 1992

Satz: Reproduktionsfertige Vorlage vom Autor/Herausgeber

33/3140-543210 – Gedruckt auf säurefreiem Papier

Vorwort

Im Februar 1991 wurde ein Arbeitskreis "Formale Beschreibungstechniken für verteilte Systeme" am Rande der Fachtagung "Kommunikation in verteilten Systemen" in Mannheim in der GI/ITG Fachgruppe 3.3.1/4.4 "Kommunikation und verteilte Systeme" gegründet.

Der Arbeitskreis wurde mit dem Ziel gegründet, einen regelmässigen wissenschaftlichen Gedankenaustausch in organisierter Form zu ermöglichen. Die folgenden Themen wurden als Interessensschwerpunkte im Arbeitskreis erkannt:

- FDT-basierte Werkzeuge zur Unterstützung des Systementwurfs
- Konformitätstesten basierend auf formalen Beschreibungen
- Leistungsanalyse basierend auf formalen Beschreibungen
- Verifikation basierend auf formalen Beschreibungen
- Ergonomie des Arbeitens mit FDTs
- Anwendung von FDTs in der ODP-Standardisierung
- theoretische Modelle
- Datenmodellierung
- FDTs im System-Life-Cycle
- FDT-basierte Implementierung

(ODP = Open Distributed Processing, Normungsaktivität ISO)
(FDT = Formal Description Technique)

Die obige Liste ist selbstverständlich offen für Veränderungen.

Der vorliegende Band enthält Arbeiten aus dem Themengebiet des Arbeitskreises. Einerseits sind dies Arbeiten, die anlässlich des ersten Fachgesprächs im Juni 1991 in Darmstadt vorgestellt wurden. Andererseits sind es Arbeiten, die nachträglich von Autoren eingereicht wurden, denen es nicht möglich war, am Fachgespräch teilzunehmen. Unter den zahlreichen Arbeiten, die eingereicht wurden, ist hier nur eine Auswahl zusammengestellt, die nach einem zweistufigen Begutachtungsprozess übrig blieben: Begutachtung - Revision ausgewählter Beiträge - erneute Begutachtung.

In meiner Eigenschaft als Sprecher des Arbeitskreises möchte ich an dieser Stelle allen Autoren für Ihr Engagement im Erstellen der Arbeiten danken. Meinen besonderen Dank verdienen aber auch die Gutachter für ihre mühevolle und gewissenhafte Arbeit: Heinz Burkhardt, Manfred Broy, Andre Danthine, Oswald Drobnik, Wolfgang Effelsberg, Kokichi Futatsugi, Reinhard Gotzhein, Jens Grabowski, Günter Karjoth, Heiko Krumm, Luigi Logrippo, Rüdiger Olderog, Juan Quemada, Wolfgang Reisig, Harry Rudin, Wilhelm Stoll, Richard Tenney, Sebastiano Trigila, Ken Turner und Georg Zörntlein.

Besonders möchte ich auch Herrn Amardeo Sarma vom Forschungsinstitut der Telekom für die Organisation des Fachgesprächs in Darmstadt und meinem Assistenten Andreas Spichiger für die Abwicklung der organisatorischen Arbeit beim Zusammenstellen dieses Bands danken.

Bern, im März 1992   Dieter Hogrefe

Inhaltsverzeichnis

Toward the Integration of Formal Description Techniques with Performance Evaluation *E. Heck, B. Müller-Clostermann*	1
Generating Parallel Code from Estelle Specifications *B. Hofmann, W. Effelsberg*	22
CDM – korrekter Entwurf von Kommunkationssoftware *J. Freudenmann*	35
LOTOS Design Methodology Based on ODP-Viewpoints *A. Vogel*	48
An introduction to compositional methods for concurrency and their application to real-time *J.J.M. Hooman, W.P. de Roever*	66
Rapid Prototyping von Estelle-Spezifikationen *R. Födisch, H. König*	110
Formale Konzepte zur Lokalisierung von Funktionen in räumlich verteilten Systemen *R. Prinoth*	125
A Simple Toy Example of a Distributed System: On the Design of a Connecting Switch *T.F. Gritzner*	144
Testfallgenerierung aus Petri-Netzen – Probleme, Konzepte, Systeme *B. Baumgarten*	177
Die Offene Petrinetz-Methode zur Analyse und Darstellung des funktionalen Verhaltens verteilter Systeme *H. Dibold*	195
Differences between Estelle and LOTOS Descriptions of a Protocol *J. Grobholz, R.L. Tenney*	222
Autorenverzeichnis	229

Toward the Integration of Formal Description Techniques with Performance Evaluation

Elke Heck, Bruno Müller-Clostermann
Informatik IV, Universität Dortmund
Postfach 50 05 00, D-4600 Dortmund 50

Abstract

Formal methods for the investigation of a future system's correctness and performance are of growing importance. Here we proceed toward an integration of state-oriented formal description techniques with model-based performance evaluation. After brief tutorial introductions to Finite State Machines and Queueing Networks an approach for the integration of the two worlds is presented. Enhancing a formal system specification with performance submodels finally yields a combined model that is quantitatively assessible and encloses the original formal description. The presented framework provides foundations for the development of new techniques and tools to support distributed systems design.

1. Introduction

1.1. Motivation and Overview

Today, computing and communicating systems form the basis for local up to worldwide information exchange. Due to their growing importance, they have to meet increasing requirements as, e.g. availability, correctness and performance. These requirements must be fulfilled either by technology improvements concerning the hardware or by powerful software implementing distributed applications. Since communication protocols form the cornerstone upon which distributed systems are built, international standardization bodies like ISO and CCITT have spent a lot of effort to simplify distributed information processing. A main result has been the development and standardization of Formal Description Techniques (FDTs) like Estelle, LOTOS and SDL. Although the main application area of FDTs has been telecommunication networks and protocols, their importance for general computing and communicating systems is growing. This growing importance is partly due to the development of FDT-based validation techniques, which have mainly been developed to cope with the correctness problem. Besides the functional correctness of a system another essential item is its quantitative correctness, usually addressed as performance, reliability and/or dependability. In real environments a functional correct system with a very bad performance, e.g., a real-time voice system with an average response time greater than some seconds, is practically incorrect.

Since formal specifications have to be as implementation-independent as possible, language constructs covering implementation issues have not been considered during the design of the FDTs. Unfortunately, performance aspects such as impediments to smooth data flow, effects due to processing overheads, response time behavior or packet loss rates cannot be investigated in an implementation-independent way [RUDI88]. Design and implementation decisions have to be made before such phenomena can be considered during the evaluation. Therefore performance analyses directly driven from an FD (in the following for Formal Definition or Formal Description) via simulation are of restricted value. Extra implementation-dependent information on the resource requirements of the processes governing the system, as well as information on the underlying machinery, has to be supplied before such analyses can be undertaken in a meaningful way. Such implementation-dependent information may cover issues involving parallelism in the execution of processes/protocol functions, scheduling strategies, processing speeds, transmission speed of the communication medium, size of memory/buffers and queues as well as timeout values. Of course the performance characteristics of the hardware devices for computing and communication that form the executing machine also must be taken into account carefully.

Advanced tools for performance evaluation cover the above mentioned issues but do not incorporate an FD directly in the modeling process. Hence two different models of the same system must be established, the (standardized) FD and the PE-model (Further on PE stands for performance evaluation). This situation brings up many questions, such as validation of the PE model with respect to the FD. In addition part of the effort spent for modeling the FD will have to be invested again for the PE model. Building two completely consistent models of the same system is a task nearly infeasible in practice. Hence, we strive for a methodology, which makes use of the formal specification for the purpose of performance evaluation. From the viewpoint of advanced performance evaluation techniques the suggestions made until today appear to be unsatisfactory. In this paper we begin to build a bridge between the two disciplines.

Note the similarity to the vision of protocol engineering [RUDI86, IEEE91, LIU91], which may sketched as follows: Formal methods and software engineering methodologies allow a designer to express his system under development formally, test this specification for correctness (validation and verification), obtain early indications of how it would perform, compile major parts of the implementation directly from the formal specification, and finally test the resulting implementation to assure that it conforms to the specification.

The rest of the paper is organized as follows. After a short review of recent work on combined FD/PE-approaches, a survey of FDTs is given. In chap. 2 in particular finite state machines, which have been chosen as the basic description technique throughout the paper, are considered in some more detail. Principles of model-based performance evaluation, and in particular queueing networks, which are common performance models, are introduced in chap. 3. Requirements for the combined FD/PE-approach are summarized in chap. 4. The central part of the paper is presented in chap. 5. Here a combined PE/FD methodology is introduced, which makes use of the original FD, enhances it by the necessary implementation-dependent PE information, and combines it with modules which serve as "machines" to execute the load imposed on the system. Parts of the methodology are illustrated by means of a textbook example. The presented results have partly been worked out in the ESPRIT project ATMOSPHERE [HECK91].

1.2. Short History of FDT-Based Performance Evaluation
For an overview on the work done in the past, see [RUDI86, RUDI88, CONW89] and [GURU89]. In the area of communication protocols for example it has usually been proposed to introduce performance aspects by attaching time durations, either deterministic or probabilistic, directly to the state transitions occurring in protocol models, cf. [RUDI83]. We shortly survey some recent contributions.

In [RUDI83] Rudin has proposed a technique based on the reachability tree to calculate the response time for an action. In Rudin's approach time durations are attached to the edges of the reachability tree such that each path in the tree is associated with a duration. The model of Rudin is time discrete, i.e., each step takes the same deterministic time. The time elapsed per edge is a single time unit. There are infinitely many cycles which are touching a certain state, but only cycles of non-neglibile weight are considered. Hence, the determination of the mean average delay for returning to the state is a nontrivial task. In [GURU89] an extension of this approach has been presented, where the considered protocol model allows incoming messages to be queued before transmission to the receiver.

In [DICH88] a technique for probabilistic validation combined with quantitative analysis has been introduced, including both performance evaluation and reliability aspects. Although the starting point of the algorithm is not a high level FDT but the so-called Reduced Global Reachability Graph (RGRG), which is a low level structure, this approach deserves attention. A tool implementing a mapping from SDL to the RGRG has been described in [BABU90]. In [BOVA88] ideas have been presented on how to introduce so-called resources into the FDT-language Estelle. Resources must be held for a certain amount of time in order to execute a transition. This idea is a step in the right direction, but from the viewpoint of advanced performance evaluation there remains much to be done. Kritzinger [KRIT86] showed how to map protocol specifications given by finite state machines to multiclass queueing networks treatable by a queueing network analysis tool. He exemplified this approach by employing the queueing network tool MICRO-SNAP. Another tool worth mentioning here is the Estelle-based PEW, the "performance engineers workbench", which is due to the work of Kritzinger and Wheeler [KRWH89]. Amongst means for specification, debugging and testing, PEW offers a performance analyzer that is based on the analytical solution of queueing networks.

Other tools that either provide or promise performance evaluation from FD are GreatSPN, a tool based on Timed Petri Nets [MBCC84, CHIO87], and the CCS-based ANALYST [BANY86]. The performance evaluation of layered OSI communication architectures is discussed extensively in [CONW89], where the specifications of the OSI-BRM are used to derive an analytical queueing network model. A recent paper of Heck, Hogrefe and Müller-Clostermann [HEHM91] describes an approach to combine FDTs with hirarchical performance evaluation for protocol design. The authors propose a PE framework to allow the inclusion of modules derived from formally specified protocols. The method is sketched by use of the performance modeling tool HIT and the FDT language SDL.

2. Formal Description Techniques

2.1. Survey on FDTs

The great steps of progress that have been made in computer networks during the last years and the growing complexity strengthen the international care for standards. An important result has been the development of concepts for Formal Description Techniques, to describe these standards in a clear and precise manner, and improve therefore the quality of specifications. Standardization bodies like CCITT and ISO have developed Formal Description Techniques for data communication services and protocols, and distributed systems. Surveys on FDTs have been given recently in [IEEE91, BOCH90, VEPI86]. Here we recall just a few essentials [ISO89, HOGR89]. It is widely accepted that the key for the success of a system is a thorough system specification and design. This requires a suitable formal specification language, satisfying the following needs:

- A well-defined set of concepts;
- Unambiguous, clear, precise and concise specifications;
- A basis for analyzing specifications for completeness and correctness;
- A basis for determining conformance of implementations to specifications;
- A basis for determining consistency of specifications relative to each other;
- Use of computer-based tools to create, maintain, analyze and simulate specifications.

A specification may serve as a basis for deriving implementations. Implementation-independent specifications should abstract from implementation details to ease an overview of a complex system, postpone implementation decisions, and not exclude valid implementations. The use of formal specification languages makes it possible to analyze and simulate the correctness and functionality of alternative system solutions. Without use of formal specification techniques this task would be nearly impossible in practice .

While in the past the development of the FDTs per se was a main subject of research, the work now focusses on applications, the development of tools for the FDTs, and the integration of the FDTs into the system life cycle. There is a wide range of various FDTs, which are all used to represent basic concepts such as abstraction, modularity, information-hiding, structuring, and synchronization, and additional architectural concepts. But they differ in their technical aspects and in the goal of their application. The FDTs Estelle, LOTOS and SDL have come to a stage where they are an International Standard by ISO (Estelle and LOTOS) or a Recommendation by CCITT (SDL). This means that they are accepted by a wide community of organizations, e.g., research institutes, network administrators and industry. Estelle and LOTOS have been developed particularly for the OSI Reference Model, whereas SDL has been developed for the wide area of telecommunication systems.

In general, two basic approaches to FD are in widespread use, the state-oriented techniques and the transition-oriented techniques:
- Using transition-oriented techniques, systems are described by the declaration of all possible sequences of transitions. Transition-oriented techniques are essentially the process algebra methods, which are based on Milner's CCS (Calculus of Communicating Systems, [MILN80]). Closely related to CCS are CSP (Communicating Sequential Processes, [HOAR78, HOAR85]) and the specification language LOTOS [TURN87]. LOTOS, developed and standardized by the ISO [ISO87b] for the design of distributed and concurrent systems, is a mathematically-defined FDT, based on two formal methods, a process calculus derived from CCS and a modification of the algebraic specification language ACT ONE [EHFH83] for the structured specification of abstract data types.
- Using state-oriented techniques, systems and their processes are described by the sequence of states each process can reach. Examples of the state-oriented techniques are finite state machines and their extensions, specification languages based on these concepts and specification by Petri Nets. The concept of states and state transitions as introduced for extended finite state machines (EFSMs) is the foundation for many models and notations, e.g., programming languages, communicating finite state machines

(CFSM), Markov chains, and Petri Nets. Each technique was designed for a special-purpose application. The CFSM model is widely used for protocol specification in practice, because most engineers are familiar with states and state transitions. Hence, CFSM will be used as basic description technique. Chapter 2.2 gives an introduction to this field. A very important FDT based on the extended finite state machine model is SDL, the Specification and Description Language [BEHO89, BEHS90], developed and standardized by CCITT [CCITT88]. SDL is supplemented by capabilities for Abstract Data Types based on the initial algebra model (the same as used in the ACT ONE part of LOTOS). This combination is supported by a well-defined formal semantic. The basic idea is to describe a system as communicating processes, where each process is an extended finite state machine, that works autonomously and concurrently with other processes. Another FDT based on extended finite state machines is Estelle, developed and standardized by the ISO [ISO87a, BUDE87].

Besides the state and transition-oriented techniques mentioned above, many other specification methods have been developed, e.g., grammars, temporal logics, and programming languages.

2.2. Finite State Machines as a Foundation of Formal Designs

In this section some principles of finite state machines (FSM) are summarized shortly. The usual approach is to subdivide the system into a number of communicating components, such that each component is modelled by a finite state machine.

A finite state machine [DANT80] is a 5-tuple $\{S, I, O, \delta, \lambda\}$ where
- S is a finite set of states;
- I is a finite set of inputs;
- O is a finite set of outputs;
- δ is a state transition function ($\delta: S \times I \rightarrow S$);
- λ is an output function ($\lambda: S \times I \rightarrow O$).

Usually a starting state $s \in S$ for each FSM as well as an initiating stimulus must be defined additionally. The behavior of the machine is expressed by the functions δ and λ. If an input of the input set is received in a given state, the output function indicates the output to be generated, and the state transition function indicates the new state of the machine. FSMs-interdependence is defined implicitly by their input/output-behavior. More generally the input and output sets may be regarded as sets of events that may occur at the interface of an FSM. The state space of an overall FSM-system, consisting of several FSMs, is given by the local state spaces of all FSMs involved.

EFSM: Extended Finite State Machines

Most specification languages like SDL and Estelle are based on "extended" finite state machines (EFSM). Most protocols are so complex that it is impractical and certainly unwieldy to record all the necessary information to describe the dynamics of the protocol as different explicit states [RUDI84]. An extended FSM can use and manipulate data stored in variables local to the machine. So conceptually less important information - such as addresses and sequence numbers - are stored as auxiliary variables. Decisions can be based and actions can be performed on these auxiliary variables. Note that this simplified specification principle caused by the use and test of variables just provides an easier notation and therefore enhanced comprehensiveness. The advantage of EFSM is that the number of necessary states can be reduced since intermediate states may be described by values of the variables. On the other side, if an EFSM specification is transformed into a simple FSM notation the state space will grow considerably.

CFSM: Communicating Finite State Machines

CFSM means communicating or coupled FSM. We distinguish explicitly between directly coupled and communicating FSMs. Direct coupling means that the output of an FSM is an immediate input for another FSM, i.e., for a given FSM entity certain types of transitions are directly coupled with transition types in other FSM entities. Hence, a transition can only be executed in parallel with a corresponding transition in another entity. One way of realizing direct coupling is by identifying certain output symbols of one entity with certain input symbols of the other entity, without any intermediate buffering [BOCH78]. In summary, direct coupling can be used to describe synchronous communications between two or more processes.

In contrast to the directly coupled FSMs there are other definitions of CFSMs, henceforth called communicating FSMs, which allow for the description of asynchronous communication. For this purpose, some authors introduce channels in order to define admitted communication routes, e.g., the existence of the channel C_{ij} allows the transmission of messages from entity i to entity j [BRZA83]. The cooperation between CFSMs is performed asynchronously by discrete messages. The term "asynchronous" implies that an entity usually continues its operation after sending out a message. It must not await the reaction caused by this particular message. Moreover, an FSM should be capable of receiving and storing more than one message at a time. This implies that an FSM needs either an input queue for storing incoming messages or channels which should be able to hold messages.

Formal description of an example protocol

We present an extension of the alternate bit protocol which, although a toy example, has been often used in the literature to explain basic principles of formal specification techniques [MERL79]. It may be seen as a simple data link layer protocol where the sender waits for a positive acknowledgement before it transmits the next message. Protocols of this kind are called PAR-protocols, where PAR stands for Positive Acknowledgement with Retransmission [TANE89, p. 222 and pp. 244-249]. Data are transmitted only from sender to receiver and acknowledgements only in the opposite direction. Since messages may be lost or damaged due to an unreliable channel the protocol employs a timeout mechanism. If the timer expires the sender repeats the last message. If the sender times out too early, while the acknowledgement is still on its way, it sends a duplicate. Hence timer setting is critical both for correctness and for performance. A formal description of sender S and receiver R by means of Finite State Machines is given below. Note that an abstract unreliable channel is assumed, which is not depicted explicitly. It may be modelled by another FSM which serves as a link between sender and receiver.

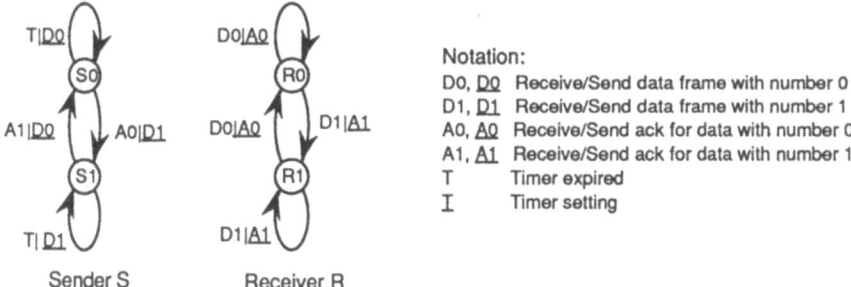

Fig. 1: Example-Protocol specified by Finite State Machines

S forwards data units D0 and D1 to R which reacts with acknowledgements A0 and A1 respectively. The data units carry a sequence number, here given by 0 or 1. The following interaction diagram shows the type of messages (events) and their destination. In this way the interface of an entity may be defined with the help of messages it can send and that it is ready to receive.

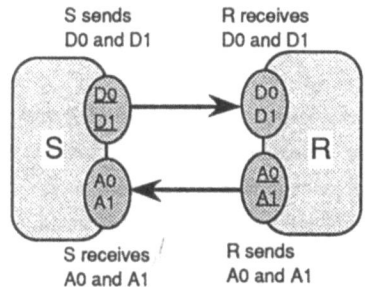

Fig. 2: Interaction diagram for the entities S and R

The connections between the interfaces describe an abstract communication relation. They do not imply concrete paths or transmission routes. Note that the messages T and $\underline{T}$ are not depicted in the interaction diagram, since they represent internal events.

3. Model-Based Performance Evaluation

3.1. Principles of Performance Evaluation

Performance evaluation and design of computing and communication systems are important and difficult problems. In particular the area of distributed systems will be a major issue for the current decade. Once resources and users are distributed, we are faced with a number of conflict resolution problems that are usually solved by queueing and blocking. Additional access problems and delays beyond queueing show up. In particular, the synchronization among coupled processes may be a major source of performance problems. So a methodology for distributed systems performance evaluation should incorporate the synchronization restrictions between coupled processes and the phenomena of queueing and blocking.

System performance is determined by the hardware and software architecture and the workload in form of application processes, which are executed on the system. Therefore performance evaluation methodology often explicitly distinguishes "machine" and "load" to be executed by the machine. Quantitative measures may be determined dependent on the underlying machine and the load both reflecting implementation decisions. Of course, intensity and type of traffic are essential parameters of the workload. Examples for performance measures of interest are throughput, utilization, response time, queue length, and wait time.

To summarize, performance evaluation may be considered to realize a mapping: f(machine, load) --> performance values. For existing systems under executable workloads such a mapping may be obtained by measurements, or alternatively by modeling and analysis. Model-based performance evaluation is the only way to investigate a future system's behavior. It is usually associated with stochastic modeling using queueing networks, Markov chain techniques, Timed Petri Nets or event driven simulation. To liberate the user from the need to be an expert in stochastic modeling, software tools for model based performance evaluation have been developed. One of the most convenient paradigms in tool building is to adopt queueing networks as underlying specification technique for performance evaluation models. Tools following this paradigm are RESQ2 [SAMN85] and QNAP2 [VEPO85]. Another approach is pursued by the tool HIT employing the familiar view of hierarchies of functional levels and function-realizing layers [BEMW88].

3.2. Queueing Networks

Because queueing networks are a well-known and widely used formalism for performance analysis, they have been chosen to explain some aspects of building performance models. We emphasize that the following considerations are not restricted to queueing networks. Queueing networks may serve as a technique to describe the congestion of multiple requests to restricted resources. Dependent on the specific features or building blocks used in the model under investigation queueing networks may be solved either by algebraical, numerical or simulative techniques. A very brief introduction into queueing networks is given subsequently. For details we refer to classical textbooks [KLEI76, LAVE83].

Queueing networks are built as a network of "stations" and "subsystems" which may themselves consist of queueing networks. Stations and subsystems represent resources, which may be summarized under the term "machine". Load is given by tasks moving through the network and requiring services. Usually the service demand of a task is assumed to be a random variable. A queueing station works as follows: ... tasks arrive from the environment, enter the queue, wait for service, receive service for a certain amount of time, leave the service station, move to the environment or to the next queueing station, ...etc. Instead of the term task also customers, processes or calls are used.

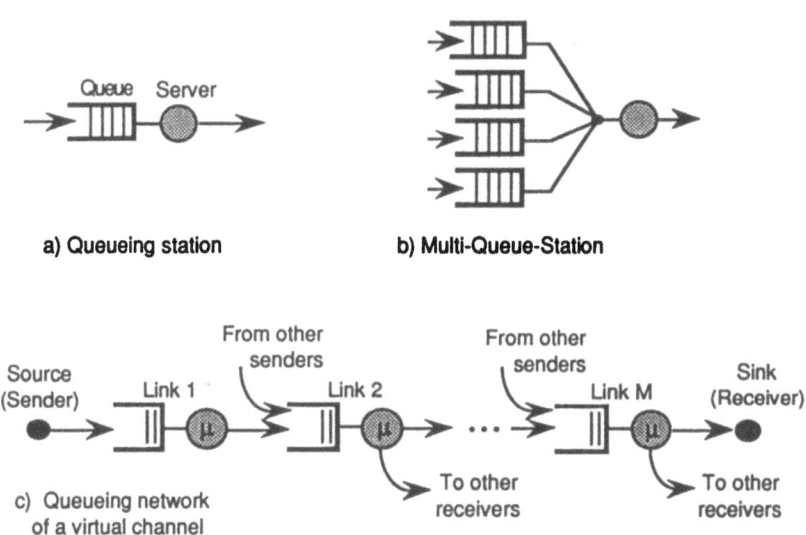

Fig. 3: Examples for queueing models

A simple queueing station is the "infinite server" abbreviated IS. Each task at an IS-station will be served immediately, without any wait delay or disturbance from other tasks being served at the same station. Usually the service demand is expressed by the parameter "mean service time". IS-stations introduce the concept of time, but they do not introduce the congestion for restricted resources due to concurrent tasks. To model congestion, queueing stations with a limited number of resources are needed. A simple example is a single server with a FIFO-queue. Congesting tasks have to queue for service until the resource represented by the single server is available, cf. Fig. 3a. In case a request (e.g., a service call) is to be executed by an elementary queueing system, the quantitative amount of the required service must be given as parameter. Typically a service demand is given by a constant real value x or by a random value, e.g., specified by negexp(1/x) for the negative exponential distribution. Since the number of requesting processes is usually greater than the number of available resources, the delay increases to the waiting time suffered in the queue. To summarize, we have to provide parameters concerning the demand of the requesting processes (i.e., the load) and we have to supply information on the performance behavior of the underlying hard- and software system to model the machine.

Apart from IS- and FIFO-stations, there are other types of stations like PS (processor sharing) or LCFS-PR (Last-Come-First-Served-Preemptive-Resume), which may be part of so-called product-form networks solvable by efficient algorithms [LAVE83]. If other stations, like multi-queue stations, multi-queue/multi-server-stations, stations with priority scheduling disciplines, degradable stations, stations with blocking and losses, or synchronizing stations (employing semaphores, counters, flags, . . .) are included in a queueing network, other solution techniques (approximate, numerical, simulative) must be used.

Queueing stations may be arranged to form a network. The stations build the vertices of a directed graph. The edges of the graph build the routes that may be taken by the tasks. In closed queueing networks a fixed number of tasks is moving permanently from station to station. In case of an open network, tasks can enter the network at sources and leave it at sinks. Usually a stochastic interarrival distribution is associated with a source, i.e., the time interval between succesive arrivals are given by a probability distribution. Figure 3c displays a model fragment that is typical for modeling packet switched communication networks. The forward buffers of a switching node are represented by a queue, the (one-way) communcation links are represented by servers. Packets enter from a source, travel through a virtual circuit and leave through the sink.

An important feature is the grouping of tasks into classes or chains, e.g., to distinguish tasks by their resource requirements (class-specific service time demands) or by their routing behavior (different paths through the network). Dependent on the application area source modeling is important to describe the pattern of traffic arriving from the environment.

Building Submodels of Queues

Queueing network models may be composed of queueing network subsystems, which are themselves composed from single stations or from lower-level subsystems. QN-subsystems are defined very similar to open queueing networks, but instead of sources and sinks a QN-subsystem provides an interface to its environment. A subsystem may be entered by tasks from the environment, provides service, and releases them finally into the environment again. According to the different paths in the subsystem it is said to provide different services to the outside. The interface for the environment may be termed as entry/exit- or input/output-port. Two examples illustrate the concepts.

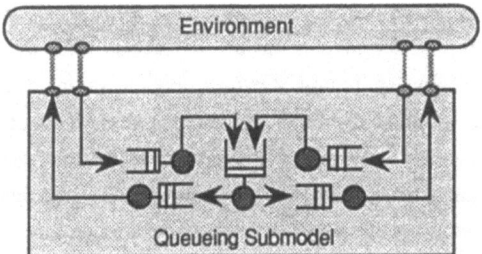
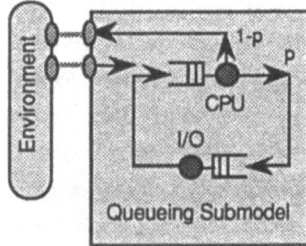

Fig. 4: Transmission medium model   Fig. 5: Computing system model

- **Example 1, Queueing submodel for a communication system**: The first example displayed in Fig. 4, shows a model of a transmission channel in a packet switched communication system. Note that the environment may be given by different (distributed) entities. Queueing stations represent links incl. buffers that are occupied for the time needed to handle and to transmit a message. The subsystem offers two directions for message transportation. Dependent on the congestion in this subsystem messages will suffer additional waiting delays. The performance behavior of the communication system, expressed in terms of throughput or delay, will be determined by the parameters of the load (like packet length, arrival intensity, arrival distribution, ...) and the parameter of the system (like band width of the channel measured in bits/sec, transmission rate, speed of computing devices).

- **Example 2, Queueing submodel for a computing system:** Figure 5 shows a computing system model and environment. From the environment calls may be submitted to the subsystem, which after some time delay returns a reply. This mechanism is known from the usual concept of "procedure call". The QN-submodel may be considered a "machine" or "module" providing a service to its environment. This service may be used (other terms are requested, called, imported) by the environment. A process will be generated and after service completion the process terminates and control returns to the requesting entity.

Other choices for PE-submodels could be Timed Petri Nets or a simulation model. In any case the interface to the environment should be given by a number of provided services accessible via a well-defined interface, whereas the specific type of the performance model used should be hidden to the outside. We stress that the major points are the interface definitions in form of provided services and their parameters.

4. Requirements for Combining FDT and PE

Model-based investigations of distributed systems requires the inclusion of time in formal designs. In the field of protocol modeling some work has been done in the last decade. The concept of a machine comprising a set of resources that is to execute the protocol has not been elaborated until now. In current FDTs there are no provisions at all for specifying any kind of machinery that is to execute a software system under time and space constraints. It is obviously indispensable to introduce these constraints in order to grasp the congestion for resources leading to wait delays, or reduced service speed due to multiple resource usage. So an important requirement in PE-modeling, namely the separate specification of "load" and "machine" has not yet been addressed in the past. The distinction between load and machine will be an important feature for the integration of FDT with PE.

FDTs supply some means for specifying the load that mainly is given by computational tasks to be performed and messages to be transported. In the context of SDL computational tasks to be performed may be specified with the help of process types, which of course must be extended by some additional information to quantify the service demands. These informations on resource requirements (e.g., number of instructions to be performed, cpu-time needed) must be provided additionally.

The machine specification should describe the resources that are capable to satisfy the service requirements of the load. Because FDTs do not supply means for the specification of "machines" we suggest to adopt ideas and concepts from the field of PE. Results obtained from research on PE-tools and techniques, presented in [BEIL85, BEMW88] may be used as starting point for the development of a conceptual framework, which aims for the supplementation of an FDT-based "load"-description with a "machine" specification tailored in PE-oriented fashion. Following these ideas a machine consists of submodels which are usable from the environment.

The connection between load and machine should be established by a particular "binding"-mechanism, which also should become a part of the combined FDT/PE-specification. The meaning is that the load's demands in form of statements, functions, tasks or processes (whatever level of detail is considered) must be referred to the services provided by the machine. In this way great flexibility can be achieved in configuring load-machine complexes for purposes of model-based PE. For the roots of these concepts see again [BEIL85, BEMW88].

Additional to the load aspects concerning the necessary computations, the sending of messages imposes load on the communication system. Whether computation or communication issues will have the greatest influence on the system's performance will dependent on many factors, in particular the kind of application and the underlying technology will play a major role. Here we consider both kinds of load.

FDTs supply means for the description of communication aspects, e.g., the sending and receiving messages is the central feature in SDL and in FSMs. The sending of a message (e.g., characterized by the amount of data to be transmitted) implies that a transmission medium must be used. In pure FDTs the transmission medium is often considered to be abstract. It also may be modeled as an unreliable channel, which is given either implicitly or is modeled explicitly. If we introduce performance aspects the connection between sender and receiver is to be described by a medium that uses a performance submodel. Parameters like band width or bit error probability are to characterize it quantitatively.

In the rest of the paper we specialize on the combination of FSM-entities and QN-submodels. For their combination we have to establish a framework that provides a consistent model world integrating both views. The semantic of the integrated world should be completely defined by the semantics of the FSM-world, the semantics of the QN-world and the semantics of the connection mechanism. A strict formalization of the new concept is not considered here, but should be feasible without difficulties. Modules of the two worlds should be preserved as far as possible. We summarize two concrete requirements.
- A basic mechanism to connect FSM and QN is to be introduced. A notation (syntax) to describe the combined world and its semantic has to be developed. The FSM-parts representing the FD of a system should be preserved completely. This is a very strong and important requirement.
- It must be explicated where and how load aspects are to be placed within the FSM-context and in particular how quantitative information has to be included. A machine composed from resources or components, must be introduced. Computational as well as communication aspects should be taken into account. Hence, additional to the concept of sending messages there should be means for performing "service calls". The connection mechanism should accomplish the binding of required services to machines (described by QN-submodels), and the passing of messages between modules.

5. A Proposal for an Integrated FD/PE-World

Our objective is a framework for the integration of state oriented FDTs with performance evaluation. We outline how a combined FSM/QN-world may be obtained that preserves the properties of both worlds. Based on the concept of FSM- and QN-modules the definition of interfaces and coupling elements is presented and illustrated by examples. Finally it is proposed how the new concept may be used.

5.1. Basic Structures for Integrating Finite State Machines and Queueing Networks

We introduce interfaces and propose how to interconnect modules, or, expressed from a different viewpoint, how QN-Submodels may be attached or bound to FSM-entities. Note that interface definitions result in a natural way from the concept of FSM-entities and QN-submodels introduced in chap. 2 and 3.

- Interface definition for Finite State Machines, Fig. 6a.:

  The proposal is to enhance an FSM with "places" that serve as interface to the environment. The addition of such places is a natural generalization of the interface definition sketched in Fig. 2. We assume input places for messages to be received and output-places for messages to be forwarded to a destination. Hence, messages may carry addresses and other necessary information. The places may be limited in capacity. An entitity may select from its input places in any order and may continue its operation while its input places are not empty. These places also support the principle of service call. Of course, a place may also be used to forward an invocation of a service call and another place may be used for receiving the result. The call mechanism implies that there is at most one outstanding call per entity. Hence the capacity of such a place is limited to one.

- Interface definition for Queueing Networks, Fig. 6b.:

  In the same spirit the interface of QN-subsystems is regarded to consist of pairs of input/output-places. For each service offered by a QN-submodel a pair of input/output-places is assumed. The input-place is to receive calls or messages, or more generally expressed: events may occur at the interface. The output-place is to offer an "event" to the environment. Whether an arrival event represents a message or a call is a matter of interpretation. In both cases the event is accepted in an entry-area, is moved through the QN-subsystem until it is placed in the exit. The services offered by the QN-submodel are accessed via the interface.

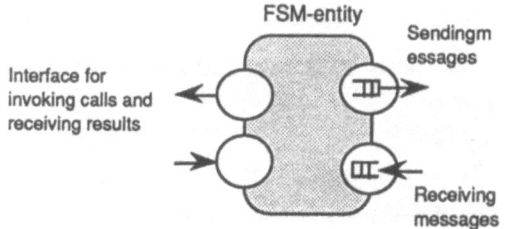

a) There are two types of "places",
 - for sending and receiving messages and
 - for making calls

b) The "places" are for receiving and sending of messages including the special case of a call and its result.

Fig. 6: Input and output places representing interfaces

To summarize, the places serve as interface to the environment. Now, coupling elements are to be introduced to connect or to bind modules together. Since the coupling elements may serve different purposes, transitions in the spirit of Petri net theory may offer many advantages. Petri nets (PN) support both intuition and rigid formalism. Hence, we introduce notions from place/transition nets and admit enhancements from other types of PN where necessary. The input- and output-places introduced above may be imagined as PN-places. Places may be connected by transitions (indicated by rectangles or bars) that serve for the exchange of calls, messages, events, tokens or whatever. From PN-theory it is known that a transition is activated if its input-place is occupied, and "fires" a token to its output-place, cf. Fig. 7a. In case of multiple input- and output-places a transition is activated if all its input-places are occupied and the firing of the transition delivers tokens to all output-places. In many cases a one-to-one relation between places will be sufficient. An example that models routing is illustrated in Fig. 7b. Multicast is shown in Fig. 7c. A two way communication that may be used to represent a call mechanism is shown in Fig. 7d.

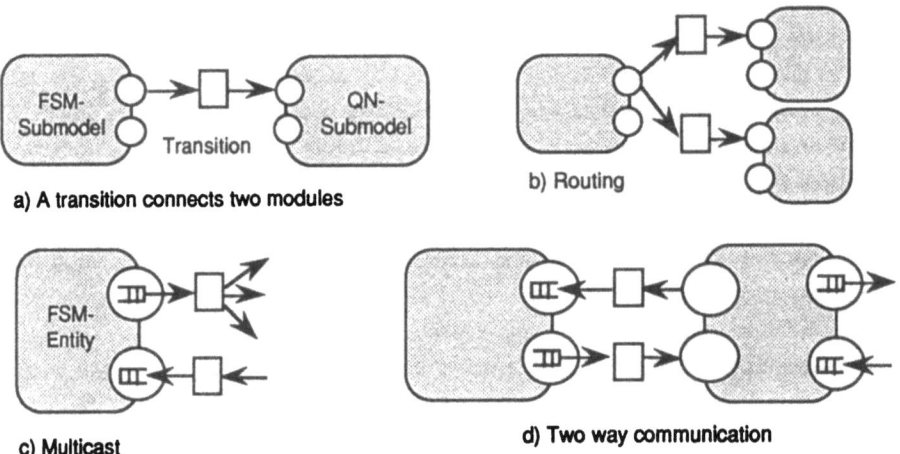

Fig. 7: Examples for module connection

For the distinction of messages individual tokens may be admitted, leading to an extension of place/transition nets. In particular the addressing information necessary to route a token may be added if necessary.

Introducing PN as a binding mechanism brings many advantages. From the abundant techniques to describe complex systems, we emphasize the most important points.
 • Routing, splitting, joining, parallelism may be decribed.
 • Refinement of transitions and places is possible, leading to hierarchical PN.
 • Performance aspects may be introduced by attaching time to transitions or to places.

The benefits of the first point is obvious, cf. the examples in Fig. 7. The last two points will be used in combination. We will assume that a transition may by refined such that a time consuming service from a QN-subsystem may be used. We take up this approach in the next section.

5.2. Principles for Building Integrated Models

As sketched in chap. 4 we assume that FSM-entities impose load on the machines represented by the QN-modules. To make the QN-modules usable by the FSMs, we need a mechanism that describes the "usage" or "binding" of required services to the services offered by QN-submodels.

The basic kind of module interaction may be viewed as "message passing", i.e., a sender forwards a message to a receiver via a medium. In pure FSM-theory the medium is abstract, i.e., it is noiseless, timeless and no concrete routes between communicating entities are specified. This situation is depicted by the interaction diagram in Fig. 8a. To introduce performance aspects the abstract medium must be made more concrete such that it can finally be associated with a QN-submodel. To this end, we do not propose to substitute the abstract medium simply by a QN-submodel. Such an approach would imply to decompose a given formal specification and reassemble it under inclusion of performance relevant modules.

Our approach is different. The FD must not be decomposed into parts, but the FD should be enhanced by attaching constituents describing performance aspects. We describe the proposal in an operational way. Starting from an interaction diagram, we select the abstract channels that should become timeful, see Fig. 8a-b. The transition symbol introduced in Fig. 8b makes the medium less abstract; it may be seen as coupling that is associated with a pure time delay, either deterministic or stochastic (e.g., in SDL there is an option to attach times to a channel). A refinement of the transition is shown in Fig. 8c-8d. The first refinement introduces a place between two transitions, afterwards the newly introduced place is refined by a QN-submodel. For techniques and formalisms concerning PN-refinements we refer to the theory of hierarchical Petri nets [JERO91].

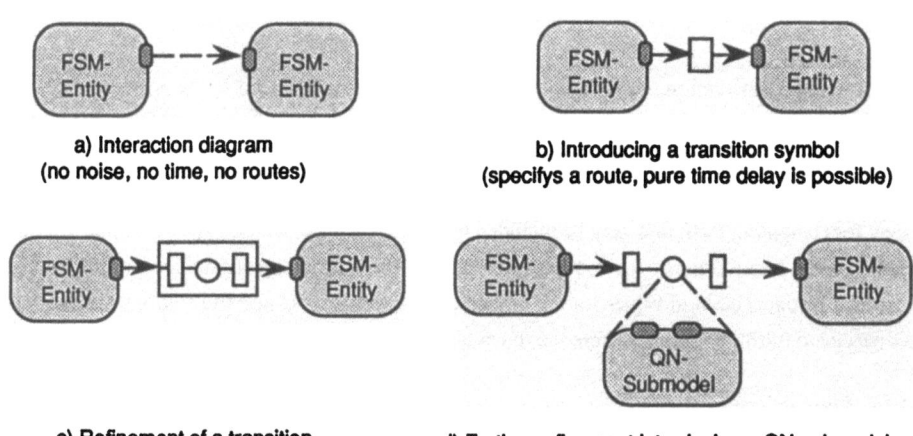

Fig. 8: Evolution of the abstract medium toward a timeful QN-submodel

Thus, finally a coupling has been achieved that models time-consuming transmission by calling a service offered by a QN-submodel. The advantage of this approach is shown in Fig. 9 where multiple entity pairs communicate over a shared medium. In this way concurrency and congestion due to resource sharing may be included in a combined FD/PE-model.

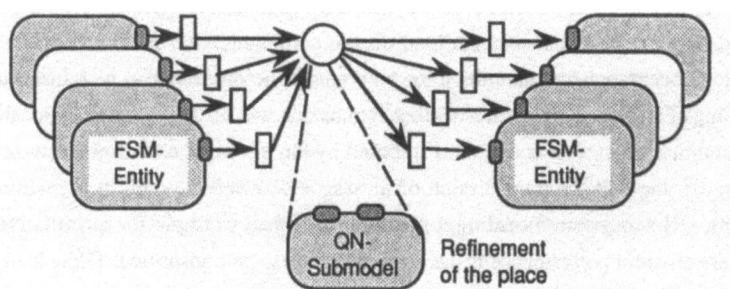

Fig. 9: Multiple entity pairs sharing a timeful medium

Another kind of interaction is given by the service call mechanism, which is well known from sequential programming languages in form of a "procedure call". The caller is blocked until the execution of the required service has been completed and control has been returned to the caller. This mechanism may be used to submit computational tasks to a "server".

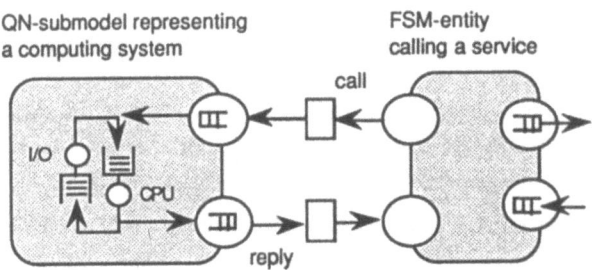

Fig. 10: Principle of service call

Figure 10 describes the situation under use of the graphical notion introduced in the previous section. The extension of FSM-entities to make calls is discussed in a later section. Provided services of a QN-submodel are requested from the FSM-entities by submitting service calls over the interface. Of course other processes, e.g., entities of the same type (or any other type) can use a QN-submodel too, such that concurreny for computing resources may be included in the quantitative evaluation. We emphasize that this possibility is very important for the modeling of parallelism and congestion. The parameters of the performance submodels include values for the processing speeds of CPU and I/O-stations and the number of cycles needed to fulfill the service offered to the outside.

5.3. Further Elaboration of the Approach

We elaborate our proposal using the protocol example introduced in chap. 2. Some of the steps necessary to achieve a combined FD/PE-model are discussed in more detail. After an overview we focus on the introduction of work load aspects into models for medium and computing resources.

We start with work load aspects and their relation to FSMs. Since a formal description of a protocol's functional behavior is implementation independent, neither the usage of resources nor any quantitative amount of service requirements is visible. Hence, the load does not show up explicitly in the formal description of the FSM-entities and needs to be introduced à posteriori. As explained both the computational load represented by "service calls" and the load on the communication medium must be considered. Computational load occurs when data units must be prepared before they can be transmitted. This may include the building of frames, encryption and other actions that will usually be done by local services. The load on the communication system is of course imposed by the messages exchanged between the entities. Here we have to assume that the transmission of messages between the entities is realized under time consumption by a QN-subsystem modeling a medium. A typical example for quantitative information necessary to assess protocol performance is the amount of data to be transmitted. These load requirements are to be fulfilled by QN-submodels.

For the example protocol model introduced in chap. 2 it is reasonable to specify two QN-submodels for computational tasks of sender and receiver respectively, and another QN-submodel that represents the medium. The performance behavior of the QN-submodels are characterized by service rates for the queueing stations. Service rates model the processing speeds of processors and the transmission capacity of the channel respectively and may depend on the type of required service.

To build a combined FD/PE-model that incorporates a formal protocol description we have to carry out the following steps, cf. Fig. 11.
- Define the interfaces of the FSM-entities
- Add information concerning the work load to be imposed
    - computation requirements, e.g. define service calls
    - communication requirements, e.g., define length of messages to be transmitted
- Specify QN-submodels providing the required services
    - build QN-submodels providing the computation services
    - build QN-submodels providing the communication services
- Integrate the QN-submodels using the introduced coupling mechanism.

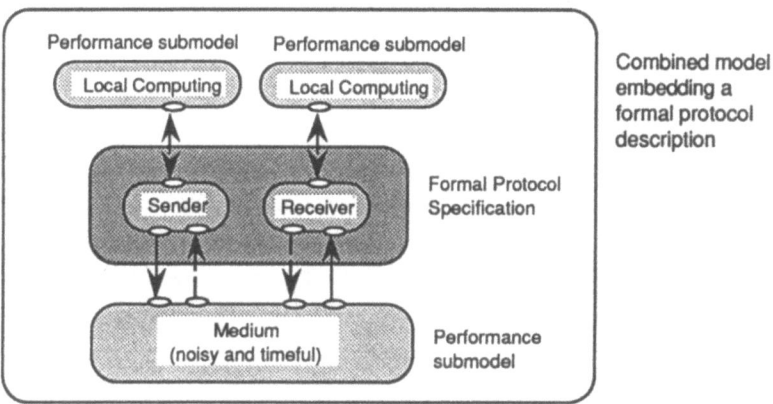

Fig. 11: Overview of the integrated FD/PE-model

5.3.1. On Computational Load and Service Calls

We may imagine that the vertical bar I used in FSM-description, e.g., in A1|D0, represents the location where service calls may have their origin. Remember, that the notion A|B is read as "after reception of message A message B is sent out, and the next state is entered". To admit service calls we introduce the graphical notion [] that may substitute the vertical bar I. The notion A [] B indicates that a service call (or even a sequence of calls) may be invoked between the receipt and the sending of a message, Fig. 12. The calling entity is blocked, i.e., it has to wait until the result has been returned. While the caller is blocked no messages can be sent out and newly arriving messages queue up in the input place. After reception of message A1 in state S1, a service is called that is executed by the machine. After some time delay the result is delivered and control returns to the calling entity. After sending message D0, the state S0 is entered.

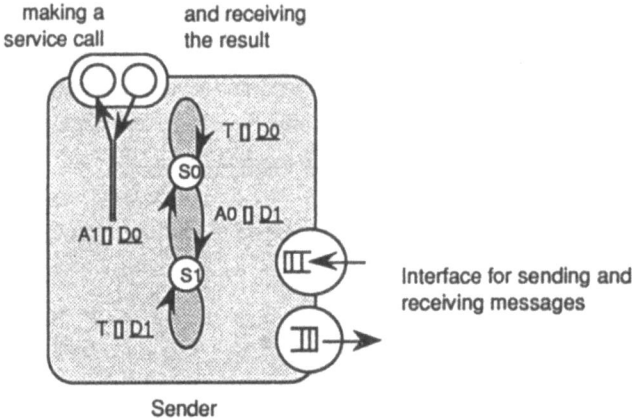

Fig. 12: Protocol entity and its interfaces

Of course, the FSM-description could be transformed into another form such that additonal intermediate states are introduced. E.g. the arc "S1 -> A1|D0 -> S0" could be substituted by "S1 -> A1 -> S1' -> D0 -> S0". In this case S1' could be used to submit a service call. For practical purposes, however, it is preferable not to change the formal description.

To model computational services required by the sender we have to specify the type and the amount of computation needed to prepare a frame for transmission. Of course we may summarize all necessary actions in a single service call, say with name "prepare_frame (...)" including operations like framing, encryption, ..., access to a file server. In a similar way the receiver may call a service "handle_frame(..)" covering all computations necessary to process an arriving frame. Also parameters describing quantitative aspects must be introduced. An example is the amount of data to be processed, e.g., measured in number of bits.

5.3.2. Modeling the Medium

Examples for quantitative parameters concerning the load imposed on the communication medium are the number of bits per data-frame and the number of bits in an ack-frame. Hence, the forwarding of a frame requires the transmission of a message of length M_D or of length M_A respectively. The parameters M_D and M_A may be added as attributes of the messages to be transmitted by sender and receiver respectively, alternatively the type of message must be added. Transmission requests will be served by the medium under consideration of this information.

The medium will be modeled under use of a QN-submodel, cf. Fig.13. To cope with performance aspects parameters like band width, frame loss probability, topology, configuration of the medium (i.e., half-duplex or full-duplex) must be included. The main parameters describing the medium quantitatively are the channel capacity C measured in bits per second (bps) and the bit error probability B. Instead of B we use the probabilities P_1 (= probability that a data-frame is lost or damaged) and P_2 (= probability that an ack-frame is lost or damaged). We assume given values for P_1 and P_2. Note that P_1 and P_2 depend on the bit error probability and the number of bits per frame. In a more detailed model damaged messsages may be passed to their destination where they are discarded.

The QN-submodel shown in Fig. 13 offers two services, namely transmission of data-frames from sender to receiver and ack-frames in the opposite direction. It is associated with a place connected to sender and receiver respectively. If a message arrives at the place a service of the QN-submodel is called and after some delay forwarded to its destination.

A promising way to formalize this proposal is the use of hierarchical Petri nets that provide concepts for refinement and substitution of transitions and places. An (arbitrarily detailed) QN-submodel may be seen as a substitute for a place that serves as coupling between entities. A formalization may be developed using recent results on hierachies in coloured Petri nets [HUJS90].

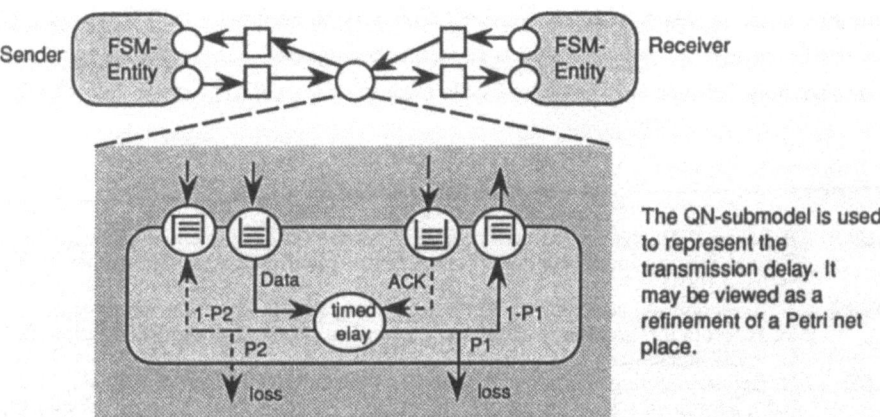

The QN-submodel is used to represent the transmission delay. It may be viewed as a refinement of a Petri net place.

Fig. 13: QN-submodel modeling a transmission medium

To cope with situations where more than one frame is on the way between sender and receiver at any time. Usually the buffer size is limited up to a certain capacity or a window mechanism is used to achieve flow control. These mechanisms should of course be part of the formal description itself, otherwise they must be specified together with the performance parameters. Note that we move on the borderline between formal description that should ideally be implementation independent and performance evaluation, which is essentially implementation dependent.

The displayed model may e.g., admit that a limited number of frames (including ack-frames) are in transit or, alternatively, that the buffer sizes are limited. So additional losses due to buffer exhaustion may occur. If frames are queueing for transmission a scheduling discipline must resolve the conflict. The time duration of a transmission is given by M_D/C for a data-frame and M_A/C for an ack-frame. This performance submodel describes a noisy and time-using channel.

5.4. Remarks on Model Evaluation

Goals of performance analysis could be the investigation of channel efficiency or throughput of data-frames as a function of system parameters like frame length or timeout duration. Guidance for an appropriate setting of these parameters may be obtained by performing experiment series with different parameter values. Here we have outlined a method to build executable models, but we did not apply techniques or tools for model solution. Due to synchronization, parallelism and other phenomena typically arising in distributed systems the models will mostly not be solvable by purely analytical techniques. Hence simulation or in some cases numerical Markovian techniques must be used.

6. Conclusion and Future Work

The paper has presented a framework for the integration of performance evaluation with formal description techniques. The proposed combination of finite state machines and queueing networks provides a consistent model world that integrates both views. As a result we obtained a foundation for the further development of techniques and tools for integrated FD/PE-modeling. To this end work should be directed toward a formalization of the approach. Also experimental tools may be eventually built by enhancing existing Estelle- or SDL-tools in a homogenous way. The term "homogeneous" means to build the performance submodels not from QN-submodels but from Estelle-modules or from SDL-processes respectively.

References

[BABU90] F. Bause, P. Buchholz, Protocol Analysis Using a Timed Version of SDL, 3rd Int. Conf. on Formal Description Techniques (FORTE'90), Madrid (Spain), November 1990, (North-Holland 1991)

[BANY86] N. Barghouti, N. Nounou, Y. Yemini, An Integrated Protocol Development Environment, in: Proc. 6th Int. Workshop on Protocol Specification, Testing and Verification, B. Sarikaya, G. v. Bochmann (eds.), Montreal 1986, (North-Holland 1987)

[BEIL85] H. Beilner, Workload Characterization and Performance Modelling Tools, in: G. Serazzi (ed.), Workload Characterization of Computer Systems and Computer Networks, (North-Holland 1986)

[BEHO89] F. Belina, D. Hogrefe, The CCITT-Specification and Description Language SDL, Computer Networks and ISDN Systems, Vol. 16, 1988/89, (North-Holland 1988), pp. 311-341

[BEHS90] F. Belina, D. Hogrefe, A. Sarma, SDL with Applications from Protocol Specification, (Prentice-Hall 1990)

[BEMW88] H. Beilner, J. Mäter, N. Weißenberg, Towards a Performance Modelling Environment: News on HIT, Proc. 4th International Conference on Modelling Techniques and Tools for Computer Performance Evaluation, Palma (Spain), 1988, (Plenum Press 1989)

[BOCH78] G. v. Bochmann, Finite State Description of Communication Protocols, Computer Networks 2, (North-Holland 1978), pp. 361-372

[BOCH90] G. v. Bochmann, Protocol Specification for OSI, Computer Networks and ISDN Systems, Vol.18, 1989/90, pp. 167-184

[BOVA88] G. v. Bochmann, J. Vaucher, Adding Performance Aspects to Specification Languages, in: S. Aggarwal, K. Sabnani (ed.), Proc. Eighth International Workshop on Protocol Specification, Testing and Verification, Atlantic City, NJ, June 7-10, 1988 (North-Holland 1988), pp. 19-31

[BRZA83] D. Brand, P. Zafiropulo, On Communicating Finite-State Machines, Journal of the ACM, Vol. 30, No. 2, April 1983, pp. 323-342

[BUDE87] S. Budkowski, P. Dembinski, An Introduction to Estelle: A Specification Language for Distributed Systems, Computer Networks and ISDN Systems Vol. 14, 1987, pp. 3-22

[CCITT88] CCITT, Specification and Description Language SDL, Recommendation Z.100, COM X-R15-E, Geneva (Switzerland), March 1988

[CHIO87] G. Chiola, GreatSPN USERS' MANUAL, Version 1.3, September 1987, University of Torino (Italy)

[CONW89] A. E. Conway, Performance Modelling of Multi-Layered OSI Communication Architectures, Proc. of the Int. Conf. on Communications 1989

[DANT80] A. Danthine, Protocol Representation with Finite State Models, IEEE Transactions on Communications, COM-28, 4 (1980), pp. 632-642

[DICH88] D. D. Dimitrijevic, M. S. Chen, An Integrated Algorithm for Probabilistic Protocol Verification and Evaluation, IBM Res. Rep. RC13901, Zürich (Switzerland), August 1988

[EHFH83] H. Ehrig, W. Frey, H. Hansen, ACT ONE - An Algebraic Specification Language With Two

	Levels of Semantics, Technical Report No. 83-03, Technical University of Berlin, 1983
[GURU89]	J. Gustafsson, H. Rudin, Including a Queue in a Formal-Description-Driven Protocol Performance Analysis, in: E. Brinksma, G. Scollo, C. A. Vissers (eds.), Proc. 9th International Workshop on Protocol Specification, Testing and Verification, Enschede (Netherlands), June 1989, (North-Holland 1989)
[HECK91]	E. Heck, Performance Evaluation of Formal Designs, in: R. Vader (ed.), ATMOSPHERE Briefings, Volume II, Esprit Project Atmosphere (EP 2565), November 1991
[HEHM91]	E. Heck, D. Hogrefe, B. Müller-Clostermann, Hierarchical Performance Evaluation Based on Formally Specified Communication Protocols, IEEE Trans. on Computers, Vol. 40, No. 4, 1991 (Special Issue on Protocol Engineering)
[HOAR78]	C. A. R. Hoare, Communicating Sequential Processes, Commun. ACM Vol. 21, No. 8, August 1978, pp. 666-677
[HOAR85]	C. A. R. Hoare, Communicating Sequential Processes, Prentice Hall, N.J., 1985
[HOGR89]	D. Hogrefe, SDL and OSI: On the Use of CCITT-SDL in the Context of OSI, Postdoctoral Thesis, Universität Hamburg, 1989
[HUJS90]	P. Huber, K. Jensen, R.M. Shapiro, Hierarchies in Coloured Petri Nets, in: [JERO91]
[IEEE91]	IEEE Trans. on Computers, Vol. 40, No. 4, 1991 (Special Issue on Protocol Engineering)
[ISO87a]	ISO: ESTELLE: A Formal Description Technique Based on an Extended State Transition Model, International Standard ISO/IS 9074, 1987
[ISO87b]	ISO: LOTOS: Language for the Temporal Ordering Specification of Observational Behaviour, International Standard ISO/IS 8807, 1987
[ISO89]	ISO: Guidelines for the Application of Estelle, LOTOS and SDL, ISO PDTR 10167, 1989
[JERO91]	K. Jensen, G. Rozenberg (Eds.), High-level Petri Nets, (Springer 1991)
[KING91]	P. W. King, Formalization of Protocol Engineering Concepts, IEEE Trans. on Computers, Vol. 40, No. 4, 1991 (Special Issue on Protocol Engineering)
[KLEI76]	L. Kleinrock; Queueing Systems, Vol.1 and 2, Wiley 1976
[KRIT86]	P. Kritzinger, A Performance Model of the OSI Communication Architecture, IEEE Trans. on Communication, Vol. 34, No. 6, June 1986, pp. 554-563(also published as IBM Research Report RZ 1346, Zürich (Switzerland), 1984
[KRWH89]	P. Kritzinger, G. Wheeler, A Protocol Engineering Workstation, 2nd Int.Conf. on Formal Description Techniques (FORTE'89), (North-Holland 1990)
[LAVE83]	S. S. Lavenberg, Computer Performance Modelling Handbook, (Academic Press 1983)
[LIU91]	M. T. Liu, Introduction to Special Issue on Protocol Engineering, IEEE Trans. on Computers, Vol. 40, No. 4, 1991 (Special Issue on Protocol Engineering)
[MBCC84]	M. A. Marsan, G. Balbo, G. Chiola, G. Conte, A Software Tool for the Automatic Analysis of Generalized Stochastic Petri Net Models, Proc. International Conference on Modelling Techniques and Tools for Performance Analysis, Paris (France), (North-Holland 1984)
[MERL79]	P. M. Merlin, Specification and Validation of Protocols, IEEE-Transactions on Communications, Vol. COM-27, No. 11, November 1979, pp. 1617-1680
[MILN80]	R. Milner, A Calculus of Communicating Systems, LNCS 92, (Springer 1980)
[RUDI83]	H. Rudin, From Formal Protocol Specification Towards Automated Performance Prediction, Proc. Third Workshop on Protocol Specification, Testing and Verification, Rüschlikon, Switzerland, May 1983 (North-Holland 1983), p. 257-269
[RUDI84]	H. Rudin, An Informal Overview of Formal Protocol Specification, IBM Research Report RZ 1344, Zürich (Switzerland), October 1984
[RUDI86]	H. Rudin, Tools for Protocols Driven by Formal Specifications, IBM Research Report RZ 1525, Zürich (Switzerland), September 1986
[RUDI88]	H. Rudin, Protocol Engineering: A Critical Assessment, in: S. Aggarwal, K. Sabnani (eds.), Proc. 8th Int. Workshop on Protocol Specification, Testing and Verification, Atlantic City, NJ, June 1988, (North-Holland 1988)
[SAMN85]	C. H. Sauer, E. A. MacNair, The Evolution of the Research Queueing Package, in: D. Potier (ed.), Modelling Techniques and Tools for Performance Analysis, (North-Holland 1985)
[TANE89]	A. S. Tanenbaum, Computer Networks (2nd Edition), (Prentice-Hall 1989)
[TURN87]	K. Turner, An Architectural Semantics for LOTOS, in: H. Rudin, C. H. West (eds.), Proc. 7th Int. Workshop on Protocol Specification, Testing and Verification, Zürich (Switzerland) 1987, (North-Holland 1987)
[VEPI86]	R. C. Venkatraman, T. F. Piatkowski, A Formal Comparison of Formal Protocol Specification Techniques, in: Proc. 5th Int. Workshop on Protocol Specification, Testing and Verification, M. Diaz (ed.), (North-Holland 1986)
[VEPO85]	M. Veran, D. Potier, QNAP2: A Portable Environment for Queueing Systems Modelling, in: D. Potier (ed.), Modelling Techniques and Tools for Performance Analysis, (North-Holland 1985)

Generating Parallel Code from Estelle Specifications

Bernd Hofmann and Wolfgang Effelsberg

Praktische Informatik IV
University of Mannheim
6800 Mannheim
Germany
hofmann@pi4.informatik.uni-mannheim.dbp.de

Abstract: High-speed networks, such as FDDI, DQDB or B-ISDN, require efficient communication software not only for layers 1 and 2, but also in the upper layers. Compared to today's implementations, a speedup of at least a factor of 50 must be achieved. A promising appoach is *parallel execution of communication software* for the upper layers. In the paper, a model for a parallel communication subsystem is presented. The C code for the concurrent processes is generated automatically from an Estelle specification of the layers.

1 Introduction

High-speed networks are becoming reality. At the level of the physical medium, fiber optics are widely used. Emerging standards, such as FDDI [RHF90], DQDB [NBH88] and Broadband-ISDN based on ATM [BKNS89], are proposing bandwidths of 100–150 MBit/s. Fast electronics for attaching to these networks and for packet switching are also becoming available.

Today's use of these new network technologies is mainly as a backbone for interconnecting slower networks. A typical example is an FDDI ring interconnecting Ethernet LANs [Spa91]. For this use, only layers 1 and 2 (and perhaps layer 3) need to be implemented, usually in hardware or in microcode. It would be very desirable to also use high-speed technology for connecting end-user workstations directly, thus enabling new applications such as multi-media (including audio and video data streams [IEE90, Hop91]). However, this would require a complete protocol stack, including higher layers, performing at the data rate of 100 MBit/s or more.

Most of the recent publications on high-speed networks have concentrated on layers 1 to 4 of the Reference Model (e.g., [SD89, DS90, DDK+90]). Very little work was reported on upper layers for this environment, in particular on layers 6 and 7. This is in contrast to the fact that upper layers consume most of the path length in today's protocol stacks, in particular in OSI. Upper layer functionality cannot be ignored in high-speed networks: as long as computers are heterogeneous in hardware and operating systems, presentation layer encoding/decoding is inherently required. Similarly, networks are useless without applications and thus an application layer is essential. In today's LANs, an application level sustained bit rate of 2 MBit/s can be considered fast [Svo89a, Svo89b]. Thus a speedup of at least a factor of 50 is necessary to reach the 100 MBit/s bit rate of high-speed networks.

In this paper, our approach to high bit rates for upper layers is **parallelism**. Parallel processing is a maturing technology in other areas of computer science [HB84, BST89]. Parallel hardware is becoming available at reasonable prices. Several processors and their memories can be placed on a network adapter card. Earlier work on the parallel execution of lower layers has resulted in a promising speedup [Zit89]. Massively parallel network adapters with RISC processors seem possible in the near future.

Most of the communication software in product use today was hand-coded from informal protocol specifications. In the meantime, however, ISO and CCITT have standardized three specification techniques: the "Formal Description Techniques (FDTs)" SDL, Estelle and LOTOS [BHT88, BD87, BB87]. Compared to specification in prose, a layer specified in an FDT is well-defined in syntax and semantics, and thus unambiguous. Whereas LOTOS is based on process algebra, SDL and Estelle are based on Extended Finite State Machines (EFSMs). EFSMs are more natural and easier to understand for the unexperienced user; also, they are less abstract and closer to an implementation. Using an FDT specification and a code generator has many advantages:

- the code generated from an FDT has fewer bugs than hand-written code
- the code is well structured and easy to understand
- protocol machine aspects are nicely separated from other aspects of a communication system node, such as buffer management, file I/O, timers, etc. The layer code is more portable than hand-written code.
- conformance to the standard is easier to achieve.

On the other hand, it is often claimed that generated code is less efficient. We don't see a good reason for *inherent* inefficiency in generated code provided that the code generator is well implemented. Obviously efficient code is of paramount importance in a high-speed environment. On the other hand, the overall speedup gained by parallel execution is much higher than a possible speedup by manual code optimization.

In [JJ89] a code generator similar to ours is described which is able to derive parallel code from Estelle specifications. The code can be generated for an Intel iPSC hypercube, a network of SUN workstations (over TCP/IP), with extensions to Transputers planned. The parallel computer is used to simulate a distributed system. This method is called "experimentation" and intended to fill the gap between simulation and test. In this approach, the main emphasis is on validation aspects rather than the production of efficient code for high performance implementations.

A similar approach for a network of Unix workstations is described in [ABH+91].

This paper is organized as follows. In Sect. 2, we present various kinds of parallelism in communication software. Section 3 describes our implementation environment. Based on the model for parallelism and the environment, we introduce the architecture of our prototype in Sect. 4. Section 5 concludes the paper.

2 Parallelism in Communication Software

As earlier work shows, several different forms of parallelism can be used for the parallel execution of protocol stacks [Zit90, JS90, UD90, GKWW89, BGK+88, Rup91]. To establish a uniform model, we define a PUnit to be the unit of execution that can be processed in parallel with other PUnits. PUnits can be assigned to the same or different operating system processes, and the processes can be assigned to the same or different processors. This allows a decomposition of communication software into parallel units independent of the target machine. In these terms, parallelism in a protocol stack can be classified as follows:

1. **PUnit-per-layer:** Often, each layer entity is implemented as a process of the operating system. Since the OS scheduler has no information on the semantics of the processes, the scheduling sequence is not optimal. E. g. it is possible that an entity is scheduled because it has been waiting for the longest time, but has nothing to do at this moment, while another entity has to process new data packets. This can be improved by using a pipeline of PUnits where each layer is implemented by a separate PUnit (Fig. 1). Similar to a hardware pipeline in vector processors, the code of subsequent protocol entities can now be executed in parallel.
2. **PUnit-per-connection:** If one host has to establish more than one connection at a time, several PUnits are created, each of them serving one connection. In conventional systems, all connections share the same processor. This results in decreasing efficiency with an increasing number of connections. Since the activities necessary for disjoint connections are independent of each other, they can easily be distributed over several PUnits, each of them serving one connection (Fig. 2).
  PUnit-per-layer and PUnit-per-connection can be combined in a matrix fashion. For example, a PUnit can be assigned to each layer in each connection.
3. **PUnit-per-packet:** Another way to improve the efficiency of a protocol stack is to integrate adjacent layers, i.e. the entities of adjacent layers are not separated into different PUnits but implemented as one PUnit [OP90]. This avoids unnecessary copying of data packets and repetition of similar work that would otherwise be done in each layer again, such as segmentation or error correction. Furthermore, this allows "bypassing" of protocol functions not needed for all data packets. The throughput can be increased by invoking several identical PUnits and executing them in parallel. Then, each data packet is associated with one PUnit through several layers (Fig. 3).
4. **PUnit-per-protocol-function:** In the case of PUnit-per-layer and PUnit-per-connection, the granularity of parallelism is an instance of a layer entity. Since the layers differ in complexity and since a pipeline is as efficient as its slowest element, internal parallelism within a complex entity can also be considered. For this purpose, functional units of an entity – so-called *protocol functions* – can be identified which

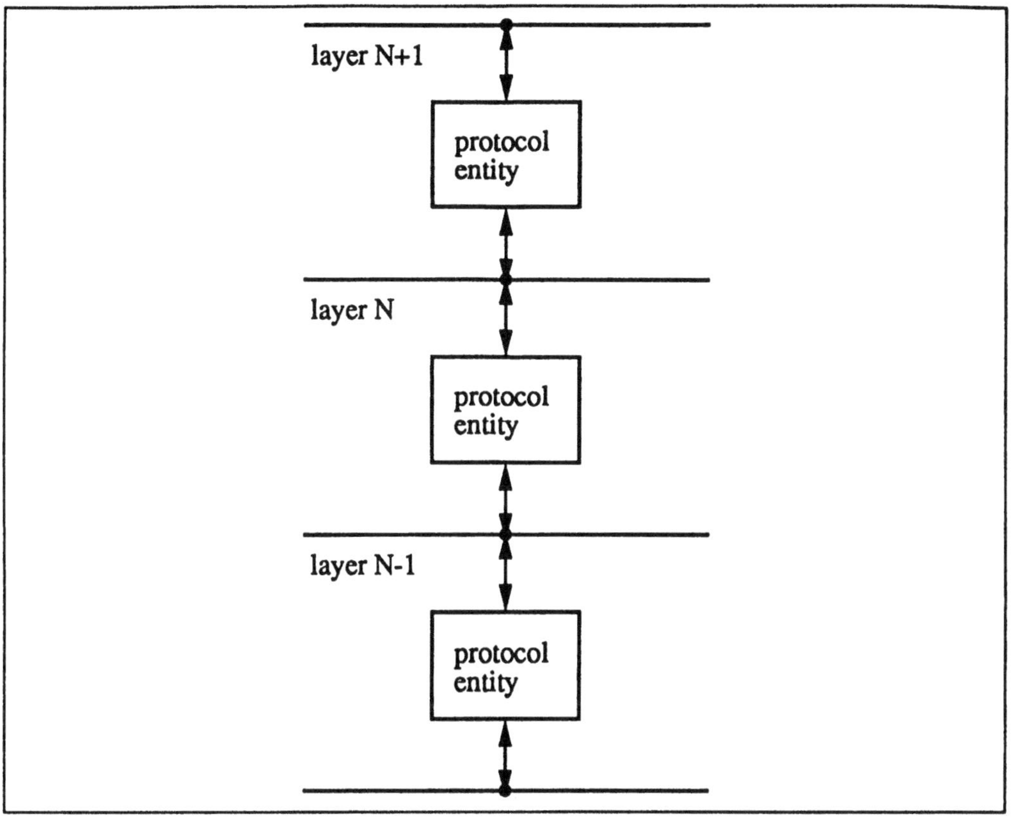

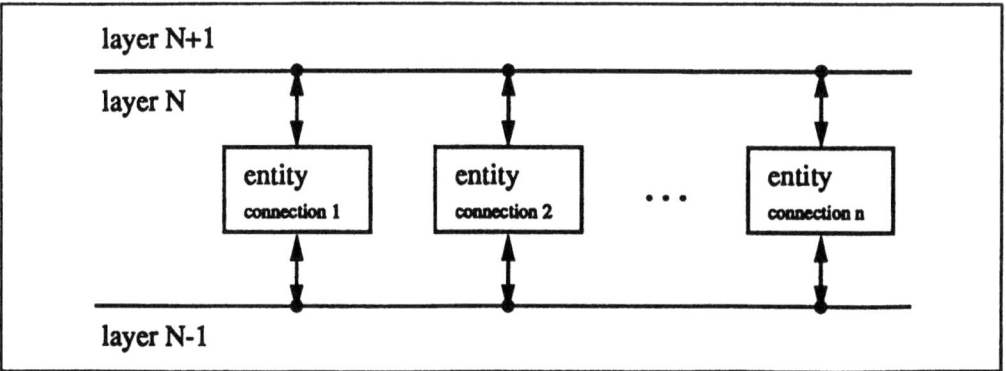

Fig. 1. PUnit-per-layer

Fig. 2. PUnit-per-connection

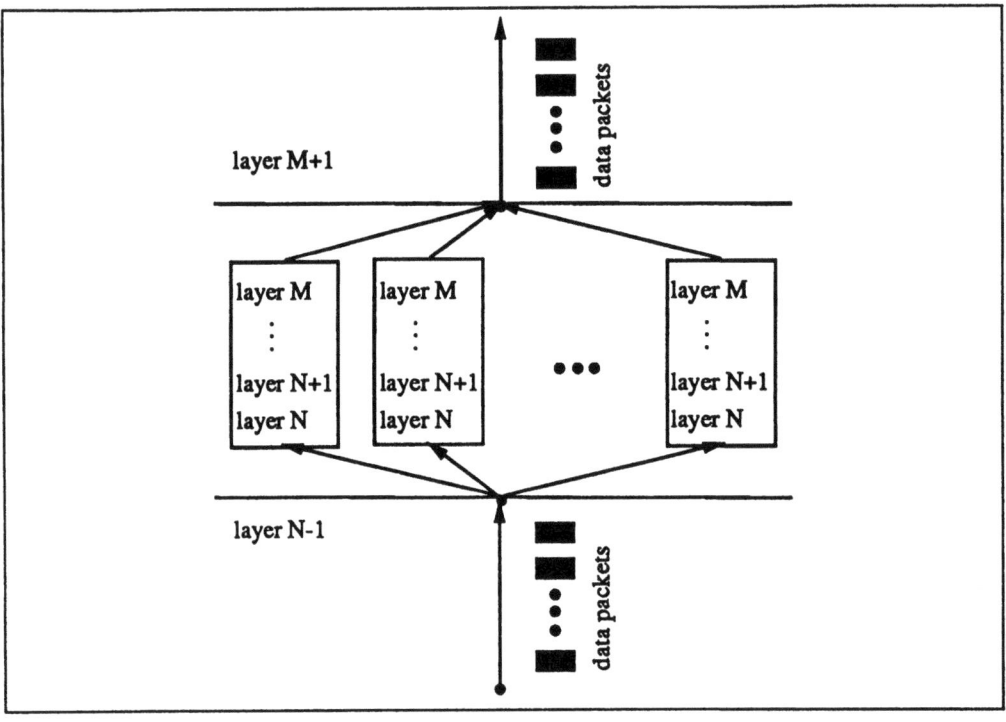

Fig. 3. PUnit-per-packet (only receiving part shown)

can be executed in parallel (Fig. 4). As a first approach it could be possible, in some full-duplex protocols, to separate the sending and the receiving part and execute them in parallel, or to do the ASN.1-encoding while generating the presentation protocol header in layer 6.

Since it is also possible to combine these methods, a generalized model comprises all different forms of parallelism. If the number of PUnits is smaller than the number of available processors, each processor can execute one PUnit. Otherwise, some processors must be shared by several PUnits.

3 Implementation Environment

To examine the different forms of parallelism as described in Sect. 2 we chose to implement a parallel version of FTAM. File transfer is one of the most important, and most frequently used functions of the application layer. Also, an FTAM file server in an OSI network can easily become a bottleneck because of the path lengths in layers 6 and 7.

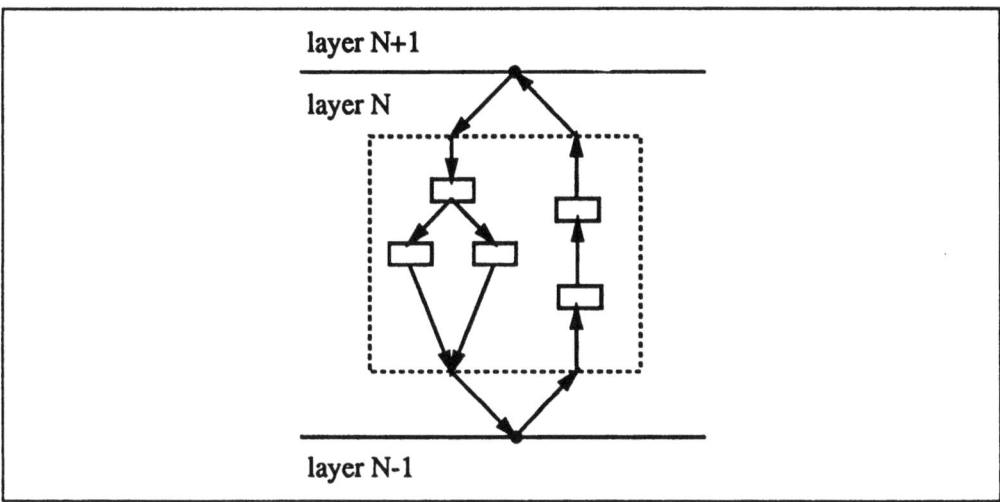

Fig. 4. PUnit-per-protocol-function

Remember that not only headers, but each user data byte must be processed in these layers.

Since we are not interested in reimplementing all other layers and the operating system environment as well, we chose to use the *ISO Development Environment (ISODE)* available under Unix [Ros90]. Our current version is ISODE Release 7.0, running on top of Ultrix on a DECSystem 5400. ISODE comprises a (sequential) FTAM implementation and presentation, session and transport layers on top of TCP/IP as well as X.25. The fact that we have a sequential implementation of FTAM available allows us to test our parallel implementation for interoperability (see Fig. 5).

Our prototype target system are Transputers. Their interconnection network is a matrix of links, four per processor. The Transputer link concept allows links between processes to be configured in hardware or in software (in the latter case, they are simply interprocess communication channels). Therefore, more than one PUnit can be assigned to a Transputer processor, and attached to its neighbors in a flexible way. Our model could also be adapted to run on another MIMD multiprocessor system.

4 Architecture and Implementation of the Parallel Prototype

For the implementation we use the generalized model described in Sect. 2, adapted to Transputers. In our first experiment shown in Fig. 6, PUnit-per-layer and PUnit-per-connection are combined. Each box represents a PUnit normally being executed on its

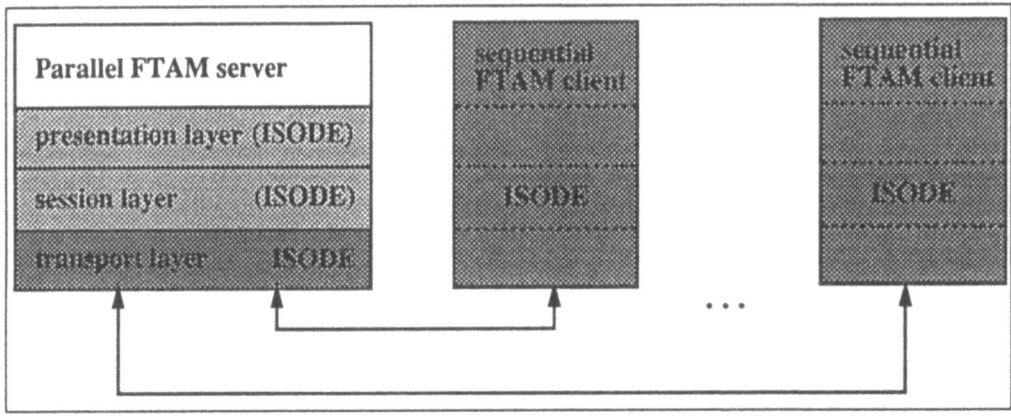

Fig. 5. Testing of the parallel FTAM entity

own processor. All *entity* processes of one layer are replicated protocol machines of an entity of that layer.

For the present, no integration of layers is done. The granularity of parallelism is one complete FTAM Application Service Element. Thus a speedup is only possible for parallel connections out of the server.

In order to allocate PUnits to processors, there is a management process called *layer manager*. The *layer manager* process accepts incoming requests for new connections and looks for free processors. If one is available, it creates a new incarnation of the entity of this layer on that processor. If all processors are in use, it creates an additional process on one of the processors, attempting to keep the load balanced. Then the connection endpoints are directly connected with that process, so further data packets need not be processed by the layer manager. At connection release, the entity process notifies the layer manager.

The communication between the processors should be asynchronous to achieve independence between them, thus enabling the greatest possible concurrency.

This architicture is particularly appropriate for network nodes with many simultaneous connections, such as file servers, WAN servers, etc. Then the advantage of increased throughput exceeds the overhead produced by the manager process and by the synchronization of the parallel processors. In any case, the processing overhead of the layer manager is not in the data path.

Our approach allows the use of existing specifications produced with formal description techniques and tools for code generation, modified for parallel use. Once the layer manager is written, new prototypes of a layer can be produced very quickly by generating code for the parallel entities from a formal specification.

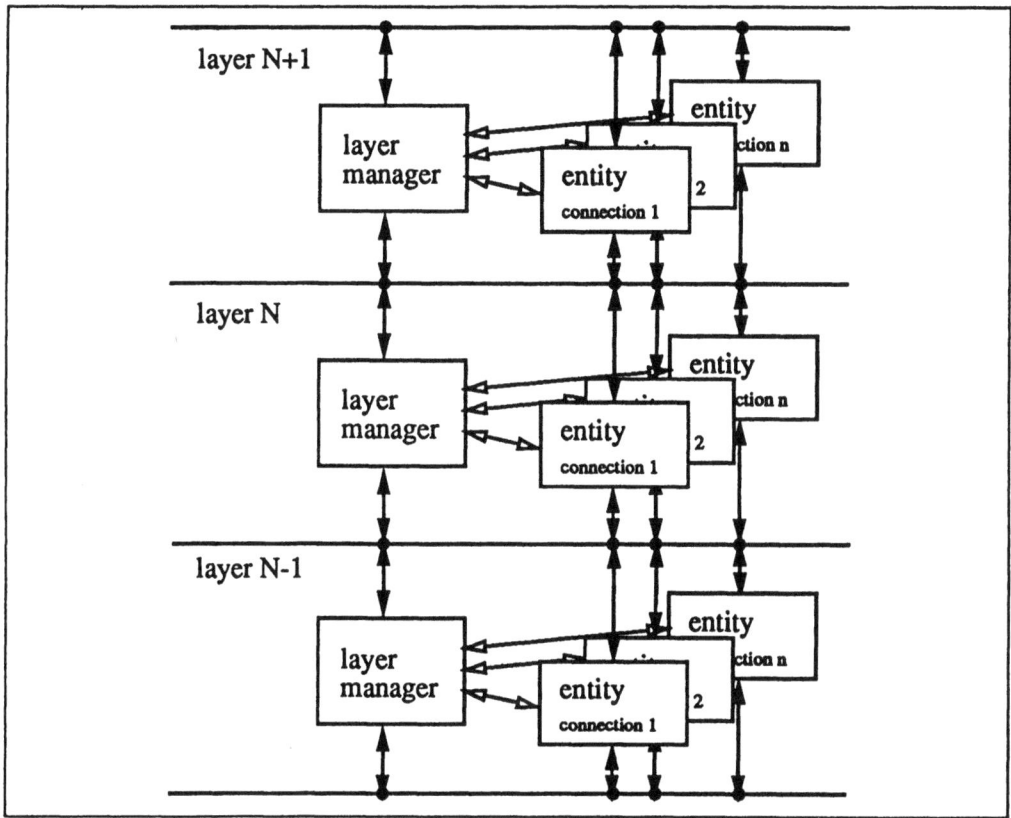

Fig. 6. Generalized model

Currently our work is concentrating on the code generator for Estelle specifications. It is based on the Estelle-Compiler of the *National Institute of Standards and Technology* (NIST) [FHSW89, NIS89]. This compiler allows to derive executable code in C from an Estelle specification very easily, and is therefore well suited for fast prototyping and validation of specifications. Not so easy is the generation of distributed executable programs. Some code has to be handwritten to implement the communication between two peer processes. Our modified code generator will do this automatically.

The code produced by our modified code generator is Parallel C. First experience shows that the code needed for parallel execution differs only slightly from the original sequential code. The main difference is that the new code generator must take into account that the communication with neighbouring layers must be mapped to interprocessor communication (i.e., Transputer links). A second difference is that we can now assume the existence of a layer manager, handling process initialization and management of all external OS resources.

The NIST Compiler creates two files of C-Code for each Estelle module: one implementing the actions and the behaviour of the finite state machine described by the module, the other is a template for implementation-dependent code. In addition, a skeleton for the main program is produced which essentially calls the Estelle scheduler. Furthermore, a runtime system provides a scheduler which determines the greatest possible set of executable transitions of the finite state machines and executes them.

What is missing and what has to be coded by hand until now, is the interface to the "outside world", i. e. other processes or the user. This code cannot be supplied by a generator suitable for different operating systems and hardware environments.

For our purposes, the environment is fixed, so the necessary code can be produced automatically. Since we want to use PUnit-per-connection and PUnit-per-layer, each top-level Estelle module has to be implemented as one PUnit. To achieve this, one main program has to be created for each module. As outlined in [FHSW89], the concept for the inter-process communication is to use *virtual interaction points* in the main program which deal with the whole communication of this module. A virtual interaction point is a data structure similar to that produced by the Estelle code generator to implement an Estelle interaction point. Thus the interaction point routines of the runtime library for message passing, connecting/disconnecting, enqueuing messages etc., can still be used.

For each external interaction point of the top-level Estelle module a virtual interaction point is created, and both are connected. The virtual interaction point serves as an internal interaction point of the main program. In addition, an array of queues is necessary, each of them associated with one external interaction point and buffering the messages sent to that interaction point.

Messages bound for other Transputer processes are enqueued in a queue of this array. To send them to the destination process, they are dequeued and sent over the hardware links or software channels connecting the Transputer processes. Incoming messages from other processes destined for an external interaction point are routed over the corresponding virtual interaction point (see Fig. 7).

The processing of the send/receive-events is done by two routines, MediumRead and MediumWrite, called immediately before and after the scheduler function call. MediumRead watches the inter-process channels for incoming events and sends them to the virtual interaction point connected to the destination interaction point. Then the Estelle scheduler is called, executing a set of possible transitions. After its completion, MediumWrite examines the array of queues, dequeues all send messages and sends them to the destination processes.

In this way we can generate most of the C code for the Transputers automatically, with only minor modifications to the Estelle code generator.

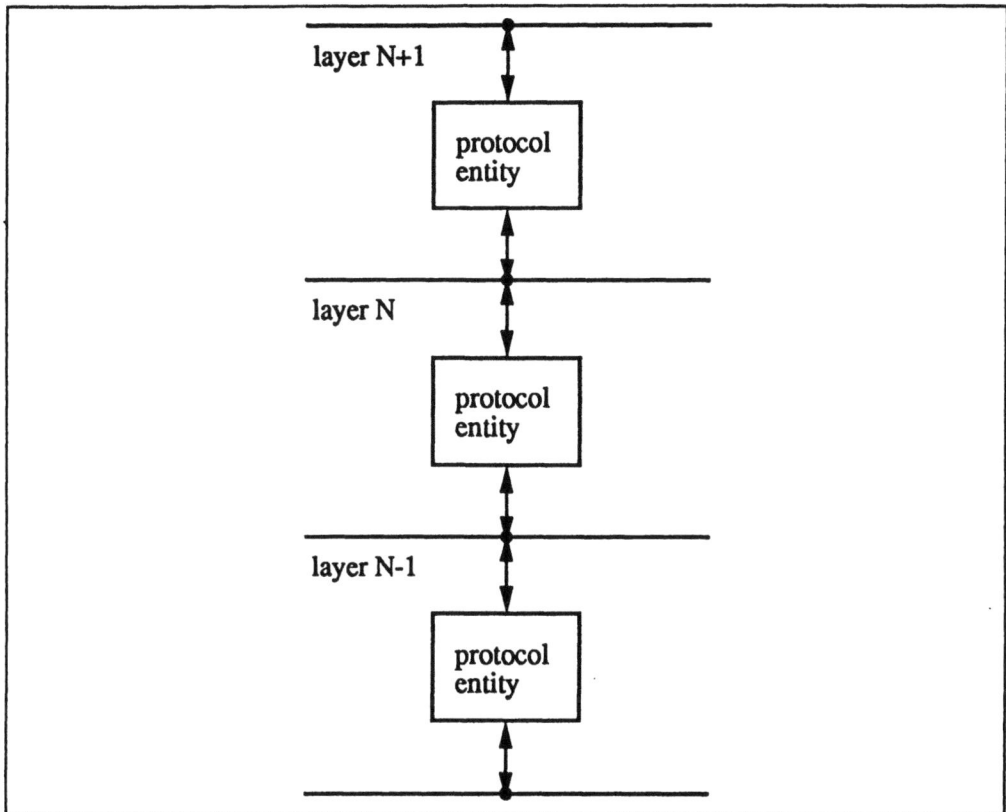

Fig. 7. Virtual interaction points

5 Conclusions and Outlook

High-speed networks will provide data rates of 100 MBit/s and more for each workstation or host in the near future. Today's network protocols and communication subsystem architectures are unable to handle such speeds in the upper layers. We are proposing to take advantage of the inherent parallism in protocol stacks to increase the performance of upper layers.

We distinguish four types of parallelism: PUnit-per-layer, PUnit-per-connection, PUnit-per-packet, and PUnit-per-protocol-function. These different kinds of parallelism can be used in combination. The software implementor is responsible for defining PUnits for a given protocol stack, and for assigning PUnits to operating system processes. At runtime, a layer manager process is responsible for assigning processes to the set of parallel processors.

For a prototype implementation, we are using ISO FTAM as an example. Our environment consists of ISODE, the Estelle compiler from NIST, and Transputer boards. First experience shows that only minor modifications to the code generated by the Estelle compiler are necessary in order to execute Estelle modules in parallel. In this way, the preparation of parallel protocol code is straight-forward if a specification is available in Estelle.

Currently, we are only investigating PUnit-per-layer and PUnit-per-connection for the FTAM example. The Estelle compiler is modified to generate parallel C code which can then be compiled for Transputers. In the next steps, we will integrate the generated code with the layer manager for layer 7, and the lower layers of ISODE. This will allow us to do conformance testing as well as performance analysis and comparison with sequential implementations.

In a next phase of the project we intend to include internal parallelism within one layer entity (PUnit-per-protocol-function) into our studies. A major problem will be to optimize the granularity of the parallel tasks; there is a trade-off between speed-up by parallelism and synchronization overhead.

Later on we intend to include a presentation layer prototype into our studies. Here the ASN.1 encoding/decoding can be done in parallel to the presentation protocol processing. Once we have two layers operational, we can study the effects of layer integration (i.e. the avoidance of an explicit crossing of inter-layer boundaries) and PUnit-per-packet on overall performance. In particular, layer integration will allow encoding/decoding to be invoked by layer 7 routines rather than being concentrated in layer 6.

This is an ongoing project, and no results on speedup or overhead are available yet. In the long run we believe that massive parallelism on communication adapters will be feasible; perhaps this is the most promising approach to high-speed networking while still guaranteeing conformance with existing international standards.

References

[ABH+91] C. Andrae, J. Bredereke, C. Hille, D. Peter, T. Reimer, U. Schüler, R. Gotzhein, and F. H. Vogt. Praktischer Einsatz und Weiterentwicklung von Estelle. In J. Encarnação, editor, *Telekommunikation und multimediale Anwendungen der Informatik*, pages 265–275, Darmstadt, 1991. GI-21. Jahrestagung, Springer-Verlag.

[BB87] T. Bolognesi and E. Brinksma. Introduction to the ISO specification language LOTOS. *Computer Networks and ISDN Systems*, 14(1):25–59, 1987.

[BD87] S. Budkowski and P. Dembinski. An introduction to Estelle: A specification language for distributed systems. *Computer Networks and ISDN Systems*, 14(1):3–23, 1987.

[BGK+88] J.E. Boillat, P.K. Goode, P.G. Kropf, D. Bärtschie, and A. Spichiger. Communication protocols and concurrency: An OCCAM implementation of X.25. In *Proceedings International Zurich Seminar on Digital Communications*, Zürich, März 1988.

[BHT88]   F. Belina, D. Hogrefe, and S. Trigila. Modelling OSI in SDL. In Kenneth J. Turner, editor, *Formal Description Techniques*, pages 135–142, Amsterdam, 1988. North Holland.

[BKNS89]   W. R. Byrne, T. A. Kilm, B. L. Nelson, and M. D. Soneru. Broadband ISDN technology and architecture. *IEEE Network*, 3(1):23–28, January 1989.

[BST89]   H. E. Bal, J. G. Steiner, and A. S. Tanenbaum. Programming languages for distributed computing systems. *ACM Computing Surveys: Special issue on Programming Language Paradigms*, 21(3):261–322, September 1989.

[DDK+90]   W. Doeringer, D. Dykeman, M. Kaiserswerth, B. Meister, H. Rudin, and R. Williamson. A survey of light-weight transport protocols for high-speed networks. *IEEE Trans. on Communications*, 38(11):2025–2039, November 1990.

[DS90]   A. Danthine and O. Spaniol, editors. *High Speed Local Area Networks II*. North Holland, Amsterdam, 1990.

[FHSW89]   J. Favreau, M. Hobbs, B. Strausser, and A. Weinstein. Internals guide for the NIST prototype compiler for Estelle. Technical Report No. ICST/SNA - 87/3, Institute for Computer Science and Technology, National Institute of Standards and Technology, September 1989.

[GKWW89]   D. Giarrizzo, M. Kaiserswerth, T. Wicki, and R. C. Williamson. High-speed parallel protocol implementation. In H. Rudin and R. Williams, editors, *Protocols for High-Speed Networks*, pages 165–180. IFIP, North-Holland, 1989.

[HB84]   K. Hwang and F. A. Briggs. *Computer Architecture and Parallel Processing*. McGraw-Hill, New York, 1984.

[Hop91]   A. Hopper. Design and use of high speed networks for multimedia applications. In *Proc. 3rd IFIP Conference on High-Speed Networking*, pages 25–38, Amsterdam, 1991. North Holland.

[IEE90]   Multimedia communications. IEEE Journal on Selected Areas in Communications, 8(3), April 1990.

[JJ89]   C. Jard and J. M. Jezequel. A multi-processor Estelle to C-compiler to prototype distributed algorithms on parallel machines. In Ed Brinksma, Giuseppe Scollo, and Chris A. Vissers, editors, *Protocol Specification, Testing, and Verification, IX*, pages 161–174. IFIP WG 6.1, North-Holland, 1989.

[JS90]   M. N. Jensen and M. Skov. VLSI-architectures implementing lower layer protocols in very high data rate LANs. In A. Danthine and O. Spaniol, editors, *High Speed Local Area Networks, II*, pages 187–205. Elsevier Science Publishers B.V. (North Holland), 1990.

[NBH88]   R. M. Newman, Z. L. Budrikis, and J. L. Hullett. The QPSX MAN. *IEEE Communications Magazine*, 26(4), 1988.

[NIS89]   User guide for the NIST prototype compiler for Estelle. Report No. ICST/SNA - 87/4, U.S. Department of Commerce, National Institute of Standards and Technology, February 1989.

[OP90]   S. W. O'Malley and L. L. Peterson. A highly layered architecture for high-speed networks. In *Second IFIP WG6.1/WG6.4 International Workshop on Protocols For High-Speed Networks, Participant's Proceedings*, 1990.

[RHF90]   F. E. Ross, J. Hamstra, and R. Fink. FDDI - a LAN among MANs. *ACM Computer Communication Reviews*, 20(3):16–31, 1990.

[Ros90]   M. T. Rose. *The ISO Development Environment: User's Manual*. Performance Systems International, Inc., February 1990.

[Rup91]   M. Rupprecht. *Implementierung und parallele Verarbeitung von Kommunikationssoftware*. PhD thesis, RWTH Aachen, 1991.

[SD89]   O. Spaniol and A. Danthine, editors. *High Speed Local Area Networks*. North Holland, Amsterdam, 1989.
[Spa91]   O. Spaniol. Measurements and experiences of a large FDDI installation. In *Proc. 3rd IFIP Conference on High-Speed Networking*. North Holland, 1991.
[Svo89a]   L. Svobodova. Implementing OSI systems. *IEEE Journal on Selected Areas in Communications*, 7(7):1115–1130, September 1989.
[Svo89b]   L. Svobodova. Measured performance of transport service in LANs. *Computer Networks and ISDN Systems*, 18(1):31–45, 1989.
[UD90]   R. Ulrich and H. Dietsch. A Transputer-based communication controller for FDDI stations. In *EFOC/LAN 90*, München, Juni 1990.
[Zit89]   Martina Zitterbart. High-speed protocol implementations based on an multiprocessor-architecture. In *Protocols for High-Speed Networks*, pages 151–163. IFIP WG 6.1/WG 6.4 International Workshop on Protocols for High-Speed Networks, Zürich, North Holland, May 1989.
[Zit90]   Martina Zitterbart. *Funktionsbezogene Parallelität in transportorientierten Kommunikationsprotokollen*. PhD thesis, Universität Karlsruhe, 1990.

CDM - korrekter Entwurf von Kommunikationssoftware

J. Freudenmann
Universität Karlsruhe
Institut für Telematik
Zirkel 2, 7500 Karlsruhe 1

E-Mail: Freudenmann@
telematik.informatik.uni-karlsruhe.[dbp.]de

1. Einleitung

Bedingt durch fallende Hardwarepreise und höheren Vernetzungsgrad der eingesetzten Rechner steigt der Bedarf an Software zur Unterstützung der Kooperation und Kommunikation dieser Rechner untereinander stark an. Dabei müssen an diese Software, als Eckpfeiler des gesamten verteilten Systems, besondere Anforderungen gestellt werden. Sie muß vollständig und präzise beschrieben sein und die Implementierungen müssen ebenfalls korrekt arbeiten um die gesamte Funktionalität des verteilten Systems zu erhalten.

Aus diesen Gründen ist beim Entwurf von Kommunikationssoftware besondere Sorgfalt und damit der Einsatz von formalen Methoden notwendig. Im Gegensatz zum Bereich der sequentiellen Programmierung (vgl. VDM /BJJ82/, CIP /BMP89/), sind im Bereich der nebenläufigen Systeme noch keine industriell einsetzbaren Entwurfssysteme bekannt. In der Forschung werden mehrere Ansätze, z.B. zur automatischen Implementierung von Protokollen /SIB90/ oder zur Verfeinerung von LOTOS-Spezifikationen /MMT89/ verfolgt.

Innerhalb der nebenläufigen Systeme bieten sich Kommunikationssysteme als Forschungsobjekt deshalb an, weil hier bereits, bedingt durch die Notwendigkeit internationaler Zusammenarbeit, Vorarbeit bei der formalen Beschreibung der Systeme geleistet wurde, s. LOTOS, Estelle und SDL /HOG89/. Im vorliegenden Ansatz sollen auf der Basis solcher formalen Beschreibungen formale Entwurfsmethoden angewandt werden. Ausgangspunkt des Entwurfs bilden Spezifikationen von einzelnen Protokollinstanzen, die entsprechend dem ISO/OSI-Basisreferenzmodell /EFF86/ zu Kommunikationssystemen strukturiert und zusammengefaßt werden. Das Ziel des Entwurfs bildet eine effiziente, korrekte Implementierung dieser Spezifikation.

Die dazu eingesetzten Konzepte und die implementierten Komponenten werden unter dem Kürzel CDM-System (Constructive Design Method) zusammengefaßt. Der vorliegende Artikel gibt einen Überblick, eine detaillierte Beschreibung findet sich in /FRE91/. Ein früher Abriß findet sich in englischer Sprache in /FRE87/.

Grundlage des hier vorzustellenden Ansatzes ist die formale Beschreibungssprache CSM (Constructive Specification Method). Sie basiert auf erweiterten endlichen Automaten. Dem Entwurf liegt das Konzept der schrittweisen Verfeinerung zugrunde. Ausgehend von Protokollspezifikationen einzelner Instanzen wird Schritt für Schritt eine Implementierung abgeleitet. Die Korrektheit der Implementierung wird garantiert durch die Definition von Regeln zur Durchführung der Verfeinerungsschritte. Diese Regeln garantieren die Einhaltung der sogenannten Verfeinerungsrelation "v=>". Eine Implementierung ist dann korrekt gegen die Spezifikation, wenn die beiden Beschreibungen die Verfeinerungsrelation erfüllen.

Im folgenden werden die Beschreibungssprache CSM, die Verfeinerungsregeln und das Werkzeugsystem zur Unterstützung der Anwendung der Methode vorgestellt.

2. Die Beschreibungssprache CSM

Der Definition von CSM liegen folgende Überlegungen zugrunde:

- Es muß eine durchgängige Beschreibung von abstrakten Protokollspezifikationen bis hin zu Implementierungen möglich sein.

- Die Sprache soll sich an allgemein gebräuchlichen Konzepten orientieren.

CSM erfüllt diese Anforderungen durch Anlehnung an das Konzept der erweiterten endlichen Automaten, eine prozedurale Schnittstelle und durch Steuerkonstrukte zur Beschreibung des Steuerflusses in Implementierungen. Den Anforderungen bzgl. der Korrektheit des Entwurfs wird dadurch Rechnung getragen, daß bei der Definition von Datentypen und Operationen ein fester Satz als Basis zur Verfügung gestellt wird und die Konstruktoren für komplexere Typen und Operationen eingeschränkt werden. Z. B. sind als Operationen nur seiteneffektfreie Funktionen erlaubt. Als Konstruktoren sind Reihungen, Felder und Listen vorgesehen. Die Zustandsübergänge sind bei CSM verknüpft mit der Bearbeitung von Schnittstellenprozeduren. Zu jeder Prozedur sind vier Übergänge definiert, einer zum Aufruf der Prozedur, einer zur Bearbeitung der Prozedur (Aktivierung), einer zum Ende der Bearbeitung der Prozedur (Terminierung) und einer zur Rückkehr der Prozedur.

Zur Verdeutlichung des Konzepts ist in Abbildung 1 die abstrakte Protokollspezifikation der Sendeinstanz des 'alternating bit'-Protokolls schematisch dargestellt.

Die Instanz AB-Send bietet Instanzen höherer Schichten eine Operation *Upp*-Send mit einem Eingabeparameter Pm, einer Nachricht, an. Sie verwendet zwei Operationen der unterliegenden Schicht, i.e. *Lop*-Sm, eine Operation zum Senden einer Nachricht Plm mit Kontrollbit Plb und *Lop*-Ra, eine Operation zum Empfang eines Kontrollbits Plb mit Gestörtanzeige Plcr. Der unterliegende Dienst ist unsicher, garantiert aber das Erkennen von Fehlern und schließt den völligen Verlust einer Nachricht aus.

Der Zustandsraum der Instanz wird gebildet durch die drei Variablen *Var*-Cb, S und M. Cb speichert das aktuell gültige Kontrollbit, S den Bearbeitungsstand und M speichert die aktuell zu übertragende Nachricht. Die Variablen werden mit den angegebenen Werten initialisiert.

Nach einem Aufruf der Operation Send wird die Aktivierungsaktion *Act* bearbeitet, wenn die Bedingung 'M=Empty' erfüllt ist. Dabei wird der Eingabewert Pm dem Nachrichtenpuffer zugewiesen. Nach der Aktivierung kann die Terminierungsaktion *Term* der Operation bearbeitet werden, die zugehörige Bedingung ist erfüllt. Die Bearbeitung verändert jedoch den Zustand nicht.

M ist nun nicht mehr leer, damit ist die Bedingung der Aufrufaktion *Call*-Sm erfüllt. Die Aufrufaktion wird bearbeitet, die Parameter werden mit Nachricht und Kontrollbit belegt, der Bearbeitungsstand S wird aktualisiert und die Operation Sm wird aufgerufen. Während der Bearbeitung von Sm sind keine weiteren Aktionen in der Instanz AB-Send möglich, keine Anwendbarkeitsbedingung ist erfüllt. Nach der Rückkehr der Operation Sm wird der Bearbeitungsstand durch die Rückkehraktion aktualisiert. S erhält den Wert Waitansw.

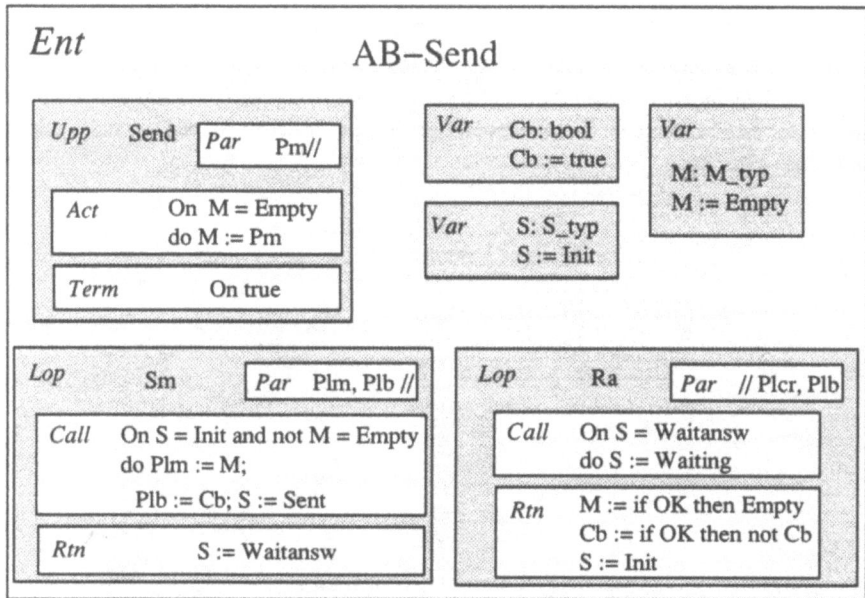

Abb. 1:   Abstrakte Protokollspezifikation

Damit ist die Bedingung der Aufrufaktion *Call*-Ra erfüllt, S erhält den Wert Waiting und die Operation Ra wird aufgerufen. Wieder sind keine weiteren Aktionen zu bearbeiten. Bei der Rückkehr wird, abhängig davon ob die Übertragung erfolgreich war, entweder die Nachricht M gelöscht und das Kontrollbit alterniert, oder die beiden Variablen werden nicht verändert. Im zweiten Fall wird ein neuer Übertragungsversuch angestoßen, denn die Variable S erhält den Wert Init.

Die Beschreibung von Implementierungen erfolgt in der Form einer Spezifikation des Softwaresystems. Sie legt die realisierungsorientierte Struktur der Instanzen fest. Im einzelnen sind zusätzlich zu definieren.

- die Struktur des Systems

- die Aufgaben der einzelnen Komponenten der Struktur

- die Schnittstellen der einzelnen Komponenten

- Algorithmen für die einzelnen Komponenten

Die Struktur eines Softwaresystems wird bestimmt durch die definierten Prozeduren, Prozesse und die Datenbasis auf denen diese arbeiten. In der CSM-Softwarespezifikation wird jede dieser Prozeduren und Prozesse durch einen Automaten repräsentiert. Die Datenbasis setzt sich aus einem globalen, allen Automaten gemeinsamen, und aus, den einzelnen Automaten exklusiv zugeordneten, lokalen Zustandsräumen zusammen.

Die Aufgaben der einzelnen Komponenten werden durch die jeweils definierten Zustandsübergänge beschrieben.

Die Schnittstellen der Komponenten können unterteilt werden in Schnittstellen zu Objekten außerhalb der Instanz, diese sind operational wie bei Protokollspezifikationen, und interne Schnittstellen zu anderen Komponenten der gleichen Instanz. Diese internen Beziehungen bestehen nur bezüglich des Zugriffs auf dieselben globalen Variablen.

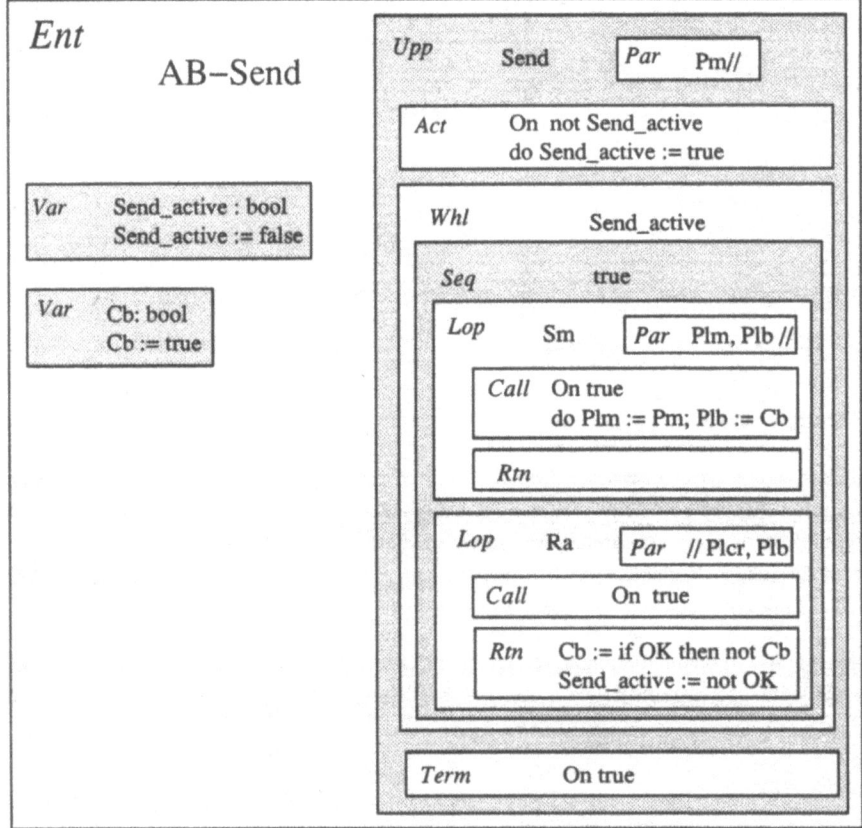

Abb. 2: Softwarespezifikation

Ein Algorithmus wird normalerweise in erweiterten Automaten nicht explizit festgelegt. Er ergibt sich als Folge von möglichen Zustandsübergängen. Damit wird im allgemeinen nicht ein einzelner Ablauf, sondern eine Menge möglicher sequentieller Abläufe beschrieben. CSM erlaubt es, diese Abläufe, den

Steuerfluß, für die einzelnen Komponenten explizit durch sogenannte Steuerkonstrukte zu beschreiben. Ein Steuerkonstrukt besteht aus einer Bedingung und einer Menge von weiteren Steuerkonstrukten oder Aktionen. Drei Typen von Steuerkonstrukten wurden definiert, das Schleifen-, das Auswahl- und das Sequenzkonstrukt. Diese Konstrukte entsprechen in ihrer Semantik den in Programmiersprachen üblichen Anweisungen ('while', 'case', Anweisungsfolge) zur Steuerung des Ablaufs. Weitere Konstrukte werden nicht definiert. Andere von Programmiersprachen her bekannte Anweisungen lassen sich aus den definierten Grundtypen zusammensetzen. In Abbildung 2 ist eine Variante einer Softwarespezifikation zu der in Abbildung 1 vorgestellten Instanz dargestellt. Andere Varianten unterscheiden sich z.B. durch Einführung eines Prozesses zur Übertragung der Nachricht.

Im Abbildung 2 wird die Übertragung der Nachricht von der Operation Send direkt mit übernommen. Ist keine andere Inkarnation der Prozedur Send aktiv, so wird die Bearbeitung der Prozedur begonnen. In einer Schleife (Whl-Konstrukt) wird eine Folge (Seq-Konstrukt) von Zustandsübergängen bearbeitet, die jeweils einer vollständigen Übertragung der Nachricht mit Erwarten der Quittung entspricht. War die Übertragung erfolgreich so wird die Bearbeitung der Operation abgeschlossen, andere Inkarnationen können nun auch bearbeitet werden (Send_active = false). Die aktuelle Inkarnation terminiert.

3. Entwurfsmethode CDM

Die Entwurfsmethode CDM (Constructive Design Method) beruht auf der Spezifikationssprache CSM und auf dem Konzept der schrittweisen Verfeinerung. Bei der schrittweisen Verfeinerung wird ausgehend von einer abstrakten Protokollspezifikation Schritt für Schritt durch Einführung von Implementierungsdetails eine entsprechende Implementierung, beschrieben durch eine Softwarespezifikation, entworfen. Jeder einzelne Schritt erfüllt einen bestimmten Zweck im Entwurf, i.e. entspricht einer Entwurfsentscheidung. Diese einzelnen Schritte lassen sich klassifizieren und zu jeder Klasse kann eine Transformationsregel definiert werden, die die Durchführung dieses Verfeinerungsschrittes beschreibt. Die Menge der Regelklassen kann aus beispielhaften Entwürfen und aus der Gegenüberstellung von abstrakten Protokollspezifikationen und Softwarespezifikationen abgeleitet werden. Als wesentliche Klassen von Entwurfsentscheidungen und damit Regelklassen ergeben sich

- Festlegen der Programmstruktur

- Zuordnung von Protokollfunktionen

- Verteilung des Protokollzustands

- Definition des Steuerflusses

In der abstrakten Spezifikation wird das Verhalten der Instanz an ihren Schnittstellen zur Umgebung festgelegt, von inneren Strukturen wird abstrahiert. Im ablauffähigen Softwaresystem entsprechen dieser Instanz eine Menge von Prozeduren und Prozessen, die zusammen die Aufgaben der Instanz bearbeiten. Je nach den Fähigkeiten des zugrundeliegenden Systems (Hardware, Betriebssystem, usw.) ergeben sich unterschiedliche Randbedingungen beim Entwurf und damit unterschiedliche Softwaresysteme. So sollte z.B. bei einer Implementierungsumgebung die Anzahl der Prozesse gering gehalten werden, während bei anderen effiziente Mehrprozeß- oder Mehrprozessorumgebungen und Interprozeßkommunikation genutzt werden können.

Entsprechend müssen auch die Protokollfunktionen auf eine unterschiedliche Zahl von Prozeduren und Prozessen verteilt werden. Beim Entwerfer stehen hier strukturelle Gesichtspunkte (z.B. Zusammenfassung gleicher Funktionen) im Vordergrund. Durch die definierten Regeln wird die Einführung von Verklemmungen vermieden.

Durch die Strukturierung und die Zuteilung der Funktionen wird hier im wesentlichen die Verteilung des Protokollzustands beeinflußt. Hier wird eine möglichst starke Dezentralisierung, i.e. Verteilung auf Prozeduren und Prozesse, angestrebt, da der Zugriff auf globale Variable in der Implementierung synchronisiert werden muß. Dies kann zu langen Wartezeiten und im Extremfall sogar zu Verklemmungen führen.

Sind für die einzelnen Prozeduren und Prozesse der lokale Zustandsraum und die zu bearbeitenden Protokollfunktionen definiert, so kann der interne Steuerfluß festgelegt werden. Hier werden zum einen implizit festgelegte Abläufe durch entsprechende Konstrukte explizit festgelegt. Zum anderen müssen Indeterminismen in der Bearbeitungsfolge der Protokollfunktionen aufgelöst werden. Die eingeführten Steuerkonstrukte werden (bei konventionellen Rechnern) in der Implementierung auf Anweisungen abgebildet, die den Programmzähler beeinflussen.

Zur Umsetzung dieser Entscheidungen wurde eine Reihe von Regeln definiert. Jede solche Regel besteht aus drei Komponenten: einem Eingangsschema, einem Ausgangsschema und einer Anwendbarkeitsbedingung. Die Beschreibung der Schemata erfolgt in der CSM-Syntax. Dabei werden Teile durch Schemavariablen repräsentiert, die bei der Anwendung der Regel mit konkreten syntaktischen Strukturen instantiiert werden. Die Bedingungen werden informell, zum besseren Verständnis des Anwenders, und formal in Form von logischen Ausdrücken angegeben.
Im Eingangsschema werden die syntaktischen Strukturen festgelegt, auf die die Regel anwendbar sein soll. Das Ausgangsschema definiert die Struktur, die sich als Ergebnis der Anwendung der Regel ergeben soll. In der Anwendbarkeitsbedingung sind Anforderungen formuliert, die erfüllt sein müssen, damit sich eine korrekte Verfeinerung ergibt. Diese Anforderungen betreffen syntaktische Strukturen der Spezifikation, Eigenschaften der statischen Semantik der Spezifikation (z.B. Variablensichtbarkeiten) und den Ablauf des spezifizierten Protokolls.

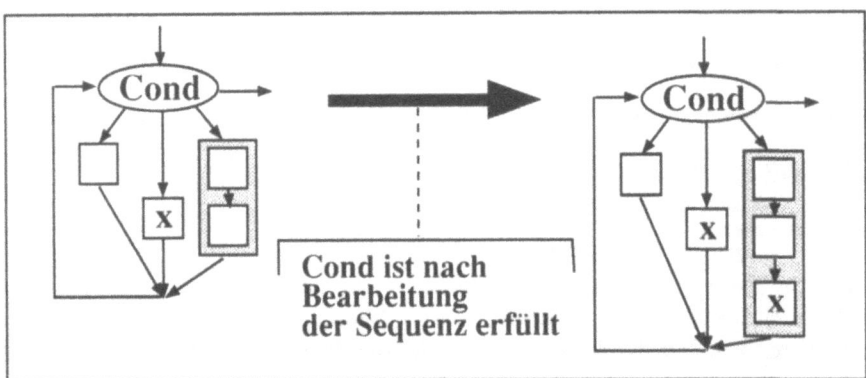

Abb. 3: Sequenzregel

Als Beispiel ist in Abbildung 3 eine Regel zur Definition des Steuerflusses schematisch dargestellt. Sie dient dazu bereits festgelegte Sequenzen zu verlängern. Im Eingangsschema der Regel ist ein Schleifenkonstrukt mit Bedingung Cond aufgeführt. Wahlweise kann eines der drei eingezeichneten Konstrukte bearbeitet werden. Rechts ist dabei ein bereits existierendes Sequenzkonstrukt hervorgehoben. Das mit x markierte Konstrukt ist zur Erweiterung der Sequenz vorgesehen.

Im Ausgangsschema ist die Erweiterung durchgeführt. Nach Bearbeitung der ersten beiden Komponenten der Sequenz wird sofort das Konstrukt x bearbeitet, ohne daß eine weitere Auswertung der Bedingung Cond notwendig wäre.

Die Anwendbarkeitsbedingung fordert dementsprechend, daß Cond in diesen Situationen immer erfüllt ist. Das Konstrukt x tritt jetzt zweimal auf. Aber durch entsprechende Regeln kann das alte Konstrukt , wenn es überflüssig geworden ist, gelöscht werden.

4. Korrektheit

Dem Aspekt der Korrektheit kommt im Entwurf große Bedeutung zu. Korrektheit bedeutet, daß das Ziel des Entwurfs nachgewiesenermaßen bestimmte Eigenschaften des Ausgangspunkts des Entwurfs immer noch besitzt. Der Nachweis solcher Eigenschaften setzt eine formale Grundlage voraus. Die Semantik der verwendeten Beschreibungssprache muß festliegen und die Entwurfssemantik muß formal faßbar sein und formal vorliegen. Die Semantik der Beschreibungssprache wird üblicherweise axiomatisch, denotational oder operational festgelegt. Bei der Entwurfssemantik kann man mehrere Klassen unterscheiden (vgl. auch /BSS86/), die Reduktionssemantik, die Erweiterungssemantik und die Verfeinerungssemantik.

Bei der Reduktionssemantik geht man davon aus, daß die Spezifikation alle möglichen Verhaltensweisen repräsentiert und daß zur Implementierung eine der Möglichkeiten gewählt wird. Sieht die Spezifikation etwa mehrere Alternativen vor, so wird eine davon ausgewählt. Das nach außen gezeigte Verhalten des speziellen Systems wird im Laufe der Verfeinerung eingeschränkt, d.h. bestimmte Verhaltensmuster treten nicht mehr auf.

Bei der Erweiterung geht man davon aus, daß im Verlauf des Entwurfs zusätzliche Verhaltensweisen eingeführt werden. Dies wird beispielsweise notwendig, wenn ein System nur teilweise spezifiziert wurde. Die zusätzlichen Verhaltensweisen müssen allerdings konsistent sein gegenüber den bereits vorhandenen. Das nach außen gezeigte Verhalten einer speziellen Spezifikation wird erweitert, d.h. es treten zusätzliche Verhaltensmuster auf.

Bei der Verfeinerung bleibt das nach außen sichtbare Verhalten unverändert. Lediglich die Strukturen und Abläufe im System werden verändert. Die beiden Beschreibungen, Spezifikation und Implementierung, legen ihren Schwerpunkt auf unterschiedliche Aspekte des Systems.

Im vorliegenden Ansatz wird die Reduktionssemantik als Basis für den Entwurf gewählt. Das Ziel des Entwurfs ist es zwar, eine Verfeinerung im Sinne der Verfeinerungssemantik zu erreichen, diese schränkt aber den Spielraum zu sehr ein. Insbesondere wäre hier die Auflösung von Indeterminismen durch Auswahl genau einer Bearbeitungsfolge nicht möglich.

Bei der Reduktion sind allerdings bestimmte Bedingungen zu beachten, so dürfen zum einen Bearbeitungsfolgen nicht einfach abgebrochen werden (Einführung von deadlocks), zum anderen muß garantiert sein, daß alle von extern an die Systemschnittstelle herangetragenen Ereignisse (Stimuli) bei der Bearbeitung auch berücksichtigt bleiben.

Basis der Definition der Semantik von CSM und der Semantik des Entwurfs bilden Historien. Historien sind definiert über Folgen von Ereignissen die an der Schnittstelle des Systems beobachtet werden können. Ereignisse sind dabei CSM-Aktionen, d.h. z.B. Aufruf und Rückkehr einer Prozedur. Für jede CSM-Spezifikation läßt sich die Menge der möglichen Historien angeben. Diese Menge repräsentiert die an der Schnittstelle relevante Semantik der Beschreibung.

Die Entwurfssemantik wird ebenfalls auf dieser Basis durch Definition der Verfeinerungsrelation "v=>" festgelegt. Zur Definition von "v=>" wird eine Historie h_S eines Systems S durch ein Zwei-Tupel dargestellt:

$$h_S = (h1; f)$$

h1 repräsentiert die Folge der aufgetretenen Ereignisse; f hat den Wert "ϵ", falls das System bei Bearbeitung der Folge h1 einen Zustand erreicht in dem kein weiteres Ereignis mehr auftreten kann. Sonst hat f den Wert "*". Bei der Berechnung von f wird auch das Verhalten der Umgebung berücksichtigt (dieses Verhalten wird in CSM-Dienstspezifikationen mit beschrieben).

Die Verfeinerungsrelation ist nun definiert über der Menge H_S aller möglichen Historien h_S eines Systems S.

$$S \text{ v=> } I \quad :<==> \quad H_I \subseteq H_S$$

Eine Implementierung I ist eine korrekte Verfeinerung einer Spezifikation S, wenn die Menge der Historien von I eine Teilmenge der Historien von S ist.

Die Verfeinerungsrelation ist transitiv, so daß sich im Fall der schrittweisen Verfeinerung aus der Korrektheit aller einzelnen Schritte auch die Korrektheit des gesamten Entwurfs ergibt.

Werden wie im vorliegenden Ansatz die einzelnen Verfeinerungsschritte nach vorgegebenen Regeln durchgeführt, so genügt es auch für die Regeln allgemein die Erhaltung der Verfeinerungsrelation nachzuweisen. Dieser Nachweis wurde geführt /FRE91/. Bei Einhaltung und ausschließlicher Verwendung der bewiesenen Regeln ergibt sich so insgesamt ein korrekter Entwurf. Daraus ist allerdings nicht zu schließen, daß nun überhaupt keine Überprüfungen mehr notwendig wären. Durch dieses Vorgehen werden lediglich allgemeine Aspekte behandelt; Aspekte, die mit dem speziellen Entwurfsschritt im speziellen Entwurfsstadium an einem bestimmten Ort zusammenhängen, werden vom Eingangsschema, dem Ausgangsschema und der Anwendbarkeitsbedingung der Verfeinerungsregel erfaßt und berücksichtigt. Die Anwendbarkeitsbedingungen müssen entwurfsbegleitend überprüft werden. Sie lassen sich drei Klassen zuordnen:

- syntaktische Anforderungen

- Anforderungen an die statische Semantik

- Anforderungen bezüglich des Ablaufverhaltens einer Instanz.

Syntaktische Anforderungen, betreffen Eigenschaften wie das Vorhandensein oder die Abwesenheit bestimmter syntaktischer Konstrukte. Ist die Beschreibung in der Form des Eingangsschemas einer Transformationsregel zu komplex, so können solche Eigenschaften in einer Bedingung formuliert werden.

Semantische Anforderungen betreffen das Vorhandensein von Definitionen für Variablen und Typen, sowie den Sichtbarkeitsbereich von Bezeichnern.

Anforderungen bezüglich des Ablaufverhaltens einer Instanz betreffen Eigenschaften des spezifizierten Systems, wie die Erreichbarkeit bestimmter Zustände und die Durchführbarkeit von bestimmten Zustandsübergängen.

Beim Entwurf eines Werkzeugsystems ist zu berücksichtigen, daß die Überprüfung dieser Anforderungen weitgehend automatisch erfolgen muß.

5. Werkzeugsystem

Zur Unterstützung der Anwendung der CDM-Methodik wurde ein System von Werkzeugen entwickelt und implementiert. Das System ist in Abbildung 4 dargestellt.

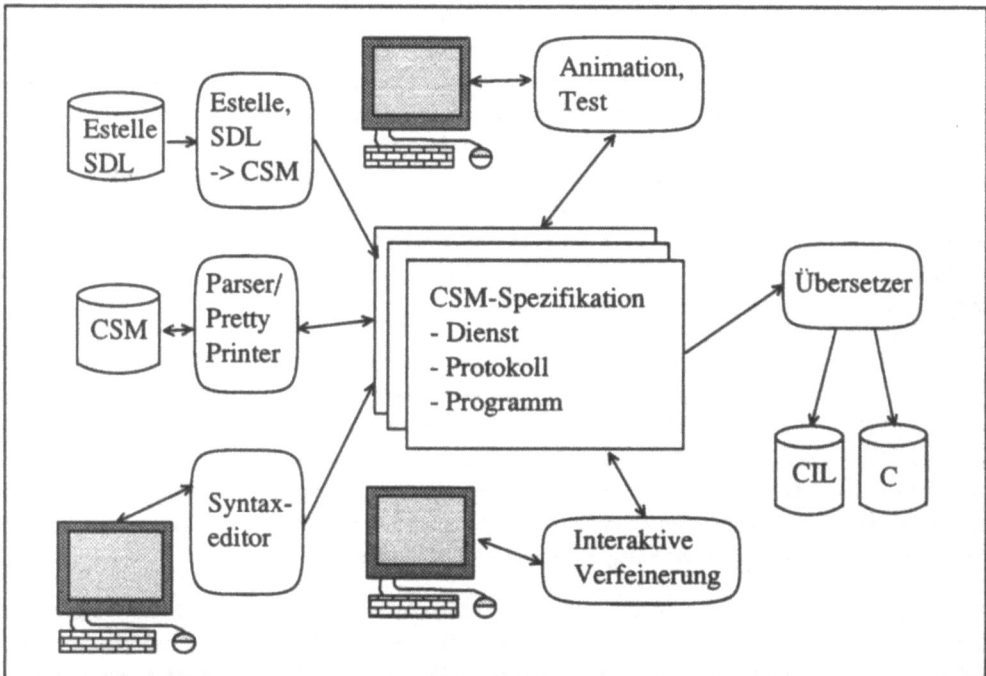

Abb. 4:   Komponenten des Werkzeugsystems

Im Mittelpunkt stehen die verschiedenen Typen von Spezifikationen. Hier werden Dienst- und Protokollspezifikationen von CSM unterschieden. Programm- oder Softwarespezifikationen sind spezielle Ausprägungen von Protokollspezifikationen. Diese Spezifikationen werden rechnerintern in einer Baumstruktur repräsentiert, auf welche die rundum gruppierten Werkzeuge aufbauen. Links sind Werkzeuge zur Erstellung von Spezifikationen aufgeführt; oben das Werkzeug zur Überprüfung von Spezifikationen; unten das Werkzeug zur interaktiven Verfeinerung (der Kern des ganzen Systems); rechts das Werkzeug zur Erzeugung von C oder CIL-Code (Comm. Services Implementation Language /KRU84/).

Zur Erstellung von CSM-Spezifikationen können bereits existierende formale Spezifikationen in Estelle oder SDL verwendet werden. Ein Generatorprogramm erzeugt automatisch eine äquivalente Darstellung in CSM. Probleme ergeben sich dabei hauptsächlich bei der Umsetzung der Datentypen. Für abstrakte Datentypen in SDL läßt sich z.B. nicht immer eine praktikable Darstellung in CSM automatisch ableiten, hier muß der Entwickler eingreifen.

Bereits erstellte CSM-Spezifikationen können auf externen Speichermedien gesichert werden. Es steht ein entsprechender Parser zum Einlesen und Analysieren dieser Spezifikationen zur Verfügung. Ein Pretty Printer dient zur strukturierten Ausgabe dieser Spezifikationen sowohl auf externe Speicher als auch am Bildschirm.

Zur direkten Erstellung von CSM-Spezifikationen steht ein syntaxorientierter Editor zur Verfügung, der beim Editieren bereits für die Einhaltung der Syntax sorgt und damit einige Überprüfungsdurchgänge einspart.

Ein wichtiges Problem stellt die Korrektheit der ersten formalen Beschreibung dar. Neben der Einhaltung gewünschter Eigenschaften /KRU90/, wie z.B. der Verklemmungsfreiheit, ist die Übereinstimmung der Beschreibung mit der Intention des Entwerfers eine wichtige Forderung. Hierzu und auch zur Prüfung des Entwurfsergebnisses auf Übereinstimmung mit der Intention des Entwicklers wurde ein Werkzeug zur Animation von CSM-Spezifikationen entworfen. Der Ablauf von CSM-Spezifikationen im verteilten System wird dabei simuliert und detailliert am Bildschirm einer Workstation dargestellt. Die Steuerung des Ablaufs erfolgt dabei automatisch oder durch Benutzervorgaben. Mit in das Werkzeug integriert sind umfangreiche 'debugging'-Möglichkeiten. So kann der interne Zustand jederzeit abgerufen und verändert werden. Das Verhalten der Systemumgebung kann vorgegeben, ergänzt und während der Bearbeitung abgeändert werden. Konsistenzbedingungen können vorgegeben werden; ihre Einhaltung wird während des Ablaufs überprüft.

Das Verfeinerungswerkzeug erlaubt die interaktive Durchführung von einzelnen Verfeinerungsschritten. Das Werkzeug stellt dem Entwickler dabei die einzelnen Verfeinerungsregeln über Kommandos zur Verfügung. Nach Vorgabe des Anwendungsortes wird die Anwendbarkeitsbedingung der Verfeinerungsregel vom Werkzeug geprüft. Verläuft die Prüfung zufriedenstellend, so wird die Spezifikation, entsprechend dem Ausgangsschema der Verfeinerungsregel, abgeändert. Zusätzliche Funktionen des Werkzeugs dienen der Protokollierung und der Information des Anwenders. Ebenso sind Funktionen zur automatischen Auswahl und Anwendung von einigen Regeln vorhanden.

Entsprechend den drei Klassen von Anwendbarkeitsbedingungen werden unterschiedliche Verfahren zu deren Überprüfung angewandt. Syntaktische Anforderungen werden durch eine Untersuchung des Strukturbaums geprüft. Anforderungen an die statische Semantik werden durch Führen von Sichtbarkeitslisten und andere aus dem Übersetzerbau bekannte Techniken überprüft. Die Anforderungen bezüglich des Ablaufverhaltens der Instanz werden auf der Basis einer vereinfachten Erreichbarkeitsanalyse und von algebraischen Techniken zur Reduktion von Termen geprüft /FRE91/.

Nach erfolgter Verfeinerung wird die resultierende CSM-Softwarespezifikation in C-Code übersetzt. Hierzu steht das Übersetzungswerkzeug zur Verfügung. Diese Umsetzung erfolgt vollständig automatisch. Der erzeugte Code ist betriebssystemspezifisch (DEC/VMS). Ein weiteres Übersetzerwerkzeug erzeugt CIL-Code. CIL steht für Communication Software Implementation Language; CIL wurde am Institut für Telematik der Universität Karlsruhe entwickelt /KRU84/ und stellt eine flexible Basis zur weiteren Bearbeitung der CSM-Softwarespezifikationen dar. Insbesondere ist die Erzeugung unterschiedlicher Zielsprachen möglich. Das Übersetzerwerkzeug ist nicht beschränkt auf CSM-Softwarespezifikationen, allgemein ist die Bearbeitung von CSM-Protokollspezifikationen möglich. Damit kann, im Sinne eines 'Rapid Prototyping', bereits früh im Entwurf ein erster Prototyp erstellt werden.

Im Laufe eines Projekts fallen viele verschiedene Dokumente, Spezifikationen und Protokolle an. Mehrere Entwickler arbeiten evtl. an einem Projekt und müssen koordiniert werden. Zur übersichtlichen Verwaltung dieser Dokumente und der Entwickler wurde eine Projektumgebung für CSM entworfen. Der Entwicklungsstand wird graphisch an einer Workstation aufgezeigt. Die jeweils möglichen, bzw. notwendigen Operationen werden ebenfalls graphisch angezeigt. Durch Menüauswahl oder durch 'Anklicken' eines 'Button' werden die beschriebenen Werkzeuge zur Durchführung der entsprechenden Operationen aktiviert.

6. Ausblick

Die vorstehend beschriebene Entwurfsmethode und die zugehörigen Werkzeuge wurden an einigen Protokollen aus der Praxis der Telematik erprobt. Die Untersuchungen konzentrierten sich auf die Flexibilität des Verfahrens, Qualität und Aufwand des Entwurfs und auf das Zusammenspiel mehrerer Entwürfe.

Speziell an 'alternating bit' und HDLC wurde ausführlich die Realisierung von unterschiedlichen Implementierungsvarianten untersucht. Hier hat sich die universelle Einsetzbarkeit gezeigt. Alle erdachten Implementierungskonzepte lassen sich auch realisieren. Insbesondere sind z.B. Unterscheidungen 'Realisierung durch eine Prozedur' oder durch einen eigenständigen Prozeß oder auch unterschiedliche Optimierungsstrategien (z.B. Effizienz beim Speicherverbrauch versus Effizienz bei der Bearbeitungszeit) möglich.

Am Beispiel des ISO-internetwork Protokolls wurden herkömmlicher Entwurf und die CDM-Methodik miteinander verglichen. Es zeigt sich dabei, daß der Entwurfsaufwand durchaus vergleichbar ist. CDM schneidet jedoch in Bezug auf die Qualität deutlich besser ab. Der Entwurf erfolgt in CDM mehr geregelt. Entwurfsentscheidungen werden bewußt gemacht, dadurch werden Entwurfsalternativen auch deutlich.

Am Beispiel eines Nachrichtenübermittlungssystems (X.400) wurde die Entwicklung eines komplexeren Systems, das sich auch über mehrere Schichten erstreckt, untersucht. Die unterschiedlichen Instanzen des Systems können hier unabhängig voneinander entwickelt werden. Ihr Zusammenspiel bei Animation und endgültiger Implementierung ist gewährleistet.

Während all dieser Untersuchungen hat sich auch das Animationswerkzeug als sehr hilfreich erwiesen. Es trug entscheidend zum Verständnis der Beschreibungssprache CSM und der definierten Protokolle und Dienste bei. Insbesondere konnten hierdurch bereits sehr früh einige Fehler in der Spezifikation erkannt werden.

In einigen Bereichen haben sich Ansätze zur Verbesserung des Systems gezeigt. Hier sind insbesondere die Datentypen, Optimierungen über Schnittstellen hinweg und die Auswahl der Verfeinerungsregeln zu nennen.

Für CSM wurde, zur Erleichterung der Überprüfung der Anwendbarkeitsbedingungen, nur ein kleiner Satz von Datentypen und Operationen vorgesehen. Für größere Spezifikationen ist dieser Satz nicht ausreichend und führt hier zu unnötig komplexen Datentypdefinitionen, Operationsdefinitionen und Ausdrücken. Dadurch und durch die notwendige abstrakte Beschreibungsform wird auch der erzeugte Zielcode relativ umfangreich. Hier kann zur Zeit nur durch manuellen Eingriff eine effiziente Realisierung der Operationen erreicht werden. Für weitere Arbeiten muß dieser Satz deshalb (z.B. um 'Strings' und 'Pointer') erweitert werden.

Wünschenswert wäre eine größere Flexibilität im Entwurf derart, daß einmal anfangs festgelegte Schnittstellen auch direkt durch bestimmte Regeln verändert werden können. So könnten Komponenten, die über Schnittstellen hinweg stark kommunizieren, in einer Komponente zusammengeführt werden, ohne die evtl. konzeptionell sinnvolle Trennung auf der Ebene der Protokollspezifikation aufzuheben.

Die für CDM definierten Verfeinerungsregeln sind, um den Satz an Regeln möglichst gering zu halten, allgemein gefaßt. Dies führt dazu, daß im Entwurf einige einfache Schritte durchgeführt werden, deren Durchführung eigentlich auf der Hand liegt oder sich direkt aus der vorher angewandten Regel ergibt. Zudem lassen sich auch bestimmte Entwurfsheuristiken angeben, die auch automatisch angewandt werden können. Arbeiten in dieser Richtung wurden bereits in Angriff genommen /BEY91/. Hier kommen Techniken der Expertensysteme zum Einsatz.

Literatur

/BJJ82/   D. Bjoerner, C.B. Jones; Formal Specification and Software Development, Prentice-Hall, 1982.

/BEY91/   P. Beyer; Zielorientierter Entwurf korrekter Kommunikationssoftware; in /KOM91/, pp. 509-523.

/BMP89/   F.L. Bauer, B. Möller, H. Partsch, P. Pepper; Formal Program Construction by Transformations - Computer Aided, Intuition-Guided Programming; IEEE Trans. on Software Engineering, 15,2(1989), pp. 165-180.

/BSS86/   E. Brinksma, G. Scollo, C. Steenbergen; LOTOS Specifications, their Implementations and their Tests; in /PST86/, pp. 349-360.

/EFF86/   W. Effelsberg, A. Fleischmann; Das ISO-Referenzmodell für offene Systeme und seine sieben Schichten; Informatik-Spektrum, 9(1986), pp. 280-322.

/EVD89/   P.H.J. van Eijk, C.A. Vissers, M. Diaz (eds.); The Formal Description Technique LOTOS; North-Holland, Amsterdam, 1989.

/FRE87/   J. Freudenmann; Development of Communication Software by Stepwise Refinement; in /PST87/, pp. 391-404.

/FRE91/   J. Freudenmann; Transformation von Protokollspezifikationen in Kommunikationssoftware; VDI-Verlag, Düsseldorf, 1991.

/HOG89/   D. Hogrefe; Estelle, LOTOS und SDL, Springer-Verlag, Heidelberg, 1989.

/KOM91/   W. Effelsberg, H.W. Meuer, G. Müller (Hrsg.); Kommunikation in verteilten Systemen; GI/ITG-Fachtagung, Mannheim, Februar 1991, Springer, Heidelberg, 1991.

/KRU84/   H. Krumm; Spezifikation, Implementierung und Verifikation von Kommunikationsdiensten für verteilte DV-Systeme, Dissertation, Univeristät Karlsruhe, 1984.

/KRU90/   H. Krumm; Funktionelle Analyse von Kommunikationsprotokollen; Springer, Heidelberg, 1990.

/MMT89/   J.A. Manas, T. de Miguel, H. van Thienen; The Implementation of a Specification Language for OSI Systems; in /EVD89/, pp. 409-421.
/PST86/   Protocol Specification, Testing and Verification VI; Montreal, Canada, Juni 1986, North Holland, Amsterdam (1986).
/PST87/   H. Rudin, C. West (eds.); Protocol Specification, Testing and Verification VII; Zürich, Schweiz, Mai 1987, North Holland, Amsterdam (1987).
/SIB90/   D.P. Sidhu, T.P. Blumer; Semi-automatic Implementation of OSI Protocols; Computer Networks and ISDN Systems 18 (1989/90), pp. 221-238.

LOTOS Design Methodology
Based on ODP - Viewpoints

Andreas Vogel
Gesellschaft für Mathematik und Datenverarbeitung
FOKUS
Hardenbergplatz 2
D 1000 Berlin 15
vogel@fokus.berlin.gmd.dbp.de

Kurzinhalt

Im Open Distributed Processing - Basic Reference Model (ODP BRM) werden fünf Gesichtspunkte (viewpoints) zur Beschreibung verteilter Systeme definiert. Eine neuer Ansatz für eine Entwurfmethodik wird vorgestellt. Er umfaßt LOTOS als Ausdrucksmittel und eine Methodologie, die auf den ODP - Gesichtspunkten beruht.

Die vorhandenen Methoden für den Entwurf mit LOTOS und insbesondere die unterstützenden Algorithmen und Werkzeuge werden diskutiert. In Hinblick auf die Ableitung von abarbeitfähigen Code aus einer Spezifikation erscheinen diese allerdings nicht befriedigend.

Es werden Klassen von LOTOS Spezifikationen für die ODP Gesichtspunkte computation, engineering und technology definiert, die Beziehungen zwischen ihnen herausgearbeitet und mögliche Transformationsalgorithmen entworfen.

Weiterhin wird eine Abbildung vom Gesichtspunkt technology in C Code definiert und in einer Fallstudie durchgeführt. Ein Werkzeug ist dafür vorhanden und seine Arbeitsweise wird erläutert.

Abstract

The Open Distributed Processing - Basic Reference Model (ODP BRM) defines five viewpoints for the description of distributed systems. A new approach for a design methodology is outlined using the formal description technique LOTOS and the concepts of ODP viewpoints. The available methodology for the design with LOTOS and supporting algorithms and tools are investigated. With respect to the code generation from a specification they do not appear satisfying.

Classes of LOTOS specifications are defined, each of them related to a certain viewpoint. As a result, relationships between the different viewpoints are identified, serving to transform the design process from computation viewpoint to engineering viewpoint and from engineering viewpoint to technology viewpoint.

Beside this, a mapping from the technology viewpoint into code is defined and illustrated in a case study for which tool support is available.

1. Introduction

LOTOS is one of the Formal Description Techniques (FDTs) [1,16,17] standardized by ISO. It is based on process algebras [2] and abstract data types [3].

The Basic Reference Model of Open Distributed Processing (ODP BRM) [5,6] is intended to create an international standard for the design and realization of open distributed systems by both ISO and CCITT.

The aim of this paper is to combine the advantages of both the FDT LOTOS and the viewpoint approach of ODP. The benefits of LOTOS are seen as the following:

- its formal semantics
- its expressive power
- available tool support,

The advantages of ODP appear as:

- defined concepts used for the modelling of distributed applications, as given in the viewpoint approach
- the implicit design method given by this approach
- standardization of these concepts.

The suitability of LOTOS and the ODP concepts has already been investigated [7,10].

LOTOS appears to be a very powerful description technique. It is suitable for both the specification of abstract requirements and of implementations. For the design, the refinement process and the use of different styles [7,8] have been proposed, but it seems impossible to arrange a LOTOS specification in this scheme without additional knowledge. For example, the identification of a LOTOS process as the abstraction of a requirement, a resource, an object, etc. is not given. However, such kind of knowledge is required for both the code generation from LOTOS specifications and the combination of LOTOS and ODP viewpoints. This observation was made during the investigation and development of algorithms transforming LOTOS specifications into implementation-oriented representations, and is discussed in more detail in section 3.1.

The idea is to suit LOTOS to the ODP viewpoint of computation, engineering and technology by restricting its syntax (section 2). For these classes of LOTOS specifications the corresponding ODP concepts are related to behaviour expressions. In the scope of this paper are only the computational, engineering and technology viewpoint. The other two viewpoints should be objects for further study. They are overlapped by the question of how to transform informal requirements into formal ones. Furthermore, the possibility of transformations between specifications from different viewpoints is investigated in section 3.2.

In section 4, code generation from specifications in the technology viewpoint is described in a case study and the tool realizing this is explained.

Finally, an overview of further work is given. The paper is appendixed by EBNF definitions of the LOTOS restrictions according to the viewpoints and by the rules for the code generation given in the case study.

2. Restriction of LOTOS with respect to the ODP viewpoints

It is suggested to restrict LOTOS with respect to the three viewpoints of computation, engineering and technology. This specialization is only a restriction of the LOTOS syntax and leaves the LOTOS semantics unchanged.

Existing LOTOS styles (specifically constraint-oriented, resource-oriented) [7,8] are thus related to the different viewpoints, and additionally for each of them a particular syntax is proposed. These restrictions are the basis for the transformations in sections 3.

The different specification styles used for the various viewpoints are illustrated by an example. It is influenced by the proposed case studies of the ODP-BRM [8]. It shows the read and write access to a distributed database. The distribution of the database is given by two singular databases, one containing the entries from a to k and the other containing the ones from l to z.

2.1. The Computation viewpoint

The computation viewpoint is defined as the one where *"the programming functions (instantiation, assignment, invocation, synchronization, communication, etc.) and data types are visible. From this viewpoint the structuring of applications is independent of the computer systems and networks on which they run."* [5]

For implementation-independent specifications the constraint-oriented style is recommended [4,8]. In order to obtain a logical structured specification it is suggested to identify a constraint with a process, and compose them by the parallel operators (||| ||). Thereby the full synchronization parallel operator can be interpreted as the logical "and" - all of the communication partners have to synchronize - and the interleaving operator as the logical "or" - only one of the communication partners has to synchronize. A single constraint should be specified in a monolithic style by use of action prefix, the choice operator and recursion, corresponding to one of ODP's aspects.

A proposal for this kind of limitation on LOTOS is given in EBNF in Appendix A 1.

From the computation viewpoint the specification of the example appears as the conjunction of the constraints *user, manager* and the disjunction of the subconstraints *identification* and *data_base*.

    specification example[login, logout, read, write]: **noexit**
        library . . . **endlib**
        type global_type **is** . . . **endtype** (* global_type *)
    behaviour
        hide check, get, put **in**
        user[login, logout, read, write]
        ||
        manager[login, logout, read, write, check, get, put]
        ||

```
        (    identification[check]
   |||
             data_base[ get, put ](empty_db) )
```

The specification of the constraint *database* is given in more detail. The relationship to the ODP aspect storage is annotated in a special comment. From the computational viewpoint the distribution of the database is invisible. Its structure is organized monolithically.

```
process data_base[ get, put ](db: db_sort): noexit :=
(* ODP aspect storage *)
        reading[ get, put](db)
        []
        writing[ get, put](db)
where
        process reading[ get, put](db: db_sort): noexit :=
                get !request ?to_seek: ident;
             get !response !seek(to_seek, db);
                data_base[ get, put ](db)
        endproc (* reading *)
        process writing[ get, put](db: db_sort): noexit :=
                put !request ?to_write: entry;
             ( let res: write_sort = write( to_write, db ) in
             put !response !err(res);
                data_base[ get, put ](db(res)))
        endproc (* writing *)
endproc (* data_base *)
```

2.2. The Engineering viewpoint

The engineering viewpoint is defined as follows : *"The objects visible from the engineering viewpoint are transparency mechanisms, processors, memory and communications networks, that together enable the distribution of programmes and data."* [5]

For specifications containing information about the distribution of resources (objects) the resource-oriented style is recommended [4,8]. Resources can be identified by their structure as processors, memory and communications networks. A better identification of objects seems possible by using ODP aspects, but the relationship of aspects and viewpoints is still undefined in the current version of the ODP BRM. In a forthcoming ODP BRM a proper definition of their relation can be expected.

The transparency mechanism can be realized for the communication structure by use of the LOTOS hide-operator. The data types are split into interface specifications, realized as global data definitions, and definitions local to an object, represented by a LOTOS process. Global definitions must be visible for all communicating processes and they contain only syntax, i.e. sorts and operations. The local definition contains the semantics of the operations syntactically specified in the interface, and also the specification of operations which are not visible outside the object.

A single resource should be specified in a monolithic style (see above). The communication between resources should be expressed by the general parallel operator, |[g_1, ..., g_n]|. The resources can communicate directly:

    resource_1[...]
    |[...]|
    ...
    |[...]|
    resource_n[...]

or via an abstract medium:

    (resource_1[...]
    |||
    ...
    |||
    resource_n[...])
    |[g_1, ..., g_n]|
    medium[...]

where the medium manages the communication between the resources.

The multiple rendezvous mechanism should not be allowed.

A proposal for this kind of restriction of LOTOS is given in EBNF in Appendix A 2.

From the engineering viewpoint communicating objects are visible. Consequently the global data types are limited to the these sorts and operations which are visible at the interfaces of the objects.

    type global_type **is**
        sorts   ident, entry, operation
        opns request, response: -> operation
    endtype (* global_type *)

The database is now visible as an object which contains three subobjects namely the two singular databases and an interface managing the access to them.

    process data_base[get, put]: **noexit** :=
        hide g1, g0, p1, p0 **in**
        (**let** db_0: db_sort = empty_db, db_1: db_sort = empty_db **in**
        db[g0, p0](0, db_0)
        |||
        db[g1, p1](succ(0), db_1))
        |[g1, g0, p1, p0]|
        db_interface[get, put, g1, g0, p1, p0]
    where

```
type db_type is
    sorts db_sort
    opns empty_db: -> db_sort
endtype (* db_type *)
```

The singular databases themselves consist of objects managing the reading and writing, respectively. The definition of the object for reading is given here, the object for writing is analogous.

```
process db[ g, p ](n: Nat, db: db_sort): noexit :=
    reading[ g, p](n, db)
    []
    writing[ g, p](n, db)
where
    process reading[ g, p](n: Nat, db: db_sort): noexit :=
        g !request ?to_seek: ident;
        g !response !seek(to_seek, db);
        db[ g, p ](n, db)
    where
        type reading_type is db_type, global_type
            opns seek: ident, db_sort -> entry
        endtype (* reading_type *)
    endproc (* reading *)
```

The interface gets the requests and controls their transmission to the databases via selecting functions.

```
process db_interface[ get, put, g1, g0, p1, p0 ]: noexit :=
    get !request ?to_seek: ident;
    ( [ a_to_k( to_seek ) ] ->
        g1 !request !to_seek; g1 !response ?res: entry;
        get !response !res; db_interface[ get, put, g1, g0, p1, p0 ]
    []
    [ l_to_z( to_seek ) ] ->
        g0 !request !to_seek; g0 !response ?res: entry;
        get !response !res; db_interface[ get, put, g1, g0, p1, p0 ])
    []
        (* analogous for writing *)
where

    type db_interface_type is global_type, Boolean
        opns a_to_k, l_to_z: ident -> Bool
    endtype (* db_interface_type *)
endproc (* db_interface *)
```

2.3. The Technology viewpoint

The technology viewpoint is defined by *"the technical artifacts (realized components) from which the distributed system is built are visible. ... It must include the hardware and software that comprise the local operating systems, the input/output devices, storage, points of access to communications, etc."* [5]

Consequently, the specification from the engineering viewpoint has to be enriched by a description of the hardware and software used. Therefore the following concepts are introduced:

LOTOS Implementation
restricted LOTOS specification using an **implementation model**.

Implementation Model
the LOTOS description of the process abstraction and the communication between instances of processes given by the **implementation environment**.

Implementation Environment
the environment which the implementation is made for. It is given by a programming language, a process concept (given by operating systems, the programming language or workbenches) and communication facilities (predefined protocols or library routines).

Code
text of the programming language (given by the implementation environment).

First, it must be ensured that the use of events, their structure (value-variable-list), the composition of guards and choice-operators, etc. fit the structures given by the programming language.

Second, a model of the communication software/hardware used must be specified in LOTOS. Such LOTOS constructs are used to replace the communication expressed by the synchronization at gates in the specification from the engineering viewpoint. There exist two way to realize this.

The first way defines a process which is working as a channel between communicating objects. This communicating structure can be given by the following skeleton:

```
    obj_1[ send1, rec1 ]
|[ send1, rec1 ]|
    channel[ send1, rec1, send2, rec2 ]
|[ send2, rec2 ]|
    obj_1[ send2, rec2 ]
```

Alternatively, the communication is enclosed in a complex interaction point, according to the skeleton:

process object_with_communication_interface[*send_gates, receive_gates*]: *func* :=

    hide *in_gates, out_gates* **in**
    send_unit[*out_gates, send_gates*]
    |[*out_gates*]|
    obj[*in_gates, out_gates*]
    |[*in_gates*]|
    receive_unit[*in_gates, receive_gates*]
    where

```
            (* obj is the object defined in the specification  *)
            (* from the engineering viewpoint                  *)
```

    endproc (* object_with_communication_interface *)

An example of a class of specifications for the technology viewpoint is given in the Appendix A 3. The definitions depends on the implementation environment defined by the case study and on the abstraction yielding the implementation model. The structure of the specifications of the example class is not very complex with respect to the rapid development of a tool. However, the principles of the code generation process are shown.

3. Approaches to the Transformations between the ODP Viewpoints

The problem of such kind of transformations can be generalized to the question: how to transform requirements on a system into the implementation of the system? Algorithms in the context of LOTOS answering this question exist already.

3.1. Discussion of "Implementation Algorithms"

Design methods based on LOTOS suggest stepwise refinement for the development of an implementation-oriented specification from a more abstract one. However, this method does not solve the problem of the transformation from constraint-oriented to implementation-oriented specifications.

There is a tool-supported algorithm for the rearrangement of processes [4], but it is limited to Basic LOTOS. On the other hand, there are algorithms transforming large classes of full LOTOS specifications even into implementation-oriented representations. Examples are the algorithm behind the LOTOS compiler *topo* [4,13], mapping to the virtual code of the $\Lambda\beta$ machine, a virtual ring algorithm for distributed implementation of the multiple rendezvous [14], and a LOTOS-to-LOTOS transformation yielding an implementation-oriented specification [9].

On investigating these algorithms, the following faults were identified. The communication between parallel processes is mapped onto a protocol given by the algorithms. This protocol defines a communication structure which is fixed and can not be influenced by various implementation environments. Additionally for *topo*, the specified aspect of distribution of components and the communication structure between them is not taken into consideration. Hence the structures are flattened and the processes communicate via a "global synchronizer".

It is the author's opinion that such algorithms do not allow the generation of satisfying code, since the code no longer reflects the distribution and structural aspects once specified.

3.2. The Draft of Viewpoint Transformation Algorithms

Transformation from computation to the engineering viewpoint

In a specification from the computation viewpoint, constraints are seen as the smallest units considered for the transformation. The aim is to build out of these constraints the objects (resources), the building blocks in the engineering viewpoint. Both, constraints and objects are identified by the ODP concepts of the aspect. Consequently, an object is the composition of constraints according to the same aspect. Additionally, the algorithm needs information about the distribution of objects. A possible tool should establish this interactively with the systems designer, e.g. it could offer a menu of transformations.

Furthermore, the transformation of a constraint into a set of communicating objects can be treated as typically. This kind of transformation was used in the data base example. The behaviour equivalence of the constraint and their realization as objects communicating via an interface can be shown.

A more detailed description of the algorithm can be given according to the ongoing development of the ODP BRM. Currently the relationship of aspects and viewpoints is still open.

Transformation from the engineering to the technology viewpoint

This transformation is to be prepared by modelling the *implementation environment*, i.e. by description of the software and hardware used, especially for communication.

Subsequently, the communication expressed by simple LOTOS synchronization must be replaced by the corresponding parts of the above model (see section 2.1.).

Each *implementation environment* needs its own function *coding*. It defines the mapping from the LOTOS constructs of the technology viewpoint to the programming language given by the implementation environment and can be understood as the inverse of the abstraction process.

The complete description of the technology viewpoint is given by the composition of these three parts: the model of the software and hardware, the definition of the function *coding*, and the expanded LOTOS specification.

4. Code Generation from Specifications of the Technology Viewpoint

Code generation depends strongly on the given implementation environment. The principle are shown in a case study given by the following implementation environment:

Programming language
    C.

Process concept
    system processes of the Sun–OS® operating system [11].

Sun-OS is a registered trademark of Sun Microsystems, Inc.

Communication
sockets in the internet domain with a datagram protocol [12].

First the mapping algorithm is defined, then the tool realizing it is explained.

4.1. The Algorithm

The algorithm is given by the definition of the function *coding*. In principle, it maps from LOTOS specifications of the technology viewpoint into code of the programming language given by the implementation environment.

In this case study LOTOS behaviour expression are mapped into C code. External events are mapped into put and get operations, hidden events into read and write operations of the sockets. The data type definitions are mapped into templates of C function definitions according to their syntax. Their semantics must be hand-coded. Progress can be expected by the use of a data type compiler like DAFY [19].

The combination of guards and the choice operator is mapped into if statements. The recursion of processes is mapped into loops using the while statement. Each process instance is realized as an OS process. It defines its sockets and runs the programme actualized by its socket parameter, defined in process definition and included via a corresponding .h file.

The complete transformation rules are given in detail in Appendix B.

4.2. The Tool

The tool realizing the algorithm is deeply influenced by the toolset *lite* developed by the LOTOSPEHERE project [4]. It uses the metatool *kimwitu* and the *common representation*. This allows the connection of LOTOS behaviour expressions and the operations generating code. Such a pair of behaviour expression and operation represent a rule. Consequently, the rules for other implementation environments can be created easily.

The tool creates a lot of files, a pair of .c and .h files for the global data types, each of the local data files and each of the process definitions and furthermore a .c file for each process instance.

The data type files must be actualized, that means the function bodies of the specified operations must be handcoded. Furthermore, the makefile template must be actualized. The make creates executable files which are able to run on and to communicate from different machines.

5. Further Work

The further work will mainly consider the two aspects of refinement and expansion of the methodology, and a case study to prove the approach.

The specification of the ODP trader in LOTOS according the proposed methodology is selected as the case study. The ODP trader is specified from all five viewpoints, but only informally [15].

Obviously, the trajectory should be expanded to the enterprise and information viewpoints as well. The refinement of the methodology contains the introduction of new concepts, the detailed definition of the transformation algorithms, and the further development of tools realizing it. Currently, the relationship between the viewpoints and aspects is an open issue in the ODP BRM. However, ongoing with its development the strong use of the aspects is intended. The outlined transformation algorithms will be worked out and tools supporting them will be developed.

Furthermore, it seems to the author that the interface of LOTOS specifications is a problem for the approaches of code generation. For example, the LOTOS compiler *topo* allows only specifications without external gates, and enables the communication with the environment via special annotated comments. The compiler *l2c* has its own special interpretation for a subclass of external gates as input or output operations - that depend on "?" and "!" parameters. In order to solve the problem generally, an additional definition for the interface (external gates) to the International Standard [1] should be formulated and suggested.

Bibliography

[1] LOTOS - A formal description technique based on the temporal ordering of observational behaviour, ISO 8807, International Standard

[2] Milner,R.: Calculus of Communicating Systems, LNCS 92, Springer Verlag 1980

[3] Ehrig,H., Mahr,B.: Fundamentals of Algebraic Specifications, Springer Verlag 1986

[4] Task 1.1 - Task 3.3 2nd Year Deliverable, Lotosphere Consortium, 1990

[5] Basic Reference Model of Open Distributed Processing - Part II: Descriptive Model, ISO/IEC JTC1/SC21A N4888, Working Document

[6] Partial text of the Basic Reference Model of Open Distributed Processing - Part I: Overview & Part IV: User Requirements, ISO/IEC JTC1/SC21A N4886, Working Document

[7] Formalisms and Specification, ISO/IEC JTC1/SC21A N4887, Working Document

[8] Architectural Semantics for ODP, ISO/IEC JTC1/SC21, Revised Working Draft

[9] Vogel,A.: An Algorithm for Transforming LOTOS Specifications for Implementation, Technical Report of GMD #531, Berlin 1991

[10] Holz,E., Vogel,A.: An Example to Prove the Qualities of FDT's in the ODP-context, DIN-contribution, 1991

[11] SunOS Reference Manual, Sun Microsystems 1990

[12] Network Programming Guide, Sun Microsystems 1990

[13] Mañas,J., Salvachúa,J.: $\Lambda\beta$: a Virtual LOTOS Machine, FORTE'91 Sydney 1991

[14] Qiang Gao, von Bochmann,G.: A Virtual Ring Algorithm for the Distributed Implementation of the Multiple Rendezvous, Technical Report #675, University of Montreal, Dept. I.R.O.

[15] Working Document on Topic 9.1. - ODP Trader, ISO/IEC JTC1/SC21/WG7 N6084, CCITT SGVII.Q19 (DAF)

[16] Hogrefe,D.: Estelle, LOTOS, SDL, Springer Verlag 1989

[17] van Eijk, Vissers, Diaz (Editors): The Formal Description Technique LOTOS, North Holland 1989

[18] Lallemand,E., Leduc,G.: On LOTOS tools and their usefulness to treat large specifications, Rept. No. OSI95/Ulg/A/11/TR/R/V2, University of Liége, Dept. Systémes et Automatique

[19] Lallemand,E.: A LOTOS Data Facility Compiler (DAFY), Rept. No. OSI95/Ulg/A/06/TR/R/V1, University of Liége, Dept. Systémes et Automatique

Appendix A

The definitions are given in EBNF. Bold printed strings are word-symbols, strings included in quotes represent special-symbols and LOTOS-expression are printed in italics according to the syntax-definition of LOTOS [1].

1. Definition of the Class of Specifications from the Computation Viewpoint

The expression "comment-about-aspect" is meant as a special LOTOS-comment providing information on the ODP-aspect, accordingly to its constraint.

computation-viewpoint-spec ==
    specification *specification-identifier formal-parameter-list*
    global-type-identifier
    constraint-oriented-expression
    endspec

constraint-oriented-expression ==
    constraint
    | constraint *parallel-operator* constraint-oriented-expression
    | constraint-oriented-process
    | declaration-expression constraint-oriented-expression

declaration-expression ==
    local-definition-expression
    | *hiding-expression*

constraint-oriented-process ==
    process-ident
    where
    process *process-identifier formal-parameter-list* ":="
        constraint-oriented-expression
    [**where** *data-type-definitions*]
    endproc

constraint ==
    process-ident
    where
    process *process-identifier formal-parameter-list* ":="
        comment-about-aspect
        sequential-expression
    [**where** *data-type-definitions*]

```
        endproc
sequential-expression ==
    guard-expression
    | stop
    | exit
    | constraint
    | sequential-expression sequential-operator sequential-expression
sequential-operator ==
    ";"
    | "[]"
    | "[>"
    | ">>"
```

2. Definition of the Class of Specifications from the Engineering Viewpoint

```
engineering-viewpoint-spec ==
    specification specification-identifier formal-parameter-list
    global-type-definitions
    object
    where
    process-definitions
    endspec
object ==
    process-instance
    | object parallel-operator object
    | hiding-operator object
process-definitions ==
    process-definition [ process-definitions ]
process-definition ==
    process-symbol process-identifier formal-parameter-list definition symbol
    definition-block
    end-process-symbol
definition-block ==
    single-object
    | composed-object
single-object ==
    behaviour-expression [ data-type-definition ]
composed-object ==
    object [ local-definitions ]
```

behaviour-expression ==
    local-definition-expression
    enable-expression

enable-expression ==
    disable-expression [*enable-operator* enable-expression]

disable-expression ==
    choice-expression [*disable-operator* disable-expression]

3. Example for a Definition of the Class of Specifications from the Technology Viewpoint

The tool l2c processes only a special class of implementation-oriented specifications. This class is defined in the following.

technology-viewpoint-spec ==
    specification *specification-identifier formal-parameter-list*
    global-type-definitions
    proc-instance-composition-expression
    where
    process-definitions
    endspec

proc-instance-composition-expression ==
    process-instance
    | proc-instance-composition-expression *parallel-operator* proc-instance-composition-expression
    | *hiding-operator* proc-instance-composition-expression

process-definitions ==
    process-definition [process-definitions]

process-definition ==
    process-symbol process-identifier [gate-parameter-list] ":" **noexit** ":="
    definition-block
    endproc

definition-block ==
    behaviour-expression [*data-type-definition*]

behaviour-expression ==
    choice-expression
    | action-prefix-expression
    | atomic-action

choice-expression ==
    guarded-expression "[]" behaviour-expression

guarded-expression ==
    "[" *operation-identifier [value-expression-list]* "]" "->" action-prefix-expression

action-prefix-expression ==
    action-denotation ";" action-prefix-expression
    | atomic-action

atomic-action ==
    stop
    | process-instance

process-instance ==
    process-identifier [gate-tuple]

action-denotation ==
    gate-identifier communication

communication ==
    "?" *identifier-declaration*
    | "!" value-expression

value-expression ==
    value-identifier
    | *operation-identifier* [value-expression-list]

value-expression-list ==
    "(" value-expression { "," value-expression } ")"

Furthermore there are some restrictions on the static semantic. The recursion inside a process definition is allowed only with the identifier of the enclosing process definition and with the same gate identifiers.

Appendix B
Transformation Rules for Code Generation

actions

LOTOS-behaviour-expression	C-code
stop	exit()
external_gate !value;	puts(value);
external_gate ?var: type;	gets(var);

hidden_gate !value;	sendto(hidden_gate_write, value, BUF_SIZE, 0,
	    (struct sockaddr *)&hidden_gate_write_name,
	    sizeof hidden_gate_write_name);

hidden_gate ?var: type;	read(hidden_gate _read, value, BUF_SIZE);

sequential operators

LOTOS-behaviour-expression	C-code

[$guard_1$] -> B_1
[]
[$guard_2$] -> B_2
	if ($guard_1$)
	  { $coding(\ B_1\ )$ }
	if ($guard_2$)
	  { $coding(\ B_2\ )$ }

process definitions and instantiations

The restricted process definition is:

    process *proc-ident* [g_1, ..., g_n]: **noexit** :=
        B
    endproc

The function coding creates two files:
*proc-ident*_def.h

    #include "sock.h"
    int *proc-ident*_def();

*proc-ident*_def.c

    #include "*proc-ident*_def.h"
    #include "*proc-ident*_data.h"
    int *proc-ident*_def(g_1_read, ..., g_n_read, g_1_write, ..., g_n_write,
        g_1_read_name, ..., g_n_read_name, g_1_write_name, ..., g_n_write_name)
    /* definition of the sockets, two for each gate */
    int g_1_read, ..., g_n_read,
        g_1_write, ..., g_n_write;
    /* definition of the according addresses of the sockets */
    struct sockaddr_in g_1_read_name, ..., g_n_read_name, g_1_write_name, ..., g_n_write_name;
    {
    /* the main loop for the noexit process */
    while(1) {
    /* the behaviour B is scanned and the local variables defined by the
        let expression and the "?" operator are defined in the place
        of this comment
    */
        coding(B)
    }}

A process instance of the process *proc-ident*[h_1, ... h_n] creates the file *proc-ident*_<i>.h where <i> is the number of current process instances:

    #include "*proc-ident*_def.h"
    #include "*proc-ident*_data.h"

```
main() {
    /* definition of the actual gate parameters as sockets */
    int proc-ident_def( h_1_read, ..., h_n_read, h_1_write, ..., h_n_write,
        h_1_read_name, ..., h_n_read_name, h_1_write_name, ..., h_n_write_name )
    int h_1_read, ..., h_n_read, h_1_write, ..., h_n_write;
    struct sockaddr_in h_1_read_name, ..., h_n_write_name,
                       h_1_read_name, ..., h_n_write_name;

    /* creation of the sockets by predefined function, imported from sock.h */
    create_active_sock( &h_1_read, &h_1_read_name );
    ...
    create_active_sock( &h_n_read, &h_n_read_name );
    create_passive_sock( &h_1_write, &h_1_write_name );
    ...
    create_passive_sock( &h_n_write, &h_n_write_name );

    /* call of the function proc-ident_def, defined in the file proc-ident_def.h */
    proc-ident_def( g_1_read_name, ..., g_n_read_name,
                    g_1_write_name, ..., g_n_write_name ) }
```

data type definitions

For the global data types, the interface specification, two files are generated. The file *specification-ident*_data.h contains for each sort s_i a C-type definition:

   type *C-type* s_i;

and for each operation:

   $opn_i: s_1, ..., s_n \rightarrow s_{res}$

a C-function declaration:

   s_{res} opn_i();

The definition of equations is still ignored.

The file *specification-ident*_data.c contains for each sort the corresponding function definitions:

   s_{res} opn_i(arg_1, ..., arg_n)
       s_1 arg_1;
       ...
       s_n arg_n; {

       /* hand coding */
   }

For the local type definitions equivalent files are generated, but the filenames are called according to the *proc-ident* of the enclosing process definition.

An introduction to compositional methods for concurrency and their application to real-time*

J.J.M. Hooman

Dept. of Mathematics and Computing Science
Eindhoven University of Technology, P.O. Box 513, 5600 MB Eindhoven, The Netherlands
e-mail: wsinjh@win.tue.nl

W.P. de Roever

Institut für Informatik und Praktische Mathematik II
Christian-Albrechts-Universität zu Kiel, Preusserstrasse 1-9, 2300 Kiel 1, Germany
e-mail: wpr@informatik.uni-kiel.dbp.de

Abstract. Formal methods to specify and verify concurrent programs with synchronous message passing are discussed. We stress the development towards compositional methods, i.e. methods in which the specification of a compound program can be inferred from specifications of its constituents without reference to the internal structure of those parts. Compositionality enables verification during the process of (top-down) design—the derivation of correct programs—instead of the more familiar a-posteriori verification based on already completed program code. We sketch the transition from non-compositional towards compositional methods for concurrent programs, indicating the main principles behind compositionality. Having achieved a compositional framework based on classical Hoare tripels, we discuss extensions to achieve a convenient formalism to specify and verify reactive systems that have an intensive interaction with their environment. Next this Hoare-style framework is adapted to specify and verify real-time properties, and a compositional proof method is formulated for real-time distributed computing. Compositional reasoning during top-down development of a real-time program is illustrated by an example concerning a watchdog timer.

1 Introduction

Formal methods for the specification and verification of distributed systems can be classified from the viewpoint of expressibility (which properties can be specified), specification

---

*This work was partially supported by Esprit-BRA project 3096: Formal Methods and Tools for the Development of Distributed and Real-Time Systems (SPEC).

language (e.g., temporal logic, Hoare triples and first-order assertions), and programming features (such as time-out, various communication mechanisms, concurrency). In this paper we concentrate on the distinction between proof methods that are only applicable to complete program code and methods that can be used to verify design steps during the process of program development. We sketch the development from a-posteriori methods (requiring the complete program text) towards compositional methods (supporting verify-while-design). Compositionality can be considered as a requirement for hierarchical, structured, program derivation. A separation of concerns is desired between the use of (and the reasoning about) a module and its implementation. This leads to the following definition of *compositionality* for proof methods:

> Properties of a compound programming language construct (such as sequential composition and parallel composition) can be deduced from specifications for its constituent parts without any further information about the internal structure of these parts.

In general, compositional program specification and verification dictates, as a principle, that all aspects of program execution which are required to define the meaning of a compound statement from its constituents, must be explicitly addressed in semantics and assertion language alike. In *semantics* because, otherwise, no compositional semantics can be defined, since compositionality in semantics requires that the meaning of a compound statement is a function of the meaning of its parts (the guiding principle of denotational semantics). In *specification languages* because, otherwise, no compositional verification rules can be formulated in which the specification of a compound statement should follow from specifications of its constituent parts without knowledge about their internal structure (the internal structure often providing implicit information which has not been explicitly stated in the specification, but is used in non-compositional methods such as [OG76,AFdR80,LG81,MP82]). The rationale for this principle is that one must be able to specify the behaviour of a module in isolation, i.e., without any implicit prior assumption regarding the environment within which it ultimately functions. Hence, all assumptions which are needed regarding the environment—because these influence the behaviour of a module—must be made explicit as parameters (in the semantics and specification of that module alike) for only then one can abstract away from the remaining aspects (such as inner syntactic structure).

In case of shared variable communication this compositionality principle implies that when defining the behaviour of a module any change of a shared variable by the environment must be explicitly expressed as an assumption of that module regarding its environment. This is worked out in Aczel's model for shared variable semantics as cited in [dR85b]. Similarly, when considering distributed communication via input/output-statements, the specification of, e.g., an input statement in one module requires explicit expressibility of assumptions regarding a corresponding output statement in another module. In case one abstracts away from blocking behaviour only assumptions regarding the value communicated must be expressible. If blocking behaviour is a focus of interest, this is again an assumption regarding program execution which must be stated explicitly; i.e. one has to state the effect of no communication partner being available in the assertion language and one must be able to express the assumption that no partner is available in the assertion language.

In this paper we also discuss the compositional verification of real-time properties for distributed systems. When the timing behaviour of a statement is considered, all factors concerning the execution of this statement which influence that timing behaviour must be expressible. E.g., for real-time systems we use in this paper the maximal progress assumption with respect to distributed i/o-communication from [KSdR+88]: no input or output statement should wait for communication when its partner is also ready to communicate. This aspect of timing behaviour requires, indeed, that one must be able to express when a partner is waiting to communicate. For, otherwise, maximal progress would not be expressible within the semantics, and hence timing behaviour of i/o-statements could not be characterized. The maximal progress assumption, which represents the situation that each process has its own processor, can be generalized to multiprogramming where several processes may share a single processor. By introducing priorities for processes on a single processor, certain statements which are ready to execute will not be executed on account of their lower priority and because at most one action can be executed at a time on a uniprocessor. Modelling the timing behaviour of such statements requires that the semantics, and hence the specification language, contains primitives to state explicitly when a statement is executing and when it is requesting processor time with a certain priority. The semantic aspects of reasoning formally about real-time and scheduling by means of priorities are addressed technically in [Hoo91a].

This paper is structured as follows. A real-time programming language with synchronous message passing is defined in chapter 2. Chapter 3 contains a description of a classical non-compositional method, and we indicate how a compositional proof system can be achieved for Hoare triples (precondition, program, postcondition). For the specification and verification of reactive systems, these triples are extended in chapter 4 with assertions (called assumption and commitment) that specify the communication interface between a program and its environment. In chapter 5 we adapt this Hoare-style framework to specify real-time properties of programs, and we give the details of a compositional proof system for real-time distributed systems. The formalism of chapter 5 is illustrated by an example of a watchdog timer in chapter 6. The extension of this formalism to assumption/commitment based reasoning for real-time is described in chapter 7. In chapter 8 we sketch the development of the field, leading to a description of the state of the art and the place of our work therein.

2 Syntax

We give syntax and informal semantics of a programming language for distributed synchronous message-passing. Our language is akin to Occam [Occ88] with concurrent processes that communicate via message passing along unidirectional channels, each connecting two processes. Communication is synchronous, i.e., both the sender and the receiver have to wait until a communication partner is available.

Let $CHAN$ be a nonempty set of channel names, VAR be a nonempty set of program variables, and VAL be a denumerable domain of values. $I\!N$ denotes the set of natural numbers (including 0). The syntax of our programming language is given in Table 1, with $n \in I\!N, n \geq 1, c, c_1, \ldots, c_n \in CHAN, x, x_1, \ldots, x_n \in VAR$, and $\vartheta \in VAL$.

Table 1: Syntax Programming Language

Expression	$e ::=$	$\vartheta \mid x \mid e_1 + e_2 \mid e_1 - e_2 \mid e_1 \times e_2$
Boolean Expression	$b ::=$	$e_1 = e_2 \mid e_1 < e_2 \mid \neg b \mid b_1 \vee b_2$
Statement	$S ::=$	$\text{skip} \mid x := e \mid c!e \mid c?x \mid S_1; S_2 \mid G \mid \star G \mid S_1 \| S_2$
Guarded Command	$G ::=$	$[[]_{i=1}^{n} b_i \rightarrow S_i] \mid [[]_{i=1}^{n} b_i; c_i?x_i \rightarrow S_i]$

Informally, the statements of our programming language have the following meaning:
<u>Atomic statements</u>

- **skip** terminates immediately.
- Assignment $x := e$ assigns the value of expression e to the variable x.
- Output statement $c!e$ is used to send the value of expression e on channel c as soon as an input command $c?x$ is available. Since we assume synchronous communication, such an output statement is suspended until a parallel process executes a corresponding input statement.
- Input statement $c?x$ is used to receive a value via channel c and assign this value to the variable x. As for the output command, such an input statement has to wait for a corresponding partner before a (synchronous) communication can take place.

Henceforth we will often refer to an input or output statement as an *io-statement*.

Compound statements

- $S_1; S_2$ indicates sequential composition: first execute S_1, and continue with the execution of S_2 if and when S_1 terminates.

- Guarded command $[[]_{i=1}^n b_i \rightarrow S_i]$. If none of the b_i evaluate to true then this guarded command terminates after evaluation of the booleans. Otherwise, non-deterministically select one of the b_i that evaluates to true and execute the corresponding statement S_i.

- Guarded command $[[]_{i=1}^n b_i; c_i?x_i \rightarrow S_i]$. A guard (the part before the arrow) is *open* if its boolean part evaluates to true. If none of the guards is open, the guarded command terminates after evaluation of the booleans. Otherwise, wait until the communication of one of the open guards can be performed and continue with the corresponding S_i.

- Iteration $*G$ indicates repeated execution of guarded command G as long as at least one of the guards is open. When none of the guards is open $*G$ terminates.

- $S_1 \| S_2$ indicates parallel execution of the statements S_1 and S_2. The components S_1 and S_2 of a parallel composition are often called *processes*.

Henceforth we use $\equiv$ to denote syntactic equality. Conventional abbreviations are used, such as $true \equiv 0 = 0$, $false \equiv \neg true$, $b_1 \wedge b_2 \equiv \neg(\neg b_1 \vee \neg b_2)$, etc.
For a guarded command $G \equiv [[]_{i=1}^n b_i \rightarrow S_i]$ or $G \equiv [[]_{i=1}^n b_i; c_i?x_i \rightarrow S_i]$, we define $b_G \equiv b_1 \vee \ldots \vee b_n$.
Observe that conventional programming constructs can be defined as an abbreviation:
if b **then** S_1 **else** S_2 **fi** $\equiv [b \rightarrow S_1 [] \neg b \rightarrow S_2]$ and **while** b **do** S **od** $\equiv *[b \rightarrow S]$.

3 Compositionality

In section 3.1 we explain the principles of traditional non-compositional methods. The development towards compositional proof systems based on Hoare triples is described in section 3.2.

3.1 Non-Compositional Methods

Classical verification methods for parallel processes, such as [OG76] for shared variable communication and [AFdR80, LG81] for synchronous message passing, consist of two stages. First a *local* correctness proof is given for each of the sequential processes by associating assertions with locations in the program. In the second, *global*, stage a consistency check is applied to the local proofs.

- For shared variables this is the *interference freedom* test which verifies that assertions in the proof of one process remain valid under actions of other processes.
- For communication via message passing the *cooperation* test is applied to verify correctness of assertions attached to locations after input- and output-statements.

Such methods are not compositional because at parallel composition they require the complete program text, annotated with assertions, of the constituent processes. Moreover, they are only suited for top-level parallelism, that is, to prove correctness of programs of the form $S_1 \| \cdots \| S_n$ where $S_1, \ldots, S_n$ are sequential processes.

As an example, we consider in more detail the method of Apt, Francez & de Roever [AFdR80] for synchronous message passing. This method is based on *Hoare triples* [Hoa69], that is, on correctness formulae of the form $\{p\}\ S\ \{q\}$ which have the following meaning: if we start program S in a state satisfying assertion p (the precondition) and if program S terminates then assertion q (the postcondition) holds for the termination state. For example, $\{x = 5\}\ x := x + 1\ \{x = 6\}$ is a valid Hoare triple.

First we indicate how a proof system can be formulated in which valid Hoare triples can be derived for sequential programs. Let $q[e/x]$ denote the textual substitution of expression e for each free occurrence of variable x in assertion q. Then we have the following assignment axiom:

Axiom 3.1 (Assignment) $\{q[e/x]\}\ x := e\ \{q\}$

Example 3.1 With this axiom we can derive $\{x = 5\}\ x := x + 1\ \{x = 6\}$, because $(x = 6)[x + 1/x]$ equals $x + 1 = 6$, which is equivalent to $x = 5$. □

Furthermore the proof system contains rules for compound constructs. For instance, sequential composition is modelled by the following rule:

Rule 3.2 (Sequential Composition) $\dfrac{\{p\}\ S_1\ \{r\},\ \{r\}\ S_2\ \{q\}}{\{p\}\ S_1; S_2\ \{q\}}$

By such a rule the formula below the line can be derived from the formulae above the line. Soundness of the rule is proved by showing that validity of the formulae above the line implies that the formula below the line is valid. Note that this rule is compositional because the formula for $S_1; S_2$ is derived without using the structure of S_1 or S_2.

For iteration we have the following rule:

Rule 3.3 (Iteration) $\dfrac{\{p \wedge b_G\}\ G\ \{p\}}{\{p\}\ \star G\ \{p \wedge \neg b_G\}}$

To strengthen preconditions and weaken postconditions, the proof system contains the following consequence rule.

Rule 3.4 (Consequence) $\quad \dfrac{p \to p_1,\ \{p_1\}\ S\ \{q_1\},\ q_1 \to q}{\{p\}\ S\ \{q\}}$

To illustrate the rule for parallel composition in [AFdR80], we consider the proof of
$\{y = 3\}\ (a?x\ ;\ x := x + 1\ ;\ b!(x + 2))\ \|\ (a!y\ ;\ b?y\ ;\ y := y + 2)\ \{x = 4 \wedge y = 8\}$.

In the first stage we attach assertions to all locations in the program text of the two processes, leading to so-called *proof outlines*:
$\{true\}\ a?x\ \{x = 3\}\ ;\ x := x + 1\ \{x = 4\}\ ;\ b!(x + 2)\ \{x = 4\}$, and
$\{y = 3\}\ a!y\ \{y = 3\}\ ;\ b?y\ \{y = 6\}\ ;\ y := y + 2\ \{y = 8\}$.
In this stage only the postconditions of assignments are verified: from the assignment axiom we obtain $\{x = 3\}\ x := x + 1\ \{x = 4\}$ and $\{y = 6\}\ y := y + 2\ \{y = 8\}$. Observe that the postconditions of the input statements $a?x$ and $b?y$ express assumptions about the values sent by the communication partner.

These assumptions are verified in the second stage by means of the *cooperation test*.[1] In general, this test requires that for $\{p_1\}\ c?x\ \{q_1\}$ and $\{p_2\}\ c!e\ \{q_2\}$ in the proof outlines of two processes we have to prove $\{p_1 \wedge p_2\}\ c?x\,\|\,c!e\ \{q_1 \wedge q_2\}$, which is equivalent to proving $\{p_1 \wedge p_2\}\ x := e\ \{q_1 \wedge q_2\}$. In our example this leads to the proof obligations:
$\{true \wedge y = 3\}\ a?x\ \|\ a!y\ \{x = 3 \wedge y = 3\}$ and
$\{y = 3 \wedge x = 4\}\ b?y\ \|\ b!(x + 2)\ \{y = 6 \wedge x = 4\}$ which are easy to prove.

After the verification of the first two stages we obtain the conjunction of all preconditions from the sequential processes as the precondition of the complete program and the conjunction of the postconditions as the final postcondition. In our example this leads to the precondition $true \wedge y = 3$ and the postcondition $x = 4 \wedge y = 8$ which are equivalent to the required conditions.

3.2 Towards Compositionality

In this section we discuss how a compositional proof method can be obtained for programs which communicate via synchronous message passing. First the cooperation test from [AFdR80] is removed by not allowing implicit assumptions in the postconditions of io-statements. The local proof of a sequential program should be valid in any arbitrary environment. Since this would weaken the method, and any valid postcondition should be provable, we use a *history variable* h which denotes the communication history of the complete program. A (communication) *history* is a sequence of records (c, ϑ) where c is a channel name and ϑ a value. For example, $< (c, 5), (b, 6), (a, 8), (b, 0) >$ is a history expressing four communications: first one via channel c with value 5, then a communica-

---

[1] In the full method of [AFdR80] auxiliary variables and a global invariant are used to make the method complete, i.e., to guarantee that any valid Hoare triple can be proved.

tion via b with value 6, etc. Let <> denote the empty sequence. History variable h does not appear in the program, but it is updated implicitly in the semantics of io-statements. This leads to the following valid formulae.

- For an output command we have, e.g., $\{h =<(c,5)>\}\ b!6\ \{h =<(c,5),(b,6)>\}$.
- For an input statement $c?x$ we can only express in the postcondition that there exists a value which is communicated via c and assigned to x. For instance, $\{h=<>\}\ c?x\ \{\exists v: h=<(c,v)> \land x=v\}$.

Example 3.2 To prove $\{true\}\ c!5\ ||\ c?x\ \{x=5\}$, we use the Hoare triples $\{h=<>\}\ c!5\ \{h=<(c,5)>\}$ and $\{h=<>\}\ c?x\ \{\exists v: h=<(c,v)> \land x=v\}$. Suppose the pre- and postcondition after parallel composition are obtained by simply taking the conjunction of, resp., pre- and postconditions of the sequential programs. Then $\{h=<>\}\ c!5\ ||\ c?x\ \{h=<(c,5)> \land \exists v: h=<(c,v)> \land x=v\}$. Since the postcondition implies $\exists v: v=5 \land x=v$, and hence $x=5$, the consequence rule leads to
$\{h=<>\}\ c!5\ ||\ c?x\ \{x=5\}$.
By a so-called *substitution rule* (not given in this paper), we could substitute <> for h in the precondition, thus obtaining precondition *true*. □

Example 3.2 suggests the following rule: $\dfrac{\{p_1\}\ S_1\ \{q_1\},\ \{p_2\}\ S_2\ \{q_2\}}{\{p_1 \land p_2\}\ S_1||S_2\ \{q_1 \land q_2\}}$

Example 3.3 Consider again $S_1||S_2$, where $S_1 \equiv a?x\ ;\ x:=x+1\ ;\ b!(x+2)$ and $S_2 \equiv a!y\ ;\ b?y\ ;\ y:=y+2$. First derive the following Hoare triples:
$\{h=<>\}\ a?x\ \{\exists v_1: h=<(a,v_1)> \land x=v_1\}$,
$\{\exists v_1: h=<(a,v_1)> \land x=v_1\}\ x:=x+1\ \{\exists v_1: h=<(a,v_1)> \land x=v_1+1\}$, and
$\{\exists v_1: h=<(a,v_1)> \land x=v_1+1\}\ b!(x+2)\ \{\exists v_1: h=<(a,v_1),(b,v_1+3)> \land x=v_1+1\}$.
By two applications of the sequential composition rule we obtain
$\{h=<>\}\ a?x\ ;\ x:=x+1\ ;\ b!(x+2)\ \{\exists v_1: h=<(a,v_1),(b,v_1+3)> \land x=v_1+1\}$.
Similarly,
$\{h=<> \land y=3\}\ a!y\ ;\ b?y\ ;\ y:=y+2\ \{\exists v_2: h=<(a,3),(b,v_2)> \land y=v_2+2\}$.
Then the parallel composition rule above leads to
$\{h=<> \land y=3\}\ S_1||S_2\ \{\ \exists v_1: h=<(a,v_1),(b,v_1+3)> \land x=v_1+1 \land$
$\exists v_2: h=<(a,3),(b,v_2)> \land y=v_2+2\}$.
The postcondition implies $\exists v_1,v_2: v_1=3 \land v_2=v_1+3 \land x=v_1+1 \land y=v_2+2$, which leads to $x=4 \land y=8$. Thus, by the consequence rule, $\{h=<> \land y=3\}\ S_1||S_2\ \{x=4 \land y=8\}$. (Again $h=<>$ in the precondition can be removed by a substitution rule.) □

Although this works nicely for two processes, the next example shows that there is a problem if more than two processes are involved.

Example 3.4 Consider $S_1 \| S_2 \| S_3$, where $S_1 \equiv a!0 \, ; \, b?x$, $S_2 \equiv a?y \, ; \, c!(y+1)$, and $S_3 \equiv c?z \, ; \, b!(z+1)$. Similar to Example 3.3, we could first prove
$\{h =<>\} \, S_1 \, \{q_1 \equiv \exists v_1 : h =< (a,0), (b,v_1) > \wedge x = v_1\}$ and
$\{h =<>\} \, S_2 \, \{q_2 \equiv \exists v_2 : h =< (a,v_2), (c,v_2+1) > \wedge y = v_2\}$.
But then the conjunction of q_1 and q_2 implies *false* whereas $S_1 \| S_2 \| S_3$ terminates and hence does not satisfy postcondition *false*. □

The problem is that h denotes the global history of the complete program—e.g., consisting of three processes—whereas each of the processes in isolation can only describe the history on its own channels. A possible solution is to give each process its own history variable, and to combine these local history variables at parallel composition. This is done in [Sou84a], using a predicate *compat*. Zwiers [ZdRvEB84,Zwi89], however, shows that a concise and simple rule for parallel composition can be formulated if each process uses projections of global history variable h onto its own channels. Such a projection expresses the view of a particular process on the global history. Formally, the *projection* of h onto a set of channel names *cset*, notation h_{cset}, denotes the sequence obtained from the history denoted by h by removing all records with a channel name not in *cset*. E.g., if $h =< (a,0), (c,1), (b,3) >$ then $h_{\{c\}} =< (c,1) >$, $h_{\{b,c\}} =< (c,1), (b,3) >$, and $h_{\{d\}} =<>$. Henceforth we write h_c, h_{bc}, and h_d instead of, resp., $h_{\{c\}}$, $h_{\{b,c\}}$, and $h_{\{d\}}$.

In the rule for $S_1 \| S_2$ we require that the postcondition of S_i only refers to history h via projections on channels occurring in S_i, for $i = 1, 2$. If, moreover, the postcondition of S_i only refers to program variables of S_i, then the following rule for parallel composition is sound:

Rule 3.5 (Parallel Composition)

$$\frac{\{p_1\} \, S_1 \, \{q_1\}, \, \{p_2\} \, S_2 \, \{q_2\}}{\{p_1 \wedge p_2\} \, S_1 \| S_2 \, \{q_1 \wedge q_2\}}$$

Observe that this is a compositional rule because a Hoare triple for $S_1 \| S_2$ can be derived without knowing the internal structure of S_1 and S_2. To obtain a valid rule we only impose a simple syntactic requirement on assertions and processes (the postcondition of a process should only refer to channels and variables of the process itself). This simple syntactic check replaces the cooperation test which requires proof outlines for the complete program text of the processes.

Except for bottom-up verification such a compositional rule can be used for top-down development. Therefore, a triple $\{p\} \, S \, \{q\}$ is considered as a specification for a program S. Suppose we decide to implement S as $S_1 \| S_2$. If we can find assertions p_1, q_1, p_2, and q_2 that satisfy $p \rightarrow p_1 \wedge p_2$ and $q_1 \wedge q_2 \rightarrow q$, and moreover certain syntactic requirements on the postconditions hold, then S_1 and S_2 can be implemented independently, using specifications $\{p_1\} \, S_1 \, \{q_1\}$ and $\{p_2\} \, S_2 \, \{q_2\}$.

Since in this compositional framework programs can be considered as black boxes, and verification is done on the basis of the specifications only, we can allow nested parallelism in programs as expressed by the syntax of chapter 2.

4 Extensions of Hoare Triples

A Hoare triple is perfectly suited to describe the observable behaviour of a sequential program which is given by initial and final state. For a parallel program also the communication behaviour on its external channels is observable. Hence a specification of a parallel component should express this communication interface. Note, however, that a specification $\{p\}\, S\, \{q\}$ has an important limitation: it only specifies the behaviour of S if S terminates. All non-terminating computations of S satisfy such a specification trivially. Thus the postcondition can not be used to express the communication interface of non-terminating programs. Therefore, a Hoare triple is extended with an invariant, called *commitment* in this paper, which must hold throughout the computation. This leads to a formulae of the form $C : \{p\}\, S\, \{q\}$ where commitment C describes the communication interface of S during its execution. The success of such formulae in many applications is based on a simple rule for parallel composition in which, besides conjunctions for pre- and postconditions, also the conjunction of commitments can be taken.

Rule 4.1 (Parallel Composition)

$$\frac{C_1 : \{p_1\}\, S_1\, \{q_1\},\; C_2 : \{p_2\}\, S_2\, \{q_2\}}{C_1 \wedge C_2 : \{p_1 \wedge p_2\}\, S_1 \| S_2\, \{q_1 \wedge q_2\}}$$

provided, for $i = 1, 2$, the assertions C_i and q_i refer to h only via projections on the channels occurring in S_i, and q_i only refers to program variables of S_i.

In this formalism the influence of the environment on the communication behaviour of a program can be expressed by using implications in the commitment. The next example indicates that for so-called reactive systems [HP85], which have an intensive interaction with their environment, this style of specification often leads to proofs using inductive arguments. This motivates a final extension of the Hoare-style formulae.

In the following examples, $seq_1 \preceq seq_2$ expresses that sequence seq_1 is an initial prefix of sequence seq_2. For a history variable h, a set of channels $cset$, and a number i, we use $h_{cset}[i] = (c, \vartheta)$ to express that if i is less than or equal to the length of h_{cset} then the record (c, ϑ) is the i-th element of by h_{cset}.

Example 4.1 We verify two reactive, non-terminating, processes that have a close interaction. Consider $S_1 \equiv c!1; *[d?x \rightarrow c!(x+1)]$ and $S_2 \equiv *[c?y \rightarrow d!(y+1)]$. The aim is to prove that $S_1 \| S_2$ satisfies the commitment $h_{cd} \preceq < (c,1), (d,2), (c,3), (d,4), (c,5), \ldots >$, that is,

$\forall i \geq 1 : h_{cd}[2i-1] = (c, 2i-1) \wedge h_{cd}[2i] = (d, 2i)$.

Define $C_1 \equiv h_{cd}[1] = (c,1) \wedge (\forall i \geq 1\, \forall v : h_{cd}[2i] = (d,v) \rightarrow h_{cd}[2i+1] = (c, v+1))$.
Then for S_1 we can prove

$C_1 : \{h_{cd} = <>\}\ c!1; *[d?x \rightarrow c!(x+1)]\ \{false\}$.

Similarly, for S_2 we define $C_2 \equiv (\forall i \geq 1\, \forall v : h_{cd}[2i-1] = (c,v) \rightarrow h_{cd}[2i] = (d, v+1))$.
Then we have

$C_2 : \{h_{cd} = <>\}\ *[c?y \rightarrow d!(y+1)]\ \{false\}$.

Since the required syntactic conditions are fulfilled, we can apply the rule for parallel composition which leads to

$C_1 \wedge C_2 : \{h_{cd} = <>\}\ S_1 \| S_2\ \{false\}$.

To prove that $C_1 \wedge C_2$ implies the required commitment, we prove by induction on i that, for $i \geq 1$, $h_{cd}[2i-1] = (c, 2i-1) \wedge h_{cd}[2i] = (d, 2i)$.

- Basic step. For $i = 1$ we have, by C_1, that $h_{cd}[1] = (c,1)$ and thus from C_2 we obtain $h_{cd}[2] = (d, 2)$.

- Induction step. Assume $h_{cd}[2i-1] = (c, 2i-1) \wedge h_{cd}[2i] = (d, 2i)$. Then $h_{cd}[2i] = (d, 2i)$ implies, by C_1, that $h_{cd}[2i+1] = (c, 2i+1)$, and thus $h_{cd}[2(i+1)-1] = (c, 2(i+1)-1)$. Using C_2 this leads to $h_{cd}[2(i+1)] = (d, 2(i+1))$.

□

This example illustrates that assumptions about the environment are important in the specification of a process, and it indicates that mutual assumptions of processes about each others communication interface usually leads to correctness proofs using inductive reasoning. Based on these observations we present an extension of the correctness formulae in which a process can be specified relative to explicit assumptions about its environment, and an inductive relation is incorporated in the specification. Therefore the specification formula is extended with a second invariant, called *assumption*, which expresses assumptions about the environment and by which we can strengthen postcondition and commitment. This leads to formulae of the form $(A, C) : \{p\}\ S\ \{q\}$, where

A is an **assumption** describing the expected behaviour of the environment of S, and
C is a **commitment** which is guaranteed by process S itself, as long as the environment does not violate the assumption.

The general idea is that assumption and commitment reflect the communication interface between parallel components (and hence do not contain program variables), whereas pre- and postcondition facilitate the reasoning at sequential composition.

Example 4.2 In this formalism assumptions about the values sent by the environment can be expressed explicitly. For instance, the assertion $h_c \preceq < (c, 3) >$ can be used as an assumption:

$(h_c \preceq < (c, 3) >, true) : \{true\}\ c?x\ \{x = 3\}$.

This assumption expresses that if a communication along c takes place then the environment will send the value 3. The next formula shows that it can be used for a commitment about the next communication:

$(h_c \preceq < (c, 3) >, h_a \preceq < (a, 4) >) : \{true\}\ c?x\ ;\ a!(x + 1)\ \{x = 3\}$. □

Assumption/commitment based reasoning was introduced in [MC81]. A proof system for these formulae has been given in [ZdRvEB84]. In this paper we discuss the proof obligations for assumptions and commitments in the parallel composition rule. Consider the parallel composition $S_1 \| S_2$, and suppose we have assumption-commitment pairs (A_1, C_1) for S_1 and (A_2, C_2) for S_2. Which conditions have to be verified to obtain a pair (A, C) for $S_1 \| S_2$? Consider assumption A_2 of S_2:

- A_2 may contain assumptions about joint channels of S_1 and S_2 which connect these two processes; these assumptions must be justified by the commitment C_1 of S_1.

- A_2 may contain assumptions about external channels of S_2. These assumptions are maintained in the new network assumption A for $S_1 \| S_2$.

This leads to the following proof obligation: $A \land C_1 \to A_2$. Similarly, $A \land C_2 \to A_1$.

To obtain a sound rule with these implications, the meaning of a formula $(A_i, C_i) : \{p_i\}\ S_i\ \{q_i\}$ has to be defined carefully. A simple implication between A_i and C_i would with the implications above and $A \equiv true$ lead to circular reasoning, that is, $A_1 \to C_1 \to A_2 \to C_2 \to A_1$. Therefore in defining the meaning of $(A_i, C_i) : \{p_i\}\ S_i\ \{q_i\}$ we require that if p_i holds in the initial state then 1) C_i holds initially, and 2) C_i holds after every communication provided A_i holds after all preceding communications. This inductive step inside the meaning of formulae is sufficient to avoid circularity (see [MC81,ZdRvEB84]).

As in Rule 4.1 we can take the conjunction of preconditions, postconditions, and commitments, provided, for $i = 1, 2$, the assertions A_i, C_i, p_i and q_i of S_i refer only to h via projections on the channels of S_i, and p_i and q_i only refer to program variables of S_i. (Program variables are not allowed in A_i and C_i.) With these constraints, the following rule for parallel composition is valid:

Rule 4.2 (Parallel Composition A-C)

$$\frac{(A_1, C_1) : \{p_1\}\ S_1\ \{q_1\},\ (A_2, C_2) : \{p_2\}\ S_2\ \{q_2\}}{(A, C_1 \wedge C_2) : \{p_1 \wedge p_2\}\ S_1 \| S_2\ \{q_1 \wedge q_2\}}$$
$$A \wedge C_1 \to A_2,\ A \wedge C_2 \to A_1$$

Example 4.3 Consider $S_1 \| S_2$ where $S_1 \equiv a?x\,;\ x := x + 1\,;\ b!(x+2)$ and $S_2 \equiv c?y\,;\ a!y\,;\ b?y\,;\ y := y + 2$. Then for S_1 and S_2 we can derive
$(A_1 \equiv h_a \preceq < (a,3) >,\ C_1 \equiv h_b \preceq < (b,6) >) : \{h_{ab} = <>\}\ S_1\ \{x = 4\}$, and
$(A_2 \equiv h_c \preceq < (c,3) > \wedge h_b \preceq < (b,6) >,\ C_2 \equiv h_a \preceq < (a,3) >) : \{h_{abc} = <>\}\ S_2\ \{y = 8\}$.
Since S_1 and S_2 communicate with each other via the channels a and b, the remaining assumption A for $S_1 \| S_2$ concerns the external channel c, that is, $A \equiv h_c \preceq < (c,3) >$. Then $A \wedge C_1 \to A_2$ and $A \wedge C_2 \to A_1$, thus the parallel composition rule leads to
$(A, C_1 \wedge C_2) : \{h_{abc} = <>\}\ S_1 \| S_2\ \{x = 4 \wedge y = 8\}$. □

Example 4.4 Consider again the two reactive processes from Example 4.1:
$S_1 \equiv c!1; *[d?x \to c!(x+1)]$ and $S_2 \equiv *[c?y \to d!(y+1)]$.
We show how the assumption/commitment formalism can be used to prove for $S_1 \| S_2$ the commitment $\forall i \geq 1 : h_{cd}[2i-1] = (c, 2i-1) \wedge h_{cd}[2i] = (d, 2i)$.
Define $A_1 \equiv \forall i \geq 1 : h_{cd}[2i] = (d, 2i)$ and $C_1 \equiv \forall i \geq 1 : h_{cd}[2i-1] = (c, 2i-1)$.
Then for S_1 we can prove
$(A_1, C_1) : \{h_{cd} = <>\}\ c!1; *[d?x \to c!(x+1)]\ \{false\}$.
Similarly, for S_2 we define $A_2 \equiv \forall i \geq 1 : h_{cd}[2i-1] = (c, 2i-1)$, and $C_2 \equiv \forall i \geq 1 : h_{cd}[2i] = (d, 2i)$. Then we have
$(A_2, C_2) : \{h_{cd} = <>\}\ *[c?y \to d!(y+1)]\ \{false\}$.
Let $A \equiv true$. Then $A \wedge C_2 \to A_1$ and $A \wedge C_1 \to A_2$. Since the other conditions on the assertions are also satisfied, we can apply the rule for parallel composition, leading to
$(true, C_1 \wedge C_2) : \{h_{cd} = <>\}\ S_1 \| S_2\ \{false\}$.
Clearly $C_1 \wedge C_2$ is equivalent to the required commitment. □

Observe that in the correctness proof for the two reactive processes from Example 4.4 there is no explicit inductive argument. The required inductive reasoning is performed only once in the soundness proof of the parallel composition rule (Rule 4.2). In this respect we have obtained a rule for parallel composition which is the analogue of Hoare's while-rule for sequential programs [Hoa69].

5 Compositionality and Real-Time

In this chapter we adapt the compositional Hoare-style proof systems from the previous chapters to real-time. We describe in detail a compositional method to specify and verify timing constraints. By describing the details of a particular compositional proof method, we illustrate the general outline of such a description which should consist of the following points:

1. A description of the programming language, i.e., syntax and informal semantics.

2. A formal semantics of the programming language.

3. The definition of an assertion language in which properties of programs can be expressed. For this assertion language we also have to give syntax, informal meaning, and formal interpretation.

4. The definition of a correctness formula that relates programs and assertions. Using the semantics of the programming language and the interpretation of assertions, the validity of such a correctness formula can be defined formally.

5. A proof system in which, by rules and axioms, correctness formulae can be derived formally.

6. The proof of soundness and (relative) completeness of the proof system: show that every correctness formula that can be derived is also valid, and that every valid formula can be derived (assuming that valid assertions can be derived).

As an example, we consider in this section a compositional proof method for distributed real-time systems based on [Hoo91b]. In section 5.1 we describe the syntax and the informal semantics of a real-time programming language with nested parallelism, communication via synchronous message passing, and time-outs. A semantic model for this language, and the syntax and the interpretation of assertions and correctness formulae can be found in [Hoo91b]. A compositional proof system is presented in section 5.3. The proofs of soundness and (relative) completeness for the proof system given in this section are not given here. The reader is referred to [Hoo91b] for all details about these proofs.

5.1 Real-Time Programming Language

Our real-time programming language is based on the Occam-like language from Chapter 2 and akin to real-time versions of CSP as defined in [KSdR+88,HGdR87]. We add a real-time statement **delay** e which suspends the execution for (at least) e time units. This statement is also used in the language Ada [Ada83] and corresponds to a *wait e* statement in [KSdR+88,HGdR87]. Similar to a delay-statement in the select construct of Ada, such a delay-statement is allowed in a guard of a guarded command to enable the programming of time-outs.

5.1.1 Syntax and Informal Meaning

The syntax of our programming language is given in Table 2, with $n \in \mathbb{N}$, $n \geq 1$, $c, c_1, \ldots, c_n \in CHAN$, $x, x_1, \ldots, x_n \in VAR$, and $\vartheta \in VAL$.

Table 2: Syntax of the Programming Language

Expression	$e ::=$	$\vartheta \mid x \mid e_1 + e_2 \mid e_1 - e_2 \mid e_1 \times e_2$
Boolean Expression	$b ::=$	$e_1 = e_2 \mid e_1 < e_2 \mid \neg b \mid b_1 \vee b_2$
Statement	$S ::=$	$\mathbf{skip} \mid x := e \mid \mathbf{delay}\ e \mid c!e \mid c?x \mid$
		$S_1; S_2 \mid G \mid \star G \mid S_1 \| S_2$
Guarded Command	$G ::=$	$[[]_{i=1}^n b_i \rightarrow S_i] \mid [[]_{i=1}^n b_i; c_i?x_i \rightarrow S_i [] b_0; \mathbf{delay}\ e \rightarrow S_0]$

We give the informal meaning of new statements:

<u>Atomic statements</u>

- **delay** e suspends execution for (the value of) e time units. If e yields a negative value then **delay** e is equivalent to **skip**.

<u>Compound statements</u>

- Guarded command $[[]_{i=1}^n b_i; c_i?x_i \rightarrow S_i [] b_0; \mathbf{delay}\ e \rightarrow S_0]$. A guard is *open* if the boolean part evaluates to true. If none of the guards is open, the guarded command terminates after evaluation of the booleans. Otherwise, wait until an input statement of the open input-guards can be executed and continue with the corresponding S_i. If the delay guard is open (b_0 evaluates to true) and no input-guard can be taken within e time units (after the evaluation of the booleans), then S_0 is executed. If b_0 evaluates to true and e yields 0 or a negative value then S_0 is executed immediately after the evaluation of the booleans.

Example 5.1 This construct makes it possible to model a *time-out*, i.e., to restrict the waiting period for certain communications. Consider the guarded command $[c?x \rightarrow S_1 [] \mathbf{delay}\ 5 \rightarrow S_2]$; if there is no partner available for the input statement within 5 time units then the delay-alternative is taken and S_2 is executed. □

For a guarded command G we define
$$b_G \equiv \begin{cases} b_1 \vee \ldots \vee b_n & \text{if } G \equiv [[]_{i=1}^n b_i \rightarrow S_i] \\ b_1 \vee \ldots \vee b_n \vee b_0 & \text{if } G \equiv [[]_{i=1}^n b_i; c_i?x_i \rightarrow S_i [] b_0; \mathbf{delay}\ e \rightarrow S_0] \end{cases}$$

Note that G terminates if b_G evaluates to false.

For a guarded command $[[]_{i=1}^n b_i; c_i?x_i \rightarrow S_i [] b_0; \mathbf{delay}\ e \rightarrow S_0]$ we write $[[]_{i=1}^n b_i; c_i?x_i \rightarrow S_i]$ if $b_0 \equiv false$. An input-guard $b_i; c_i?x_i$ is written as $c_i?x_i$ if $b_i \equiv true$, and, similarly, a delay-guard b_0; **delay** e is abbreviated as **delay** e if $b_0 \equiv true$.

Let $var(S)$ be the set of program variables occurring in statement S. Recall that for $S_1 \| S_2$ we require $var(S_1) \cap var(S_2) = \emptyset$. Let $DCHAN$ be the set of channels extended with directional channels; $DCHAN = CHAN \cup \{c! \mid c \in CHAN\} \cup \{c? \mid c \in CHAN\}$.

Definition 5.1 (Channels Occurring in Statement) The set of (directional) channels occurring in a statement S, notation $dch(S)$, is defined as the smallest subset of $DCHAN$ such that if c is an output channel of S then $\{c, c!\} \subseteq dch(S)$, and if c is an input channel of S then $\{c, c?\} \subseteq dch(S)$.

For example, $dch(a?; b! \| b?; c!) = \{a, a?, b, b!, b?, c, c!\}$.

5.1.2 Basic Timing Assumptions

We express the timing behaviour of a program from the viewpoint of an external observer with his own clock. Thus, although parallel components of a system might have their own, physical, local clock, the observable behaviour of a system is described in terms of a single, conceptual, global clock. Since this global notion of time is not incorporated in the distributed system itself, it does not impose any synchronization upon processes.

In this paper we use a time domain which is dense, i.e., between every two points of time there exists an intermediate point. With such a dense time domain a communication can be represented by an interval of communication records, and we can easily model communications that overlap in time or that are arbitrarily close to each other in time. When combining the real-time specifications of independently designed components, it is convenient to have a dense time domain to avoid problems such as finding a common unit of time. Having dense time is also suitable for the description of reactive systems that interact with a time-continuous environment (see, e.g., [Koy90]). Furthermore, a dense time domain allows the refinement of a single action into a sequence of sub-actions.

For notational convenience, a special value ∞ is used with the following properties: $\infty \notin TIME$, for all $\tau \in TIME$: $\tau < \infty$, and for all $\tau \in TIME \cup \{\infty\}$: $\tau + \infty = \infty + \tau = \infty$, $\infty - \tau = \infty$, $\tau \times \infty = \infty \times \tau = \infty$, $max(\infty, \tau) = max(\tau, \infty) = \infty$, $min(\infty, \tau) = min(\tau, \infty) = \tau$, and $min\ \emptyset = \infty$. For a point $\tau_0 \in TIME$, a left-closed right-open interval $[0, \tau_0)$ is defined as $\{\tau \mid \tau \in TIME \wedge 0 \leq \tau < \tau_0\}$.

In our proof system the correctness of a program with respect to a specification, which may include timing constraints, is verified relative to assumptions about:

- The execution time of atomic statements. In general, time bounds on the execution time will be given. In this paper we assume that there is a fixed constant which gives the execution time, but the framework can be easily adapted to the more general case. Thus assume that

- there exists a constant $T_{assign} > 0$ such that each assignment takes T_{assign} time units;
- **delay** e takes exactly e time units if e is positive and 0 time units otherwise;
- there exists a constant $T_{comm} > 0$ such that each communication takes T_{comm} time units. Note that the execution of an input or output statement includes a waiting period when no communication partner is available. Since this waiting period depends on the environment of the communication statement, no assumptions can be made about its length.

- The overhead time required for compound programming constructs. Here we assume that there exists a constant $T_{exec} > 0$ such that the evaluation of guards in a guarded command takes T_{exec} time units. Note that in this way we avoid an infinite loop in zero time. There is no overhead for other compound statements.

- The execution model of parallel processes, representing assumptions about the allocation of processes on processors. We use in this paper the *maximal parallelism model* to represent the situation that each parallel process has its own processor. Then a process never waits with the execution of a local, non-communication, command. An input or output command can cause a process to wait, but only when no communication partner is available; as soon as a partner is available the communication must take place. Thus maximal parallelism implies minimal waiting. In [Hoo91a,Hoo91b] the framework is generalized to multiprogramming where several processes may share a single processor.

5.2 Modification of Hoare Triples to Real-Time

We modify the Hoare-style framework of the previous chapters by extending the first-order assertion language with primitives to specify the timing behaviour of programs. As already explained, by means of Hoare triples we can only specify partial correctness. Hence, we add a third assertion, called *commitment*, to specify real-time properties of terminating and non-terminating computations. In contrast with the previous chapters, the aim is to specify, besides safety properties (which can be falsified in finite time), also liveness properties. Therefore the commitment will not be an invariant which holds at any point during a computation (as in chapter 4), but it should hold for complete, possibly infinite, computations.

To extend a Hoare triple $\{p\}$ S $\{q\}$ to real-time, a special variable *time* is introduced. Consider, for instance, the formula $\{time = 3\}$ **delay** 2 $\{time = 5\}$. In the precondition the variable *time* specifies the starting time of the program, whereas in the postcondition *time* denotes the termination time.

Since our aim is to specify and verify timing properties of open systems that communicate with their environment by synchronous message passing along channels, the logic contains the following primitives to express this communication behaviour:

- *comm* (c, exp_1) *at* exp_2 with $c \in CHAN$, exp_1 and exp_2 expressions yielding a value in, respectively, VAL and $TIME \cup \{\infty\}$. This predicate expresses that value exp_1 is communicated along channel c at time exp_2.

Recall that our maximal progress assumption implies minimal waiting for communications. To express this assumption in our compositional framework, we include the following primitives in the logic:

- *wait c!* *at* exp and *wait c?* *at* exp to express that a process is waiting to send, resp. waiting to receive, a message along channel c at time exp.

As usual in Hoare-style formalisms, logical variables are used to relate pre- and postcondition. In this chapter we have logical variables ranging over $TIME \cup \{\infty\}$, and quantification over these variables. For instance, with logical variable t, the specification $\{time = t\}\ S\ \{t + 4 < time < t + 7\}$ expresses that if S terminates then it takes between 4 and 7 time units.

Recall that a formula $\{p\}\ S\ \{q\}$ can only express the behaviour of terminating computations, and hence such a specification is trivially satisfied by non-terminating programs. Therefore we extend a Hoare triple $\{p\}\ S\ \{q\}$ with a third assertion, called a *commitment*, which expresses the real-time communication behaviour of all executions of S, including the non-terminating ones. This leads to a correctness formula of the form $C : \{p\}\ S\ \{q\}$. In general, commitment C reflects the real-time communication interface between parallel components, whereas the pre- and postcondition facilitate the reasoning for sequential composition and iteration.

Finally, we argue that termination should be expressible in commitments. Consider, e.g., the statements $S_1 \equiv c!0$ and $S_2 \equiv [c!0 \rightarrow \text{skip} [] c!0 \rightarrow \star[\text{delay } 1 \rightarrow \text{skip}]]$. Then the programs $S_1; d!1$ and $S_2; d!1$ cannot be distinguished with classical Hoare triples, but in our extended framework we can use the commitment

$$\forall t_0 : (comm\ (c, 0)\ at\ t_0 \rightarrow \exists t_1 \geq t_0 : [wait\ d!\ at\ t_1 \vee comm\ d\ at\ t_1])$$

(which is satisfied by $S_1; d!1$ but not by $S_2; d!1$). Since we aim at a compositional proof system, the difference between $S_1; d!1$ and $S_2; d!1$ implies that S_1 and S_2 must also be distinguishable. This means that we have to express termination in the commitment. This can be done conveniently, without introducing new primitives, by allowing the special variable *time* to occur in commitments. Observe that the commitment can be seen as an extension of the postcondition to non-terminating computations. Hence, by interpreting *time* similar as in postconditions, *time* in commitments expresses the termination time of terminating computations. For non-terminating computations we use the special variable ∞ and such computations satisfy the commitment $time = \infty$. In the example above, S_1 and S_2 can be distinguished by using the commitment $\forall t_0 : (comm\ c\ at\ t_0 \rightarrow time < \infty)$.

The syntax of the assertion language is given in Table 3, with $\tau \in TIME \cup \{\infty\}$, $\vartheta \in VAL$, t and v logical variables ranging over, respectively, $TIME \cup \{\infty\}$, VAL, $c \in CHAN$, and $x \in VAR$.

Table 3: Syntax of the Assertion Language

Expression	$exp ::=$	$\tau \mid \vartheta \mid t \mid v \mid time \mid x \mid$
		$exp_1 + exp_2 \mid exp_1 - exp_2 \mid exp_1 \times exp_2$
Assertion	$p ::=$	$comm\ (c, exp_1)\ at\ exp_2 \mid wait\ c!\ at\ exp \mid wait\ c?\ at\ exp \mid$
		$exp_1 = exp_2 \mid exp_1 < exp_2 \mid exp \in I\!N \mid$
		$\neg p \mid p_1 \vee p_2 \mid \exists t : p$

Let $var(p)$ denote the set of program variables occurring in assertion p. $dch(p)$ denotes the set of (directed) channels occurring in assertion p. E.g., $dch(comm\ (c, exp_1)\ at\ exp_2) = \{c\}$ and $dch(wait\ c!\ at\ exp) = \{c!\}$.

The conventional abbreviations are used, such as $p_1 \wedge p_2 \equiv \neg(\neg p_1 \vee \neg p_2)$, $p_1 \rightarrow p_2 \equiv \neg p_1 \vee p_2$, and $\forall t : p \equiv \neg \exists t : \neg p$. Relativized quantifiers are defined as usual, for instance,

- $\forall t, t_0 \leq t < time : p \equiv \forall t : (t_0 \leq t < time \rightarrow p)$
- $\exists t, t_0 \leq t < time : p \equiv \exists t : (t_0 \leq t < time \wedge p)$.

Furthermore, the following abbreviations are frequently used:

- $comm\ (c, v)\ during\ [t_0, t_1) \equiv \forall t_2, t_0 \leq t_2 < t_1 : comm\ (c, v)\ at\ t_2$
- $comm\ c\ at\ exp \equiv \exists v : comm\ (c, v)\ at\ exp$
- $end\ comm\ (c, v)\ at\ t \equiv comm\ (c, v)\ during\ [t - T_{comm}, t)$
- $end\ comm\ c\ at\ t \equiv \exists v : end\ comm\ (c, v)\ at\ t$
- $wait\ c!\ during\ [t_0, t_1) \equiv \forall t_2, t_0 \leq t_2 < t_1 : wait\ c!\ at\ t_2$
- $no\ comm\ c\ during\ [t_0, t_1) \equiv \forall t_2, t_0 \leq t_2 < t_1 : \neg\ comm\ c\ at\ t_2$
- $wait\ c!\ at\ t_0\ until\ comm\ (c, exp)\ at\ t_1 \equiv$
  $wait\ c!\ during\ [t_0, t_1) \wedge comm\ (c, exp)\ during\ [t_1, t_1 + T_{comm})$
- $await\ (c!, exp)\ at\ t_0 \equiv \exists t_1 \geq t_0 : wait\ c!\ at\ t_0\ until\ comm\ (c, exp)\ at\ t_1$
- $await\ c!\ at\ t_0 \equiv \exists v : await\ (c!, v)\ at\ t_0$

Let $cset$ be a finite subset of $DCHAN$. Then

- $no\ cset\ during\ [t_0, t_1) \equiv \forall t, t_0 \leq t < t_1 : \bigwedge_{c!\in cset} \neg\ wait\ c!\ at\ t \wedge$
$\bigwedge_{c?\in cset} \neg\ wait\ c?\ at\ t \wedge \bigwedge_{c\in cset} \neg\ comm\ c\ at\ t$

The abbreviations above are also used with $c?$ instead of $c!$, and with other intervals such as $\langle t_0, t_1 \rangle$ and $\langle t_0, \infty \rangle$ instead of the interval $[t_0, t_1)$. It is easy to extend these definitions to general expressions instead of t_0 or t_1.

Finally we describe the informal meaning of our correctness formulae. A formula $C : \{p\}\ S\ \{q\}$ is valid if

1. $var(C) = \emptyset$, and

2. if p holds in the initial state and the communication history in which S starts its execution then

    (a) C holds in the initial communication history extended with the communication behaviour of S, and

    (b) if the computation terminates then q holds in the final state and the initial communication history extended with the communication behaviour of S.

Examples

First a few general safety properties:

- Program S does not terminate.
  $true : \{true\}\ S\ \{false\}$.
  (S will also satisfy the formula $time = \infty : \{true\}\ S\ \{false\}$.)
- S does not perform any communication along channel c.
  $(\forall t \geq 0 : \neg comm\ c\ at\ t) : \{time = 0\}\ S\ \{true\}$.

Next a number of real-time safety properties:

- If S starts its execution in a state where variable x has the value 3 and if S terminates, then S terminates within 2 time units in a state where x has the value 4.
  $true : \{x = 3 \wedge time = t\}\ S\ \{x = 4 \wedge time < t + 2\}$.
- S terminates in less than 12 time units, incrementing x by 5.
  $time < 12 : \{x = v \wedge time = 0\}\ S\ \{x = v + 5\}$.
  (Instead of starting at time 0 we could also use a general starting time t_0:
  $time < t_0 + 12 : \{x = v \wedge time = t_0\}\ S\ \{x = v + 5\}$.)
- S communicates along channel c within 25 time units.
  $(\exists t < 25 : comm\ c\ at\ t) : \{time = 0\}\ S\ \{true\}$.

- If S receives value v via channel c then it will try to send the value $v + 7$ along channel d in less than 10 time units.

  $comm\ (c, v)\ at\ t_0 \rightarrow \exists t_1, t_0 < t_1 < t_0 + 10 : await\ (d!, v + 7)\ at\ t_1$:
  $\{time = 0\}\ S\ \{false\}$

Some liveness properties:

- S terminates.
  $time < \infty : \{true\}\ S\ \{true\}$.
- S eventually communicates along channel c.
  $(\exists t : comm\ c\ at\ t) : \{time = 0\}\ S\ \{true\}$.
- Program S either communicates infinitely often via channel c, or it eventually waits forever to receive along c.
  $(\forall t_0 < \infty\ \exists t_1 > t_0 : comm\ c\ at\ t_1) \vee (\exists t_0 < \infty\ \forall t_1 > t_0 : wait\ c?\ at\ t_1)$:
  $\{time = 0\}\ S\ \{false\}$.
  Note that the program $*[c?x \rightarrow skip]$ satisfies this liveness specification.

Finally, a few examples of programs satisfying a certain specification:

- $await\ (c!, 8)\ at\ 5 : \{time = 3\}\ \textbf{delay}\ 2\ ;\ c!8\ \{true\}$.
- $(v < 5 \rightarrow time = \infty) \wedge (v \geq 5 \rightarrow time < \infty)$:
  $\{x = v\} * [x > 5 \rightarrow x := x - 1 [\!] x < 5 \rightarrow skip]\{x = 5\}$.
  This formula describes how the termination of the program depends on the initial value of x.
- $comm\ (c, 9)\ at\ 3 \wedge await\ (d!, 6)\ at\ 7$:
  $\{comm\ (c, 9)\ at\ 3 \wedge time = 5 \wedge x = 4\}\ \textbf{delay}\ 2\ ;\ d!(x + 2)$
  $\{comm\ (c, 9)\ at\ 3 \wedge await\ (d!, 6)\ at\ 7 \wedge time < \infty\}$
  Observe that the communication history before the starting time, i.e. $comm\ (c, 9)\ at\ 3$ in the precondition, is included in the commitment and the postcondition.

5.3 Proof System for Maximal Parallelism Model

We formulate a compositional proof system for our extended Hoare triples. First we formulate rules and axioms that are generally applicable to any statement. Next we axiomatize the programming language by formulating rules and axioms for all atomic statements and compound programming constructs.

A general requirement for all rules and axioms is that the commitment should not contain program variables. Further, we assume that all logical variables which are introduced in the rules are fresh.

We start with an axiom expressing general well-formedness properties of a computation. Define $WF_c \equiv \forall t < time : MW_c(t) \wedge Excl_c(t) \wedge Comm_c(t)$, with

$MW_c(t) \equiv \neg(\text{wait } c? \text{ at } t \wedge \text{wait } c! \text{ at } t)$

(*Minimal waiting*: It is not possible to be simultaneously waiting to send and waiting to receive on a particular channel.)

$Excl_c(t) \equiv \neg(\text{wait } c! \text{ at } t \wedge \text{comm } c \text{ at } t) \wedge \neg(\text{wait } c? \text{ at } t \wedge \text{comm } c \text{ at } t)$

(*Exclusion*: It is not possible to be simultaneously communicating and waiting to communicate on a given channel.)

$Comm_c(t) \equiv \text{comm }(c, exp_1) \text{ at } t \wedge \text{comm }(c, exp_2) \text{ at } t \rightarrow exp_1 = exp_2$

(*Communication*: at any time at most one value is transmitted on a particular channel.)
Then, for every channel c we have the following axiom.

Axiom 5.2 (Well-Formedness) $WF_c \; : \; \{true\} \; S \; \{WF_c\}$

We give a few lemmas concerning the use of WF_c. These lemmas will be used in the chapter 6 where we illustrate our formalism by an example of a watchdog timer.

Lemma 5.3 For all t_0, t_1, $\text{wait } c? \text{ during } \langle t_0, t_1 \rangle \wedge WF_c \rightarrow \text{no } \{c!, c\} \text{ during } \langle t_0, t_1 \rangle$

Lemma 5.4 For all t_0, $\text{await } c! \text{ at } t_0 \wedge WF_c \rightarrow \forall t_1, t_0 < t_1 < t_0 + T_{comm} : \neg \text{ wait } c? \text{ at } t_1$

Next we give two axioms to deduce invariance properties. The first axiom expresses that a precondition which satisfies certain restrictions remains valid during the execution of a program.

Axiom 5.5 (Initial Invariance) $C : \{p \wedge C\} \; S \; \{p\}$

provided *time* does not occur in p or C, and $var(S) \cap var(p) = \emptyset$.

The channel invariance axiom below expresses that during the execution of a program S no activity takes place on channels not occurring in S. Let $cset$ be a finite subset of $DCHAN$.

Axiom 5.6 (Channel Invariance)

$\text{no } cset \text{ during } [t_0, time) \; : \; \{time = t_0\} \; S \; \{\text{no } cset \text{ during } [t_0, time)\}$

provided $cset \cap dch(S) = \emptyset$.

The proof system contains a consequence rule which is an extension of the classical consequence rule for Hoare triples. There is a disjunction rule which makes it possible to split a precondition in several cases (see the rules for guarded commands below). Further, the proof system contains the rules for conjunction and substitution.

Rule 5.7 (Consequence) $\dfrac{C_0 : \{p_0\}\ S\ \{q_0\},\ p \to p_0,\ C_0 \to C,\ q_0 \to q}{C : \{p\}\ S\ \{q\}}$

Rule 5.8 (Disjunction) $\dfrac{C_1 : \{p_1\}\ S\ \{q_1\},\ C_2 : \{p_2\}\ S\ \{q_2\}}{C_1 \vee C_2 : \{p_1 \vee p_2\}\ S\ \{q_1 \vee q_2\}}$

Rule 5.9 (Conjunction) $\dfrac{C_1 : \{p_1\}\ S\ \{q_1\},\ C_2 : \{p_2\}\ S\ \{q_2\}}{C_1 \wedge C_2 : \{p_1 \wedge p_2\}\ S\ \{q_1 \wedge q_2\}}$

Rule 5.10 (Substitution) $\dfrac{C : \{p\}\ S\ \{q\}}{C[exp/u] : \{p[exp/u]\}\ S\ \{q[exp/u]\}}$

for any logical variable u, provided *time* does not occur in expression exp.

Axiom 5.11 (Skip) $C : \{p \wedge C\}\ \textbf{skip}\ \{p\}$

Rule 5.12 (Assignment) $C : \{(q \wedge C)[time + T_{assign}/time, e/x]\}\ x := e\ \{q\}$

Note that by this rule, and the consequence rule, we can derive
$time = t_0 + T_{assign} : \{time = t_0 \wedge e = v\}\ x := e\ \{time = t_0 + T_{assign} \wedge x = v\}$,
since $time = t_0 \wedge e = v \to (time = t_0 + T_{assign} \wedge x = v)[time + T_{assign}/time, e/x]$.

Axiom 5.13 (Delay) $C : \{(q \wedge C)[time + max(0, e)/time]\}\ \textbf{delay}\ e\ \{q\}$

Example 5.2 With the delay axiom we can derive the formula
$time = 5 : \{x = 3 \wedge time = 2\}\ \textbf{delay}\ x\ \{x = 3 \wedge time = 5\}$,
because $(x = 3 \wedge time = 5)[time + max(0, x)/time] \equiv (x = 3 \wedge time + max(0, x) = 5)$ is equivalent to $x = 3 \wedge time = 2$. Similarly we can derive
$time = t_0 + max(0, e) : \{time = t_0\}\ \textbf{delay}\ e\ \{time = t_0 + max(0, e)\}$. □

To obtain a compositional proof system we do not make any assumption about the environment of a statement. Thus in the rule for an output statement no assumption is made about when a communication partner is available, and hence this rule includes any arbitrary waiting period (including an infinite one). In the rule for $c!e$ this is achieved by using $\exists t \geq t_0 : wait\ c!\ at\ t_0\ until\ comm\ (c, e)\ at\ t$ where t_0 is the starting time. Observe logical variable t, ranges over $TIME \cup \{\infty\}$, and thus

$\exists t \geq t_0 : wait\ c!\ at\ t_0\ until\ comm\ (c, e)\ at\ t$

is equivalent to

$wait\ c!\ at\ t_0\ until\ comm\ (c, e)\ at\ \infty \vee$
$\exists t, t_0 \leq t < \infty : wait\ c!\ at\ t_0\ until\ comm\ (c, e)\ at\ t$,

which is equivalent to

$wait\ c!\ during\ [t_0, \infty) \vee$
$(\exists t, t_0 \leq t < \infty : wait\ c!\ during\ [t_0, t) \wedge comm\ (c, e)\ during\ [t, t + T_{comm}))$,
since $comm\ (c, e)\ during\ [\infty, \infty + T_{comm}) \leftrightarrow true$, and thus
$wait\ c!\ at\ t_0\ until\ comm\ (c, e) \leftrightarrow wait\ c!\ during\ [t_0, \infty)$.

Hence the commitment can express a non-terminating computation in which the output statement waits forever to communicate. When a communication partner is available the actual communication takes place during an interval of length T_{comm}. This leads to the following rule.

Rule 5.14 (Output)
$$\frac{p[t_0/time] \wedge \exists t \geq t_0 : wait\ c!\ at\ t_0\ until\ comm\ (c, e)\ at\ t \wedge time = t + T_{comm} \rightarrow q \wedge C}{C\ :\ \{p\}\ c!e\ \{q \wedge time < \infty\}}$$

Example 5.3 With the output rule we can derive

$await\ (c!, 5)\ at\ 1 : \{time = 1 \wedge x = 3\}\ c!(x+2)\ \{end\ comm\ (c, 5)\ at\ time \wedge time < \infty\}$

since, for the commitment,

$t_0 = 1 \wedge x = 3 \wedge \exists t \geq t_0 : wait\ c!\ at\ t_0\ until\ comm\ (c, x + 2)\ at\ t \wedge time = t + T_{comm}$
$\rightarrow await\ (c!, 5)\ at\ 1$ and for the postcondition we have
$t_0 = 1 \wedge x = 3 \wedge \exists t \geq t_0 : wait\ c!\ at\ t_0\ until\ comm\ (c, x + 2)\ at\ t \wedge time = t + T_{comm} \rightarrow$
$\exists t \geq t_0 : comm\ (c, 5)\ during\ [t, t + T_{comm}) \wedge time = t + T_{comm}$, and thus we obtain the postcondition $end\ comm\ (c, 5)\ at\ time \wedge time < \infty$. □

In the input rule we allow any arbitrary input value, since no assumption should be imposed upon the environment, and hence any value can be received.

Rule 5.15 (Input)
$$\frac{p[t_0/time] \wedge \exists t \geq t_0 : wait\ c?\ at\ t_0\ until\ comm\ (c, v)\ at\ t \wedge time = t + T_{comm} \rightarrow q[v/x] \wedge C}{C\ :\ \{p\}\ c?x\ \{q \wedge time < \infty\}}$$

Example 5.4 With the input rule we can derive

$comm\ (d, 0)\ at\ 2 \wedge \exists v_1 : await\ (c?, v_1)\ at\ 4 :$
   $\{comm\ (d, 0)\ at\ 2 \wedge time = 4\}\ c?x\ \{comm\ (d, 0)\ at\ 2 \wedge end\ comm\ (c, x)\ at\ time\}$

since, for the commitment,
$comm\ (d, 0)\ at\ 2 \wedge t_0 = 4 \wedge \exists t \geq t_0 : wait\ c?\ at\ t_0\ until\ comm\ (c, v)\ at\ t \wedge time = t + T_{comm}$
$\rightarrow comm\ (d, 0)\ at\ 2 \wedge \exists v_1 : await\ (c?, v_1)\ at\ 4$. For the postcondition we have
$comm\ (d, 0)\ at\ 2 \wedge t_0 = 4 \wedge \exists t \geq t_0 : wait\ c?\ at\ t_0\ until\ comm\ (c, v)\ at\ t \wedge time = t + T_{comm}$
$\rightarrow comm\ (d, 0)\ at\ 2 \wedge end\ comm\ (c, v)\ at\ time \equiv$
$(comm\ (d, 0)\ at\ 2 \wedge end\ comm\ (c, x)\ at\ time)[v/x]$. □

The inference rule for sequential composition is an extension of the classical rule for Hoare triples. To explain the commitment of $S_1; S_2$, observe that a computation of $S_1; S_2$ is either a non-terminating computation of S_1 or a terminated computation of S_1 extended with a computation of S_2. The commitment of $S_1; S_2$ expresses the non-terminating computations of S_1 by using the commitment of S_1 with $time = \infty$. Terminating computations of S_1 are characterized in the postcondition of S_1 which is also the precondition of S_2. Then these computations are extended by S_2 and described in the commitment of S_2.

Rule 5.16 (Sequential Composition)

$$\frac{C_1 : \{p\}\ S_1\ \{r\},\ C_2 : \{r\}\ S_2\ \{q\}}{(C_1 \wedge time = \infty) \vee C_2 : \{p\}\ S_1; S_2\ \{q\}}$$

Example 5.5 Consider the program $c?x; d!(x+1)$. We prove

$(wait\ c?\ during\ [0,\infty) \wedge time = \infty) \vee$
$(\exists t_1 < \infty\ \exists v : end\ comm\ (c, v)\ at\ t_1 \wedge await\ (d!, v+1)\ at\ t_1) :$
$\quad \{time = 0\}\ c?x\ ;\ d!(x+1)$
$\quad\quad \{\exists t_1 < \infty\ \exists t_2, t_1 < t_2 < \infty : end\ comm\ (c, x)\ at\ t_1 \wedge end\ comm\ (d, x+1)\ at\ t_2\}.$

Define $C^1_{nonterm} \equiv wait\ c?\ during\ [0, \infty)$, $C^1_{term} \equiv end\ comm\ (c, v)\ at\ t_1$, and $q \equiv \exists t_1 < \infty\ \exists t_2, t_1 < t_2 < \infty : end\ comm\ (c, x)\ at\ t_1 \wedge end\ comm\ (d, x+1)\ at\ t_2$. Then

$(time = \infty \to C^1_{nonterm}) : \{time = 0\}\ c?x\ \{\exists t_1 < \infty\ \exists v : C^1_{term} \wedge time = t_1 \wedge x = v\}.$

For $d!(x+1)$, define $C_2 \equiv await\ (d!, v+1)\ at\ t_1$, then we can derive

$(\exists t_1 < \infty\ \exists v : C^1_{term} \wedge C_2) :$
$\quad \{\exists t_1 < \infty\ \exists v : C^1_{term} \wedge time = t_1 \wedge x = v\}\ d!(x+1)$
$\quad\quad \{\exists t_1 < \infty\ \exists v\ \exists t_2, t_1 < t_2 < \infty : C^1_{term} \wedge end\ comm\ (d, v+1)\ at\ t_2 \wedge x = v\}.$

Observe that the terminating behaviour of $c?x$ is characterized by its postcondition, thus by the precondition of $d!(x+1)$, and hence can be included in the commitment of $d!(x+1)$. Since the postcondition of $d!(x+1)$ implies q, we obtain by the sequential composition rule and the consequence rule:

$(C^1_{nonterm} \wedge time = \infty) \vee (\exists t_1 < \infty\ \exists v : C^1_{term} \wedge C_2) : \{time = 0\}\ c?x\ ;\ d!(x+1)\ \{q\}.$ □

For guarded commands, we start with a simple rule for the case that none of the guards is open, i.e. b_G evaluates to false. Recall that the evaluation of b_G takes T_{exec} time units.

Rule 5.17 (Guarded Command Termination)

$$\frac{C : \{p \wedge \neg b_G\}\ \text{delay}\ T_{exec}\ \{q\}}{C : \{p \wedge \neg b_G\}\ G\ \{q\}}$$

Next consider $G \equiv [\|_{i=1}^{n} b_i \to S_i]$.

Rule 5.18 (Guarded Command with Purely Boolean Guards)

$$\frac{C_i \;:\; \{p \wedge b_i\} \text{ delay } T_{exec}\,;\, S_i \; \{q_i\}, \text{ for all } i \in \{1,\ldots,n\}}{\bigvee_{i=1}^{n} C_i \;:\; \{p \wedge b_G\} \;[\|_{i=1}^{n} b_i \to S_i] \; \{\bigvee_{i=1}^{n} q_i\}}$$

Now let $G \equiv [\|_{i=1}^{n} b_i; c_i?x_i \to S_i \,[\!]\, b_0; \text{delay } e \to S_0]$. Define

- *wait in* G *during* $[t_0, t_1) \equiv \bigwedge_{i=1}^{n}(b_i \leftrightarrow \text{wait } c_i?\text{ during } [t_0, t_1)) \wedge$
  $\text{no }(dch(G) - \{c_1?, \ldots, c_n?\}) \text{ during } [t_0, t_1), \text{ and}$
- *await comm* G *at* $t_0 \equiv \exists t \geq t_0 :$ wait in G during $[t_0, t) \wedge (b_0 \to t < max(0, e)) \wedge$
  $\exists i \in \{1, \ldots, n\} : b_i \wedge \text{comm } c_i \text{ during } [t, t + T_{comm}) \wedge$
  $\text{no }(dch(G) - \{c_i\}) \text{ during } [t, t + T_{comm}) \wedge time = t + T_{comm}.$

Then we can formulate the rule for a guarded command as follows.

Rule 5.19 (Guarded Command with Input Guards)

$$\frac{\begin{array}{l} true \;:\; \{p \wedge b_G\} \text{ delay } T_{exec} \; \{\hat{p}\} \\ \hat{p}[t_0/time] \wedge \text{wait in } G \text{ during } [t_0, \infty) \wedge time = \infty \to C_{nonterm} \\ \hat{p}[t_0/time] \wedge \text{await comm } G \text{ at } t_0 \wedge \text{end comm } (c_i, v) \text{ at time} \to p_i[v/x_i], \text{ for } i = 1, \ldots, n \\ \hat{p}[t_0/time] \wedge b_0 \wedge \text{wait in } G \text{ during } [t_0, t_0 + max(0, e)) \wedge time = t_0 + max(0, e) \to p_0 \\ C_i \;:\; \{p_i\} \; S_i \; \{q_i\}, \text{ for } i = 0, 1 \ldots, n \end{array}}{C_{nonterm} \vee \bigvee_{i=0}^{n} C_i \;:\; \{p \wedge b_G\} \;[\|_{i=1}^{n} b_i; c_i?x_i \to S_i \,[\!]\, b_0; \text{delay } e \to S_0] \; \{\bigvee_{i=0}^{n} q_i\}}$$

The rule for the iteration construct does not contain any explicit well-foundedness argument, although we deal with liveness properties. The main principle is that liveness properties can be derived from real-time safety properties, and these properties can be proved by means of an invariant.

Rule 5.20 (Iteration)

$$\frac{\begin{array}{l} C \;:\; \{p \wedge b_G\} \; G \; \{p\} \\ C_{term} \;:\; \{p \wedge \neg b_G\} \; G \; \{q\} \\ (\forall t_1 < \infty \; \exists t_2 > t_1 : C[t_2/time]) \to C_{nonterm} \end{array}}{(C_{nonterm} \wedge time = \infty) \vee C_{term} \;:\; \{p\} \; *G \; \{q\}}$$

The soundness of this rule is shown as follows:

For a computation of $*G$, starting in a state satisfying p, there are three possibilities:

- It is a terminating computation, obtained from a finite number of terminating computations from G. For all these computations of G, except for the last one, b_G is true initially. From the first condition of the rule we can then prove by induction also that p is true in the initial state of these computations. Since for the last computation $\neg b_G$ must be true, the second condition of the rule then leads to C_{term} and q for this computation.

- It is a non-terminating computation obtained from a non-terminating computation of G. Then, as in the previous point, we have $p \wedge b_G$ in the initial state of this computation. Thus, using the first condition and the fact that it is a non-terminating computation, $C \wedge time = \infty$ holds for this computation. Hence, $\forall t_1 < \infty \exists t_2 > t_1 : C[t_2/time]$, and then the third condition leads to $C_{nonterm}$.

- It is a non-terminating computation obtained from an infinite sequence of terminating computations of G. Then, similar to the first point, the first condition leads by induction to C for all these computations. Since each computation of G takes at least T_{exec} time units, we obtain C after any point of time. Thus, $\forall t_1 < \infty \exists t_2 > t_1 : C[t_2/time]$, and then by the third condition we obtain $C_{nonterm}$.

Example 5.6 Termination of program $\star[x > 0 \rightarrow x := x-1]$ can be expressed as follows:
$$time < \infty \; : \; \{x \in I\!N\} \; \star[x > 0 \rightarrow x := x - 1] \; \{true\}.$$
We indicate how this formula is proved by means of an invariant. First this liveness property is strengthened to a real-time safety property. Let $K \equiv T_{exec} + T_{assign}$. We prove $time = v \times K + T_{exec} \; : \; \{time = 0 \wedge x = v \wedge v \in I\!N\} \; \star[x > 0 \rightarrow x := x - 1] \; \{true\}.$
Define $C_{term} \equiv time = v \times K + T_{exec}$, $p_0 \equiv time = 0 \wedge x = v \wedge v \in I\!N$, and $q \equiv true$. Then we have to prove $C_{term} \; : \; \{p_0\} \; \star[x > 0 \rightarrow x := x - 1] \; \{q\}$. We apply the iteration rule with $p \equiv time = (v-x) \times K < \infty \wedge x \in I\!N$, $C \equiv time \leq v \times K$, and $C_{nonterm} \equiv false$. Observe that the three conditions of the rule are fulfilled:

1. $C \; : \; \{p \wedge x > 0\} \; [x > 0 \rightarrow x := x - 1] \; \{p\}$ can be proved as follows.
   Let $r \equiv time = (v - x) \times K + T_{exec} < \infty \wedge x \in I\!N \wedge x > 0$.
   By the delay axiom we obtain $time < \infty \; : \; \{p \wedge x > 0\} \; \textbf{delay} \; T_{exec} \; \{r\}$,
   and from the assignment axiom, $time \leq v \times K \; : \; \{r\} \; x := x - 1 \; \{p\}$.
   Then the sequential composition rule leads to
   $(time < \infty \wedge time = \infty) \vee (time \leq v \times K) \; : \; \{p \wedge x > 0\} \; \textbf{delay} \; T_{exec}; x := x - 1 \; \{p\}$,
   and thus $time \leq v \times K \; : \; \{p \wedge x > 0\} \; \textbf{delay} \; T_{exec}; x := x - 1 \; \{p\}$.
   Hence by rule 5.18 we obtain $C \; : \; \{p \wedge x > 0\} \; [x > 0 \rightarrow x := x - 1] \; \{p\}$.

2. $C_{term} \; : \; \{p \wedge x \leq 0\} \; [x > 0 \rightarrow x := x-1] \; \{q\}$, can be derived from rule 5.17; observe that $C_{term} \; : \; \{p \wedge x \leq 0\} \; \textbf{delay} \; T_{exec} \; \{q\}$ follows from the delay axiom since $(p \wedge x \leq 0) \rightarrow (time = v \times K)$.

3. $\forall t_1 < \infty \exists t_2 > t_1 : C[t_2/time] \equiv \forall t_1 < \infty \exists t_2 > t_1 : t_2 \leq v \times K$ implies $false$, i.e., $C_{nonterm}$.

Then by the iteration rule we obtain
$$(C_{nonterm} \wedge time = \infty) \vee C_{term} \; : \; \{p\} \; \star[x > 0 \rightarrow x := x - 1] \; \{q\}.$$
Since $C_{nonterm} \wedge time = \infty \rightarrow false$ and $p_0 \rightarrow p$, the consequence rule leads to

C_{term} : $\{p_0\} \star [x > 0 \to x := x - 1] \{q\}$.

Using $C_{term} \to time < \infty$, we obtain

$time < \infty$: $\{time = 0 \land x = v \land v \in I\!N\} \star [x > 0 \to x := x - 1] \{true\}$.

By the substitution rule, replacing v by x, we can derive

$time < \infty$: $\{time = 0 \land x \in I\!N\} \star [x > 0 \to x := x - 1] \{true\}$.

To obtain precondition $x \in I\!N$ one should replace $time = 0$ by $time = t_0$ and prove $time = t_0 + v \times K + T_{exec} < \infty$:

$$\{time = t_0 < \infty \land x = v \land v \in I\!N\} \star [x > 0 \to x := x - 1] \{true\}.$$

As a last step in the proof we can then replace t_0 by $time$, using the substitution rule. □

Consider the parallel composition $S_1 \| S_2$. If $time$ does not occur in commitments and postconditions of the components S_1 and S_2, then we have the following simple rule:

Rule 5.21 (Simple Parallel Composition)

$$\frac{C_1 : \{p_1\} S_1 \{q_1\}, \quad C_2 : \{p_2\} S_2 \{q_2\}}{C_1 \land C_2 : \{p_1 \land p_2\} S_1 \| S_2 \{q_1 \land q_2\}}$$

provided $dch(C_i, q_i) \subseteq dch(S_i)$, $var(q_i) \subseteq var(S_i)$, for $i \in \{1, 2\}$, and $time$ does not occur in C_1, C_2, q_1, and q_2.

Example 5.7 To illustrate the problem with the termination times at parallel composition, consider the following two (valid) formulae:

$time = 5$: $\{time = 0\}$ **delay** 5 $\{time = 5\}$, and

$time = 7$: $\{time = 0\}$ **delay** 7 $\{time = 7\}$.

Then for **delay** $5 \|$ **delay** 7 we cannot take the conjunction of commitments and postconditions, since this would imply $false$. □

The problem is that, in general, the termination times of S_1 and S_2 will be different. Therefore we substitute a logical variable t_i for $time$ in q_i, $i = 1, 2$. Then the termination time of $S_1 \| S_2$, expressed by $time$ in its postcondition, is the maximum of t_1 and t_2. A similar construction is used for the commitments. This leads to the following rule:

Rule 5.22 (Parallel Composition)

$$\frac{\begin{array}{c} C_i : \{p_i\} S_i \{q_i\}, i = 1, 2 \\ \bigwedge_{i=1}^{2} C_i[t_i/time] \land time = max(t_1, t_2) \to C \\ \bigwedge_{i=1}^{2} q_i[t_i/time] \land time = max(t_1, t_2) \to q \end{array}}{C : \{p_1 \land p_2\} S_1 \| S_2 \{q\}}$$

provided $dch(C_i, q_i) \subseteq dch(S_i)$, and $var(q_i) \subseteq var(S_i)$, for $i = 1, 2$.
(To obtain a complete system one should add $no\ dch(S_i)\ during\ [t_i, time)$ to the conjunction before the implications in the rule.)

In example 5.7 we can now obtain the commitment and postcondition $time = 7$ because $(\exists t_1, t_2 : time = max(t_1, t_2) \land t_1 = 5 \land t_2 = 7) \to time = 7$.

6 Example Watchdog Timer

The formalism from chapter 5 is illustrated by an example of a watchdog timer. Consider the network pictured in Fig. 1. Process W is a "watchdog" process: its job is to ensure

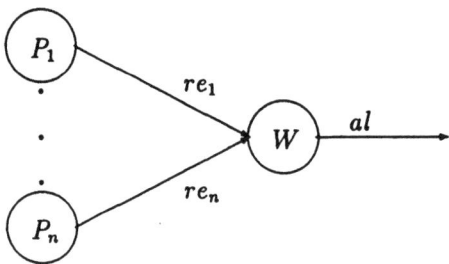

Figure 1: Watchdog Timer Network

that processes $P_1, \ldots, P_n$ are functioning properly. We abstract from the task that has to be performed by P_i, but we assume that P_i is functioning correctly iff it is ready to send (or sending) a reset signal on channel re_i to W at least once every v_i time units, for some constant v_i. So long as all processes P_i are ready to send a reset signal in time, watchdog timer W communicates on each re_i at least once every v_i time units and then it does not communicate on channel al. As soon as W has to wait for a reset signal on a particular re_k during v_k time units, then it is ready to send (or sending) an alarm message on channel al within, say, K time units.

In this section we first give a formal specification for process W. Then, given specifications for the P_i, we prove that $P_1 \| \cdots \| P_n \| W$ is ready to send (or sending) on channel al iff one of the P_i is not functioning correctly. This is verified using our proof system without knowing the implementations of $P_1, \ldots, P_n$ and W. To demonstrate program design from a specification, W is implemented as a parallel composition $W_1 \| \cdots \| W_n \| A$, and we derive the specification of W using specifications for W_i and A. Next W_i and A are, independently, implemented, and we prove that these programs satisfy the corresponding specifications.

6.1 Specification of the Watchdog Timer

We give a formal specification for the watchdog timer W and derive properties from it, using certain specifications for the processes P_i. In the specification of W we express that if there is a waiting period of v_k time units to receive input via re_k then, for some constant K, W starts waiting to send on channel al within K until the actual communication takes place. Furthermore, W tries to communicate value k via channel al at a certain point of time only if there was a previous period of at least v_k time units during which W is waiting to receive input via re_k. Let

$$C_0^W \equiv \forall t_0 < \infty : wait\ re_k?\ during\ \langle t_0, t_0 + v_k \rangle \to \exists t \le t_0 + v_k + K : await\ al!\ at\ t$$

$$C_1^W \equiv \forall t_1 < \infty : await\ (al!, k)\ at\ t_1 \to \exists t_2 \le t_1 - v_k : wait\ re_k?\ during\ \langle t_2, t_2 + v_k \rangle)$$

Then we specify W by $\quad C_0^W \wedge C_1^W : \{time = 0\}\ W\ \{true\}$.

We prove that W tries to send a message via al iff there is an error in one of the processes P_i. Therefore we assume given a specification for P_i in which we use a predicate $error_i$ representing some erroneous behaviour of P_i. Thus assume that, for all i, we have $C^{P_i} : \{time = 0\}\ P_i\ \{true\}$, where

$$C^{P_i} \equiv error_i \leftrightarrow \exists t_0 < \infty : no\ \{re_i!, re_i\}\ during\ \langle t_0, t_0 + v_i \rangle$$

This asserts that there is an error in P_i iff there exists a period of v_i time units during which P_i is not communicating via re_i and not waiting to communicate via re_i. Given our specifications for $P_1, \ldots, P_n$ and W, we try to prove that $P_1 \| \cdots \| P_n \| W$ satisfies the commitment

$(\exists k : error_k) \leftrightarrow (\exists t < \infty : await\ (al!, k)\ at\ t)$. Applying the simple parallel composition rule n times we obtain

$$C_0^W \wedge C_1^W \wedge \bigwedge_{i=1}^n C^{P_i} : \{time = 0\}\ P_1 \| \cdots \| P_n \| W\ \{true\}.$$

By the well-formedness axiom and the consequence rule we can derive

$$\bigwedge_{k=1}^n WF_{re_k} : \{true\}\ P_1 \| \cdots \| P_n \| W\ \{true\}.$$

Using the conjunction rule we obtain the following commitment:

$$C_0^W \wedge C_1^W \wedge \bigwedge_{i=1}^n C^{P_i} \wedge \bigwedge_{k=1}^n WF_{re_k}.$$

First we prove that this commitment implies
$\exists t < \infty : await\ (al!, k)\ at\ t \to error_k$.

$$\exists t < \infty : await\ (al!, k)\ at\ t$$
$\Rightarrow \quad \{C_1^W\} \quad \exists t < \infty\ \exists t_2 \le t - v_k : wait\ re_k?\ during\ \langle t_2, t_2 + v_k \rangle$
$\Rightarrow \quad \{calculus\} \quad \exists t_2 < \infty : wait\ re_k?\ during\ \langle t_2, t_2 + v_k \rangle$
$\Rightarrow \quad \{Lemma\ 5.3\} \quad \exists t_2 < \infty : no\ \{re_k!, re_k\}\ during\ \langle t_2, t_2 + v_k \rangle$
$\Rightarrow \quad \{C^{P_k}\} \quad error_k$

Next we try to prove $\exists k : error_k \to \exists t < \infty : await\ (al!, k)\ at\ t$.
From $\exists k : error_k$ we obtain, by C^{P_k}, $\exists k\ \exists t_0 < \infty : no\ \{re_k!, re_k\}\ during\ \langle t_0, t_0 + v_k \rangle$, and

thus $\exists k \, \exists t_0 < \infty$: *no comm re_k during $\langle t_0, t_0 + v_k \rangle$*. With the current specification of W, however, nothing can be derived from this. The specification of W only expresses how W should behave if it does something on any of the channels. But then W need not do anything; even the simple program skip would satisfy its specification. Therefore we modify the specification for W as follows: $C_1^W \wedge C_2^W : \{time = 0\} \, W \, \{true\}$, with

$C_1^W \equiv \forall t_1 < \infty : await \, (al!, k) \, at \, t_1 \rightarrow \exists t_2 \le t_1 - v_k : wait \, re_k? \, during \, \langle t_2, t_2 + v_k \rangle$

$C_2^W \equiv \forall t_3 < \infty : no \, comm \, re_k \, during \, \langle t_3, t_3 + v_k \rangle \rightarrow \exists t_4 \le t_3 + v_k + K : await \, al! \, at \, t_4$

Note that C_0^W follows from C_2^W, because *wait re_k? during $\langle t_0, t_0+v_k \rangle$* implies by Lemma 5.3, *no $\{re_k!, re_k\}$ during $\langle t_0, t_0 + v_k \rangle$*, and hence *no comm re_k during $\langle t_0, t_0 + v_k \rangle$*. Now the proof proceeds as follows, for all k,

$\exists k : error_k$

$\Rightarrow \{C^{P_k}\} \quad \exists k \, \exists t_0 < \infty : no \, \{re_k!, re_k\} \, during \, \langle t_0, t_0 + v_k \rangle$

$\Rightarrow \{\text{definition}\} \quad \exists k \, \exists t_0 < \infty : no \, comm \, re_k \, during \, \langle t_0, t_0 + v_k \rangle$

$\Rightarrow \{C_2^W\} \quad \exists k \, \exists t_0 < \infty \, \exists t_4 \le t_0 + v_k + K : await \, (al!, k) \, at \, t_4$

$\Rightarrow \{\text{calculus}\} \quad \exists t_4 < \infty : await \, (al!, k) \, at \, t_4$

6.2 Implementing the Watchdog Timer

Next we design a program implementing watchdog process W that satisfies the required specification. Since W has to watch all processes $P_1, \ldots, P_n$ simultaneously, our first design step is to implement W as a parallel composition, $W \equiv W_1 \| \cdots \| W_n \| A$.

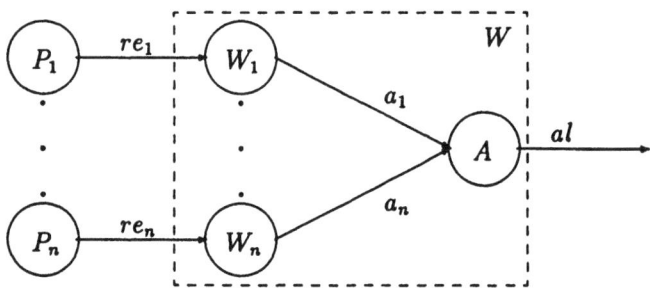

Figure 2: Implementation Watchdog Timer

Process W_i is a watchdog for P_i, and it signals process A via channel a_i as soon as there is no communication on re_i for at least v_i time units. Process A waits for a signal on any of the a_i's; after receipt of a signal it tries to send a message on al (see Fig. 2). We give specifications for W_i and A and prove that they are sufficient to derive the specification of W. The specification for W_i expresses that W_i tries to communicate via a_i only if it has been waiting to communicate via re_i during a period of v_i time units. On the other hand, if there is a period of v_i time units during which no communication via re_i occurs, then W_i will try to communicate via a_i within a certain time bound K_i. Define

$$C_1^{W_i} \equiv \forall t_1 < \infty : await\ a_i!\ at\ t_1 \to \exists t_2 \leq t_1 - v_i : wait\ re_i?\ during\ \langle t_2, t_2 + v_i \rangle$$

$$C_2^{W_i} \equiv \forall t_1 < \infty : no\ comm\ re_i\ during\ \langle t_3, t_3 + v_i \rangle \to \exists t_4 \leq t_3 + v_i + K_i : await\ a_i!\ at\ t_4$$

Then W_i is specified by $\quad C_1^{W_i} \wedge C_2^{W_i} : \{time = 0\}\ W_i\ \{true\}$.

The specification for A asserts that it tries to send a message via al only if there was a preceding communication via one of the a_k. If A is not waiting to communicate via one of the a_k at a certain point of time, then within, say, K_A time units it will wait to communicate via al until the actual communication can be performed. Define

$$C_1^A \equiv \forall t_1 < \infty : await\ (al!, k)\ at\ t_1 \to \exists t_2 \leq t_1 : comm\ a_k\ during\ [t_2, t_2 + T_{comm})$$

$$C_2^A \equiv \forall t_3 < \infty : \neg\ wait\ a_k?\ at\ t_3 \to \exists t_4 < t_3 + K_A : await\ al!\ at\ t_4$$

Then $C_1^A \wedge C_2^A : \{time = 0\}\ A\ \{true\}$.

We show that $W_1 \| \cdots \| W_n \| A$ satisfies the specification of W (using the specifications of $W_1, \ldots, W_n$, and A only). By the repeated application of the simple parallel composition rule, we obtain the conjunction of the commitments of the processes: $\bigwedge_{i=1}^n (C_1^{W_i} \wedge C_2^{W_i}) \wedge C_1^A \wedge C_2^A$. By the well-formedness axiom and the conjunction rule we can add $\bigwedge_{k=1}^n WF_{a_k}$, leading to the following commitment: $\bigwedge_{i=1}^n (C_1^{W_i} \wedge C_2^{W_i}) \wedge C_1^A \wedge C_2^A \wedge \bigwedge_{k=1}^n WF_{a_k}$. This implies C_1^W as follows, for all $t_1 < \infty$,

$\qquad\qquad\qquad await\ (al!, k)\ at\ t_1$
$\Rightarrow \qquad \{C_1^A\} \qquad \exists t_2 \leq t_1 : comm\ a_k\ during\ [t_2, t_2 + T_{comm})$
$\Rightarrow \qquad \{\text{definition}\} \qquad \exists t_2 \leq t_1 : wait\ a_k!\ at\ t_2\ until\ comm$
$\Rightarrow \qquad \{C_1^{W_k}\} \qquad \exists t_2 \leq t_1 \exists t_3 \leq t_2 - v_k : wait\ re_k?\ during\ \langle t_3, t_3 + v_k \rangle$
$\Rightarrow \{t_3 \leq t_2 - v_k \leq t_1 - v_k\}\ \exists t_3 \leq t_1 - v_k : wait\ re_k?\ during\ \langle t_3, t_3 + v_k \rangle$

Next we prove C_2^W. For all $t_3 < \infty$,

$\qquad\qquad\qquad no\ comm\ re_k\ during\ \langle t_3, t_3 + v_k \rangle$
$\Rightarrow \quad \{C_2^{W_k}\} \qquad \exists t_4 \leq t_3 + v_k + K_k : wait\ a_k!\ at\ t_4\ until\ comm$
$\Rightarrow \{\text{Lemma 5.4}\}\ \exists t_4 \leq t_3 + v_k + K_k\ \forall t_5, t_4 < t_5 < t_4 + T_{comm} : \neg\ wait\ a_k?\ at\ t_5$
$\Rightarrow \qquad \{C_2^A\} \qquad \exists t_4 \leq t_3 + v_k + K_k\ \forall t_5, t_4 < t_5 < t_4 + T_{comm}\ \exists t_6 < t_5 + K_A :$
$\qquad\qquad\qquad\qquad await\ al!\ at\ t_6$
$\Rightarrow \quad \{\text{calculus}\} \quad \exists t_4 \leq t_3 + v_k + K_k\ \exists t_6 \leq t_4 + K_A : await\ al!\ at\ t_6$
$\Rightarrow \quad \{\text{calculus}\} \quad \exists t_6 \leq t_3 + v_k + K_k + K_A : await\ al!\ at\ t_6$

Hence the specification of W can be derived provided $K_k + K_A \leq K$, for all k.

6.3 Final Implementations

Finally, we give implementations for the processes A and W_i that meet the required specifications. We implement W_i by

$$*[re_i?x_i \rightarrow \text{skip} \, [] \, \text{delay } v_i \rightarrow a_i!0].$$

This program satisfies the required specification $C_1^{W_i} \wedge C_2^{W_i} : \{time = 0\} \, W_i \, \{true\}$, provided $v_i > 0$ and $T_{guard} = 0$. Process A can be implemented as

$$[[]_{i=1}^n a_i?y_i \rightarrow a!!i].$$

provided $T_{comm} < K_A$ and $T_{guard} = 0$.

By giving programs that implement A and W_i, we have obtained an implementation that satisfies the top-level specification for a watchdog timer as given in section 6.1. To conclude this example, we analyze the requirements which have been imposed upon the constants K, K_i, K_A and v_i to prove the correctness of our implementation. To justify the refinement step from the previous section, we required $K_i + K_A \leq K$, for all $i = 1, ..., n$. The implementation given in this section has been proved to satisfy the specification for all K_A and K_i such that $T_{comm} < K_A$ and $K_i \geq 0$, for all $i = 1, .., n$. Observe that if $K > T_{comm}$ then for $K_A = K$ and $K_i = 0$ we have $T_{comm} < K_A$ and $K_i \geq 0$, and also $K_i + K_A = K \leq K$. Hence, if $v_i > 0$ for all $i = 1, .., n$, $T_{guard} = 0$, and $K > T_{comm}$, i.e., the constant in the specification must be greater than the duration of a communication, then our implementation meets the top-level specification for W.

7 Adding Assumptions to Real-Time

In general, real-time embedded systems have an intensive interaction with their environment, and usually their correctness strongly depends on assumptions about the environment. Therefore it is convenient to use correctness formulae in which these assumptions can be expressed. Hence, similar to chapter 4, we extend the formulae $C : \{p\} \, S \, \{q\}$ from chapter 5 with a fourth assertion, called *assumption*, leading to formulae of the form $(A, C) : \{p\} \, S \, \{q\}$. Assumption A should neither contain program variables nor the special variable *time*.

In this assumption/commitment formalism for real-time properties, we can now express, for instance, in the assumption when the environment of a process is waiting to communicate. With such an assumption we can determine when the communication must take place. For example,

$(A \equiv$ wait to c! at 5 until comm $\wedge$ no comm via c during $[3,5)$,
$C \equiv$ comm via c during $[5, 5+T_{comm}))$:

$$\{time = 3\}\ c?x\ \{time = 5 + T_{comm}\}$$

Note that, by using the maximal parallelism model, a communication takes place as soon as both process and environment are ready for to perform the communication.

In the remainder of this chapter we indicate how the proof system from the chapter 5 can be extended to obtain a compositional proof system for these assumption/commitment-formulae. First we consider a compositional rule for the parallel composition of S_1 and S_2. Concerning preconditions, postconditions and commitments, the rule is identical to Rule 5.22 for the commitment-formalism. For the assumptions, we have same requirements as in Rule 4.2: $A \wedge C_1 \to A_2$ and $A \wedge C_2 \to A_1$.

Rule 7.1 (Parallel Composition)

$$\frac{\begin{array}{c}(A_1, C_1) : \{p_1\}\ S_1\ \{q_1\},\ (A_2, C_2) : \{p_2\}\ S_2\ \{q_2\} \\ q_1[t_1/time] \wedge q_2[t_2/time] \wedge time = max(t_1, t_2) \to q \\ C_1[t_1/time] \wedge C_2[t_2/time] \wedge time = max(t_1, t_2) \to C \\ A \wedge C_1 \to A_2,\ A \wedge C_2 \to A_1\end{array}}{(A, C) : \{p_1 \wedge p_2\}\ S_1 \parallel S_2\ \{q\}}$$

provided $dch(C_i, q_i) \subseteq dch(S_i)$ and $var(q_i) \subseteq dch(S_i)$, for $i \in \{1, 2\}$.

A typical application of this rule can be found in the next example.

Example 7.1 Consider the following specifications (**delay** d is used to represent any internal actions which takes d time units and assume communications take one time unit, i.e., $T_{comm} = 1$):

$(A_1 \equiv$ wait to c? at 2 until comm $\wedge$ no comm via c during $[0, 2) \wedge$
wait to d? at 6 until comm $\wedge$ no comm via d during $[3, 6)$,
$(C_1 \equiv$ comm via c during $[2, 3) \wedge$ wait to d! at 3 until comm $\wedge$
no comm via d during $[0, 3))$:

$$\{time = 0\}\ c!x\ ;\ d!x\ ;\ \textbf{delay}\ 2\ \{time = 9\},\ \text{and}$$

$(A_2 \equiv$ wait to d! at 3 until comm $\wedge$ no comm via d during $[0, 3)$,
$C_2 \equiv$ wait to d? at 6 until comm $\wedge$ no comm via d during $[0, 6) \wedge$
comm via d during $[6, 7))$:

$$\{time = 0\}\ \textbf{delay}\ 6\ ;\ d?y\ \{time = 7\}$$

Take for $(c!x\ ;\ d!x\ ;\ \textbf{delay}\ 2) \parallel (\textbf{delay}\ 6\ ;\ d?y)$ the following assumption:
$A \equiv$ wait to c? at 2 until comm $\wedge$ no comm via c during $[0, 2)$.

Since $A \wedge C_1 \to A_2$ and $A \wedge C_2 \to A_1$, the Parallel Composition Rule leads to $(A, C_1 \wedge C_2) : \{time = 0\}\ (c!x\ ;\ d!x\ ;\ \text{delay } 2) \parallel (\text{delay } 6\ ;\ d?y)\ \{time = 9\}$.
Using a consequence rule we can easily derive from $C_1 \wedge C_2$ the following commitment:
comm via c during $[2, 3) \wedge$ *comm via d during* $[6, 7)$. □

Similar to chapter 4, to achieve a sound rule for parallel composition, an inductive relation between assumption and commitment is necessary to avoid circularity. In our real-time specifications we require for the validity of $(A, C) : \{p\}\ S\ \{q\}$ that there exists a $\delta > 0$ such that

1. for all t with $0 \leq t < \delta$: C holds at t, and

2. for all $t \geq \delta$: if A holds at $t - \delta$ then C holds at t.

In our examples this requirement is fulfilled if *comm via D during* $[t_0, t_1)$ is not considered as an abbreviation but as a primitive which is trivially true at all points of time before t_1.

Other rules and axioms for the assumption/commitment formalism can be obtained by adapting the commitment-based proof system of the previous section. We simply add assumption *true* to the rules and axioms for atomic statements, and the proof system is extended by a rule that allows the addition of an assumption to strengthen commitment and postcondition. To formulate this rule, let $p@t$ denote assertion p at time t, ignoring the part of p that refers to points of time after t. (By definition, $(p@t) \equiv true$ if $t < 0$.) Then, for all $\delta > 0$, we have the following rule:

Rule 7.2 (Strengthen)
$$\frac{\begin{array}{l}(A_1, C_1) : \{p_1\}\ S\ \{q_1\} \\ \forall t : (A@(t - \delta)) \wedge (C_1@t) \to (C@t) \\ \forall t : (A@(t - \delta)) \wedge (q_1@t) \to (q@t)\end{array}}{(A_1 \wedge A, C) : \{p\}\ S\ \{q\}}$$

Furthermore, the rules for compound statements have to be adapted, and in the consequence rule we require that all implications hold point-wise.

Rule 7.3 (Consequence)

$$\frac{\begin{array}{l}(A_1, C_1) : \{p_1\}\ S\ \{q_1\} \\ \forall t : (A@t) \to (A_1@t),\ \forall t : (p@t) \wedge (time < \infty) \to (p_1@t) \\ \forall t : (C_1@t) \to (C@t),\ \forall t : (q_1@t) \to (q@t)\end{array}}{(A, C) : \{p\}\ S\ \{q\}}$$

8 Related Work and State of the Art

For concurrent programs communicating via message passing as well as for shared variable concurrency, one can observe a development from non-compositional proof methods which require the (final) program text for their application, such as [AFdR80,LG81,OG76], towards compositional theories, e.g. [CH81,Sou84a,Zwi89,Sti86,Stø91] (see [dR85b,HdR86] for an overview of this development). An early Indian pioneer in compositional proof methods for concurrency is Soundararajan [Sou84a,Sou84b,SS85]. Whereas these methods verify only safety properties, with temporal logic [Pnu77,MP82] also liveness (progress) properties can be verified. Compositional proof systems for temporal logic have been given in [BKP84,NDGO86]. In [PJ91] a compositional proof system called P-A logic (for Presupposition-Affirmation logic) is described for establishing weak total correctness and weak divergence correctness of CSP-like distributed programs with synchronous and asynchronous communication. This extension allows compositional deadlock proofs and, moreover, compositional proof rules are given for *until*-properties of the form Q *until* R, where Q and R are assertions over communication traces. It seems that this paper describes how far one can go towards proving liveness in a compositional framework using a non-temporal formalism in an assumption-commitment based setting.

Interestingly, more involved programming language fragments for concurrency, such as the concurrency fragment of Ada [Ada83] or those for monitor based languages, have not been characterized until now through compositional trace-based methods, although their non-compositional characterization was possible and has been given—see [dR85a] for an overview. However, if one allows locations inside programs as observables, there exists a straightforward technique to convert non-compositional proof methods for concurrency to compositional ones, as reported in [GdR86]. Similarly, Stirling reports in [Sti86] how a basically non-compositional proof method for shared variable concurrency, such as the one by Owicki & Gries [OG76], can be reformulated compositionally within a framework based on relevance logic. Also for the parallel object oriented language POOL first non-compositional proof methods [AdB90] have been developed based on the method of [AFdR80]. The principal author, de Boer, recently reformulated his proof system along history-based compositional lines [dB91] using the work of Zwiers [Zwi89] as a starting point.

In the present paper we discuss a compositional proof system for distributed message passing in which assumptions can be made about the behaviour of the environment in the style of [MC81,ZdRvEB84]. The main idea of the method is that suitable assumptions about the environment reduce the immense number of possible behaviours of complex reactive systems. Misra and Chandy [MC81] were the first ones to demonstrate the advantages of assumptions in the hierarchical design and verification of distributed processes with message passing. They proposed a compositional rule for the parallel operator and demonstrated their method on several examples. In [ZdRvEB84] these ideas have been formalized, resulting in a compositional proof system for assumption/commitment based specifications together with its soundness and completeness proof. The examples in [Oss83] show that the Misra-Chandy method is easy to use, indeed, and that it leads to simple and natural correctness proofs. In [Pan88,PJ91] the formalism of [ZdRvEB84] is extended to asynchronous communication and progress properties. Also related is the formalism given by Stark [Sta85], who uses rely- and guarantee-conditions for deriving global liveness properties of a distributed system. Interestingly, at present new nonstandard applications of the assumption-commitment framework are burgeoning, e.g., for characterizing specifications of fault-tolerant processes or within development methods for mutual exclusion algorithms in which the final algorithm is obtained by a series of error containing approximations in which errors are gradually removed until all are absent, see e.g. [CK92]. In [JPZ91] a compositional theory for action refinement is developed, in a setting of partial orders, and this theory is applied to proving the correctness of distributed databases by formalizing the notion of serializability. It contains an example of how a general specification using an unbounded number of processors is refined to an implementation using two processors. Recently, a number of developments in assumption/commitment based reasoning have taken place; for the state of the art consult the papers by Pandya [Pan89] and Abadi & Lamport [AL89].

In the present paper we focus on concurrent processes with synchronous message passing along channels. What about compositional approaches to shared-variable concurrency and asynchronous message passing? As to shared-variable concurrency, a basic observation is made by Aczel (as reported in [dR85b]), in which in the semantics of a process a distinction is made between a so-called "component action" Π (an action of the process itself) and an "environment action" E (an update to a shared variable by another process). Then (x, Π) and (x, E) are the analogues of the communication records in the channel-based theory above. The original reference in which these notions where informally introduced is [Jon81]; [Jon83] is a more accessible reference to these ideas and contains a proposal for assumption/commitment based reasoning (called rely/guarantee reasoning) about shared variable concurrency together with the formulation of a compositional proof rule for the parallel operator. The idea is exemplified in [WD88] and formally worked out in [Stø91]. In a mixed (temporal logic)-(transition system) based approach, asynchronous communication is characterized compositionally in the work by Bengt Jonsson [Jon87a, Jon87b]. In [dBKPR91] it is shown that for a compositional description of any programming language based upon asynchronous communication a trace model is sufficient, i.e., no additional structures to encode some relevant branching information (trees, failure sets) are needed. In [HRdR90, HRdR92] a compositional axiomatization is given for the graphical specification language Statecharts which includes features like concurrency, broadcast communication, and time-out.

In [ZdR89] a dichotomy is observed in compositional proof theories for concurrency; one class of methods (including, e.g., Temporal Logic and VDM), is based on programs as predicates, and has a simple proof theory (due to the power of the consequence rule), but has trouble in characterizing sequential composition and iteration. Methods in the other class (including weakest precondition calculi, Hoare triples and dynamic logics), are based on programs as predicate transformers, and have no trouble in dealing with sequential composition and iteration, but are more complicated due to awkward implication rules. In [ZdR89] an attempt at unification is made by using adjoints.

Except for Statecharts, these methods are not designed to verify and specify real-time properties. Now an obvious approach towards a verification theory for real-time programs is to adapt and extend an already existing method which does not incorporate any notion of time. For instance, in traditional linear temporal logic safety and liveness properties are expressed by means of a qualitative notion of time (e.g. "eventually", "henceforth", "until"). In order to express real-time constraints, extensions of this logic have been proposed [Koy89,BH81,SL87] which also includes a quantitative notion of time (e.g. "eventually within 5 time units", "always after 7 time units"). These extensions have been applied to the specification of real-time communication properties of a transmission medium [KVdR83] and the verification of local area network protocols [SPE84]. A compositional proof theory for real-time distributed message passing using an assertion language based on real-time temporal logic has been given in [HW89]. In [Hoo91b] this compositional method is extended to uniprocessors implementations and priorities. Non-compositional proof methods, based on Manna & Pnueli's classical approach to linear time temporal logic [MP82], can be found in [Har88,Ost89]. They express real-time properties in Explicit Clock Temporal Logic and give decision procedures for this logic.

Similarly, real-time extensions have been formulated for other methods. There is an early paper of Haase [Haa81] in which time is introduced by a special variable in the weakest precondition calculus. Bernstein [Ber87] discusses several ways of modelling message passing with time-out in the, non-compositional, framework of [LG81]. Zwarico and Lee [ZL85] have adapted Hoare's trace model [Hoa85] (with one invariant and a satisfaction relation) to real-time. Nested parallelism is not allowed in their programming language, a restricted version of sequential composition is used, and there is no explicit mechanism for expressing time constraints. In [JM86] a real-time logic to analyze safety properties is defined based on a function which assigns a time value to each occurrence of an event. Real-time properties of sliding window protocols are verified by Shankar and Lam in [SL87] using special state variables, called timers, to measure the passage of time. The compositional proof system from [DS89, Sch90] for Timed CSP supports semantic reasoning in the framework of Reed and Roscoe [RR86]. Furthermore, Schneider [Sch90] defines a notion of time-wise refinement to transfer properties of non-timed CSP programs to their timed version, thus exploiting the hierarchy of timed and untimed models from [Ree89]. Baeten and Bergstra [BB91] have incorporated real-time aspects in the process algebra of [BK84] by adding time stamps to atomic actions. In their approach atomic actions have a positive duration, whereas in the process algebra of [NRSV90] actions have no duration in general, except a distinguished time action which models the ticks of a synchronized global clock. Furthermore [NRSV90] contains a systematic approach to delay-constructs. Milner's CCS [Mil89] is extended in [MT90, Wan90] with explicit time. To obtain a calculus for shared resources, in [GL90] a priority-based process algebra is presented.

References

[Ada83]   *The Programming Language Ada, Reference Manual*, 1983.

[AdB90]   P. America and F. de Boer. A proof system for process-creation. In *TC2 Working Conference on Programming Concepts and Methods*, 1990.

[AFdR80] K.R. Apt, N. Francez, and W.P. de Roever. A proof system for Communicating Sequential Processes. *ACM Transactions on Programming Languages and Systems*, 2:359–385, 1980.

[AL89]   M. Abadi and L. Lamport. Composing specifications. In *Stepwise Refinement of Distributed Systems*, pages 1–41. LNCS 430, Springer-Verlag, 1989.

[BB91]   J.C.M. Baeten and J.A. Bergstra. Real time process algebra. *Formal Aspects of Computing*, 3(2):142–188, 1991.

[Ber87]   A.J. Bernstein. Predicate transfer and timeout in message passing. *Information Processing Letters*, 24:43–52, 1987.

[BH81]   A. Bernstein and P.K. Harter, Jr. Proving real-time properties of programs with temporal logic. In *8th ACM Symposium on Operating System Principles*, pages 1–11, 1981.

[BK84]   J.A. Bergstra and J.W. Klop. Process algebra for synchronous communication. *Information & Control*, 60:109–137, 1984.

[BKP84]   H. Barringer, R. Kuiper, and A. Pnueli. Now you may compose temporal logic specifications. In *16th ACM Symposium on Theory of Computing*, pages 51–63, 1984.

[CH81]　　　Zhou Chao Chen and C.A.R. Hoare. Partial correctness of Communicating Sequential Processes. In *IEEE International Conference on Distributed Computing Systems*, pages 1–12, 1981.

[CK92]　　　A. Cau and R. Kuiper. Formalising Dijkstra's development strategy using Stark's formalism. In *BCS FACS Fifth Refinement Workshop, Theory and Practice of Formal Software Development*, 1992.

[dB91]　　　F. de Boer. A compositional proof system for dynamic process creation. In *Symposium on Logic in Computer Science*, 1991.

[dBKPR91]　F. de Boer, J. Kok, C. Palamidessi, and J. Rutten. The failure of failures in a paradigm for asynchronous communication. In *CONCUR '91*, pages 111–126. LNCS 527, Springer-Verlag, 1991.

[dR85a]　　W.P. de Roever. The cooperation test: a syntax directed verification method. In *Logics and Models of Concurrent Systems*, volume 13 of *NATO ASI Series, Series F*, pages 213–260. Springer-Verlag, 1985.

[dR85b]　　W.P. de Roever. The quest for compositionality - a survey of assertion-based proof systems for concurrent programs, Part I: Concurrency based on shared variables. In *IFIP Working Conference 1985: The role of abstract models in computer science*, pages 181–207. North-Holland, 1985.

[DS89]　　　J. Davies and S. Schneider. Factorizing proofs in timed CSP. In *Mathematical Foundations of Programming Semantics*, pages 129–159. LNCS 442, Springer-Verlag, 1989.

[GdR86]　　R. Gerth and W.P. de Roever. Proving monitors revisted: a first step towards verifying object oriented systems. *Fundamenta Informatica*, IX:371–400, 1986.

[GL90]　　　R. Gerber and I. Lee. CCSR: a calculus for communicating shared resources. In *CONCUR '90*, pages 263–277. LNCS 458, Springer-Verlag, 1990.

[Haa81]　　V.H. Haase. Real-time behaviour of programs. *IEEE Transactions on Software Engineering*, SE-7(5):494–501, 1981.

[Har88]　　E. Harel. Temporal analysis of real-time systems. Master's thesis, The Weizmann Institute of Science, Rehovot, Israel, 1988.

[HdR86]　　J. Hooman and W.P. de Roever. The quest goes on: a survey of proof systems for partial correctness of CSP. In *Current Trends in Concurrency*, pages 343–395. LNCS 224, Springer-Verlag, 1986.

[HGdR87]   C. Huizing, R. Gerth, and W.P. de Roever. Full abstraction of a real-time denotational semantics for an OCCAM-like language. In *14th ACM Symposium on Principles of Programming Languages*, pages 223–237, 1987.

[Hoa69]   C.A.R. Hoare. An axiomatic basis for computer programming. *Communications of the ACM*, 12(10):576–580,583, 1969.

[Hoa85]   C.A.R. Hoare. *Communicating Sequential Processes*. Prentice Hall, 1985.

[Hoo91a]   J. Hooman. A denotational real-time semantics for shared processors. In *Parallel Architectures and Languages Europe*, volume II, pages 184–201. LNCS 506, Springer-Verlag, 1991.

[Hoo91b]   J. Hooman. *Specification and Compositional Verification of Real-Time Systems*. LNCS 558, Springer-Verlag, 1991.

[HP85]   D. Harel and A. Pnueli. On the development of reactive systems. In *Logics and Models of Concurrent Systems*, pages 477–498. NATO, ASI-13, Springer-Verlag, 1985.

[HRdR90]   J. Hooman, S. Ramesh, and W.P. de Roever. A compositional axiomatisation of safety and liveness properties of Statecharts. In *Semantics for Concurrency*, Workshops in Computing, pages 242–261. Leicester, Springer-Verlag, 1990.

[HRdR92]   J. Hooman, S. Ramesh, and W.P. de Roever. A compositional axiomatization of Statecharts. *Theoretical Computer Science*, to appear, Volume 102, November, 1992.

[HW89]   J. Hooman and J. Widom. A temporal-logic based compositional proof system for real-time message passing. In *Parallel Architectures and Languages Europe*, volume II, pages 424–441. LNCS 366, Springer-Verlag, 1989.

[JM86]   F. Jahanian and A. Mok. Safety analysis of timing properties in real-time systems. *IEEE Transactions on Software Engineering*, SE-12(9):890–904, 1986.

[Jon81]   C.B. Jones. *Development Methods for Computer Programs including a notion of Interference*. PhD thesis, Oxford University, Technical Monograph 25, 1981.

[Jon83]   C.B. Jones. Tentative steps towards a development method for interfering programs. *ACM Transactions on Programming Languages and Systems*, 5(4):596–619, 1983.

[Jon87a]   B. Jonsson. *Compositional Verification of Distributed Systems*. PhD thesis, Uppsala University, Sweden, 1987.

[Jon87b]   B. Jonsson. Modular verification of asynchronous networks. In *6th ACM Symposium on Principles of Distributed Computing*, pages 152–166, 1987.

[JPZ91]   W. Janssen, M. Poel, and J. Zwiers. Action systems and action refinement in the development of parallel systems. In *CONCUR '91*, pages 298–316. LNCS 527, Springer-Verlag, 1991.

[Koy89]   R. Koymans. *Specifying Message Passing and Time-Critical Systems with Temporal Logic*. PhD thesis, Eindhoven University of Technology, 1989.

[Koy90]   R. Koymans. Specifying real-time properties with metric temporal logic. *Real-Time Systems*, 2(4):255–299, 1990.

[KSdR+88] R. Koymans, R.K. Shyamasundar, W.P. de Roever, R. Gerth, and S. Arun-Kumar. Compositional semantics for real-time distributed computing. *Information and Computation*, 79(3):210–256, 1988.

[KVdR83]   R. Koymans, J. Vytopyl, and W.P. de Roever. Real-time programming and asynchronous message passing. In *2nd ACM Symposium on Principles of Distributed Computing*, pages 187–197, 1983.

[LG81]   G.M. Levin and D. Gries. A proof technique for Communicating Sequential Processes. *Acta Informatica*, 15:281–302, 1981.

[MC81]   J. Misra and K.M. Chandy. Proofs of networks of processes. *IEEE Transactions on Software Engineering*, 7(7):417–426, 1981.

[Mil89]   R. Milner. *Communication and Concurrency*. Prentice Hall, 1989.

[MP82]   Z. Manna and A. Pnueli. Verification of concurrent programs: a temporal proof system. In *Foundations of Computer Science IV, Distributed Systems: Part 2*, volume 159 of *Mathematical Centre Tracts*, pages 163–255, 1982.

[MT90]   F. Moller and C. Tofts. A temporal calculus of communicating systems. In *CONCUR '90*, pages 401–415. LNCS 458, Springer-Verlag, 1990.

[NDGO86]   V. Nguyen, A. Demers, D. Gries, and S. Owicki. A model and temporal proof system for networks of processes. *Distributed Computing*, 1(1):7–25, 1986.

[NRSV90]   X. Nicollin, J.-L. Richier, J. Sifakis, and J. Voiron. ATP: an algebra for timed processes. In *IFIP Working Group Conference on Programming Concepts and Methods*, pages 402–429, 1990.

[Occ88]   INMOS Limited. OCCAM 2 *Reference Manual*, 1988.

[OG76]   S. Owicki and D. Gries. An axiomatic proof technique for parallel programs. *Acta Informatica*, 6:319–340, 1976.

[Oss83]   M. Ossefort. Correctness proofs of communicating processes: Three illustrative examples from the literature. *ACM Transactions on Programming Languages and Systems*, 5(4):620–640, 1983.

[Ost89]   J. Ostroff. *Temporal Logic for Real-Time Systems*. Advanced Software Development Series. Research Studies Press, 1989.

[Pan88]   Paritosh Pandya. Compositional verification of distributed programs. Technical Report CS-88/3 (Ph.D. Thesis), Tata Institute of Fundamental Research, Bombay, India, 1988.

[Pan89]   P. Pandya. Some comments on the assumption-commitment framework for compositional verification of distributed programs. In *Stepwise Refinement of Distributed Systems*, pages 622–640. LNCS 430, Springer-Verlag, 1989.

[PJ91]   P. Pandya and M. Joseph. P-A logic – a compositional proof system for distributed programs. *Distributed Computing*, 4(4), 1991.

[Pnu77]   A. Pnueli. The temporal logic of programs. In *18th Symposium on Foundations of Computer Science*, pages 46–57, 1977.

[Ree89]   G.M. Reed. A hierarchy of domains for real-time distributed computing. In *Mathematical Foundations of Programming Semantics*, pages 80–128. LNCS 442, Springer-Verlag, 1989.

[RR86]   G. Reed and A. Roscoe. A timed model for Communicating Sequential Processes. In *13th International Colloquium on Automata, Languages and Programming*, pages 314–323. LNCS 226, Springer-Verlag, 1986.

[Sch90]   S. Schneider. *Correctness and Communication in Real-Time Systems*. PhD thesis, Oxford University, 1990.

[SL87]   A.U. Shankar and S.S. Lam. Time-dependent distributed systems: proving safety, liveness and real-time properties. *Distributed Computing*, 2:61–79, 1987.

[Sou84a]   N. Soundararajan. Axiomatic semantics of Communicating Sequential Processes. *ACM Transactions on Programming Languages and Systems*, 6(4):647–662, 1984.

[Sou84b]   N. Soundararajan. A proof technique for parallel programs. *Theoretical Computer Science*, 31:13–29, 1984.

[SPE84]   D.E. Shasha, A. Pnueli, and W. Ewald. Temporal verification of carrier-sense local area network protocols. In *11th ACM Symposium on Principles of Programming Languages*, pages 54–65, 1984.

[SS85] A. Sobel and N. Soundararajan. A proof system for distributed processes. In *Logics of Programs*, pages 343–359. LNCS 193, Springer-Verlag, 1985.

[Sta85] E. Stark. A proof technique for rely/guarantee properties. In *5th Conference on Foundations of Software Technology and Theoretical Computer Science*, pages 369–391. LNCS 206, Springer-Verlag, 1985.

[Sti86] C. Stirling. A compositional reformulation of Owicki-Gries's partial correctness logic for a concurrent while language. In *13th International Colloquium on Automata, Languages and Programming*, pages 407–415. LNCS 226, Springer-Verlag, 1986.

[Stø91] K. Stølen. A method for the development of totally correct shared-state parallel programs. In *CONCUR '91*, pages 510–525. LNCS 527, Springer-Verlag, 1991.

[Wan90] Yi Wang. Real-time behaviour of asynchronous agents. In *CONCUR '90*, pages 502–520. LNCS 458, Springer-Verlag, 1990.

[WD88] J. Woodcock and B. Dickinson. Using VDM with rely and guarantee-conditions, experiences from a real project. In *Second VDM-Europe Symposium*, pages 434–458. LNCS 328, Springer-Verlag, 1988.

[ZdR89] J. Zwiers and W.P. de Roever. Predicates are predicate transformers: a unified compositional theory for concurrency. In *8th ACM Symposium on Principles of Distributed Computing*, 1989.

[ZdRvEB84] J. Zwiers, W.P. de Roever, and P. van Emde Boas. Compositionality and concurrent networks: soundness and completeness of a proofsystem. Technical Report 57, University of Nijmegen, The Netherlands, 1984.

[ZL85] A. Zwarico and I. Lee. Proving a network of real-time processes correct. In *IEEE Real-Time Systems Symposium*, pages 169–177, 1985.

[Zwi89] J. Zwiers. *Compositionality, Concurrency and Partial Correctness*. LNCS 321, Springer-Verlag, 1989.

RAPID PROTOTYPING VON ESTELLE-SPEZIFIKATIONEN

Rainer Födisch   Hartmut König

TU "Otto von Guericke" Magdeburg
Institut für Rechnerverbund und Betriebssysteme
Postfach 4120
O-3010 Magdeburg

foedisch@dmdtu11.bitnet   koenig@dmdtu11.bitnet

Der folgende Artikel gibt einen Überblick über die Rapid Prototyping-Komponente eines an der TU Magdeburg entwickelten Estelle-Arbeitsplatzes. Nach einer Einführung in das verwendete Rapid Prototyping Tool ATLANTIC wird die Estelle-ATLANTIC-Realisierung mit ihren Elementen Interface, Compiler und Laufzeitsystem erläutert. Anwendungsmöglichkeiten des Rapid Prototyping von Estelle-Spezifikationen werden vorgestellt.

1. Einleitung

Die Bedeutung formaler Beschreibungstechniken für die Spezifikation von Kommunikationsprotokollen ist allgemein anerkannt. Mit den Sprachen Estelle, LOTOS, SDL und ASN.1 sind hierfür durch internationale Gremien standardisierte Spezifikationssprachen bereitgestellt worden (/Hogrefe 89/). Durch die Spezifikation wichtiger Dienst- und Protokollstandards liegen erste Erfahrungen in der Anwendung dieser Sprachen vor. In zunehmenden Maße werden auch Werkzeuge für die Behandlung der Spezifikationen bereitgestellt (für Estelle z.B. siehe /Budko 91/,/Chari 89/,/Richard 89/).

Die Überführung der verbal formulierten Protokollstandards in formale Darstellungen ist ein aufwendiger und komplizierter Prozeß, der durchaus mehrere Monate in Anspruch nehmen kann. Dabei ist eine umfangreiche methodische Arbeit erforderlich, die Erfahrungen mit Kommunikationsprotokollen und Spezifikationstechniken voraussetzt (/König 91/). Trotzdem sind Fehler in der formalen Spezifikation unvermeidlich. Sie resultieren aus subjektiven Fehlern wie z.B.

aus Fehlinterpretationen des nichtformalen Standards und aus der eventuellen Unvollständigkeit, Mehrdeutigkeit oder gar Widersprüchlichkeit desselben. Aus diesem Grund ist eine Verifikation der Spezifikation wünschenswert. Diese Verifikation verfolgt das Ziel, nachzuweisen, daß das spezifizierte Protokoll den spezifizierten Dienst erbringt und daß die Spezifikation widerspruchsfrei ist. Zu diesem Zweck wurden eine Vielzahl von Ansätzen entwickelt, die sich jedoch nicht alle als praktikabel erwiesen. Gegenwärtig werden vor allem die Erreichbarkeitsanalyse, die Petri-Netz-Analyse und das Rapid Prototyping favorisiert. Erreichbarkeits- und Petri-Netz-Analyse liefern formale Beweise, wobei man aber rasch an Grenzen bezüglich der Komplexität einer Spezifikation gelangt. Rapid Prototyping erlaubt die simulative Abarbeitung einer Spezifikation ohne Komplexitätsgrenzen und erhöht so das Vertrauen in deren Funktionsfähigkeit. Eine Garantie über eine Fehlerfreiheit kann nicht gegeben werden, da ein formaler Nachweis nicht erbracht wird. Eine Verifikation sollte deshalb mehrere sich ergänzende Verfahren umfassen.

Der Beitrag stellt ein Rapid-Prototyping-Werkzeug für die ISO-Sprache Estelle (/ISO 9074/,/ISO 5710/,/Dembi 89/) vor. Dieses Werkzeug wurde unter Nutzung des Rapid-Prototyping-Systems ATLANTIC (/Fischer 88/) entwickelt und dient der Simulation von Estelle-Prototypen. Ausgehend von einer kurzen Vorstellung des ATLANTIC-Systems wird die Schnittstelle von Estelle zu ATLANTIC, insbesondere die Modellierung des Estelle-Laufzeitsystems, erläutert. Danach werden die Analyse- und Ausgabemöglichkeiten der Simulation dargestellt. Zum Schluß wird auf Einsatzmöglichkeiten des Tools eingegangen.

2. Rapid-Prototyping mit ATLANTIC

Das Programmpaket ATLANTIC unterstützt das schnelle Erstellen von Software-Prototypen, d.h. deren Modellierung und simulative Ausführung, für verschiedene Spezifikations- und Implementationssprachen. Das System ist in der Sprache C++ implementiert, wodurch sich die Vorteile der objektorientierten Programmierung in bezug auf die Generierung von Software-Prototypen nutzen lassen.

Das System ATLANTIC besitzt eine modulare Struktur. Aufbauend auf das Basissystem, das die nötigen Simulations- und Verwaltungskomponenten enthält, werden Interfacemodule für verschiedene Spezifikations- und Implementationssprachen bereitgestellt (siehe Abb. 2/1). Sie umfassen ein geeignetes abstraktes Laufzeitsystem für die jeweilige Sprache unter Nachbildung ihrer typischen Sprachkonstrukte in C++. Da vorher bereits Interfacemodule für die Implementationssprachen Modula 2 und Concurrent Pascal und für die Spezifikationssprachen PDL, SDL und LOTOS existierten, wird mit der neuen Estelle-Schnittstelle das Spektrum der standardisierten Spezifikationssprachen erweitert. ATLANTIC ermöglicht eine simulative Abarbeitung von Software-Prototypen der angegebenen Sprachen schon von frühen Phasen des Software-Lebenszyklusmodells an.

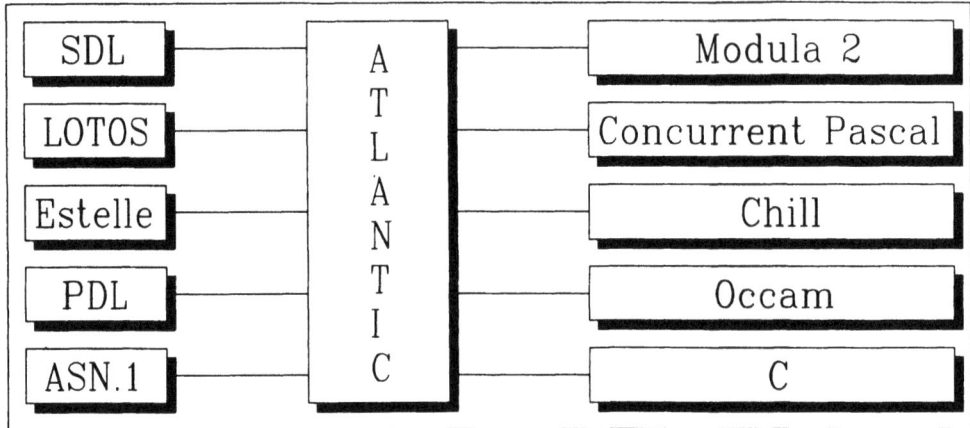

Abb. 2/1: Sprachanschlüsse von ATLANTIC

ATLANTIC basiert auf folgendem Grundkonzept:
- es wird unterschieden zwischen Typen und Instanzen,
- verschachtelte Prozesse werden nicht unterstützt,
- rekursive Prozesse sind möglich,
- Prozesse sind als Koroutinen realisiert; sie sind durch einen sequentiellen Algorithmus und eine Datenstruktur bestimmt,
- zwei Zeitmodelle werden unterstützt: diskreter Zeitverbrauch und Zeitverbrauch gleich Null,
- Puffergrößen sind unbeschränkt (Hardwaregrenze),
- Prozesse können in verschiedenen Relationen zueinander stehen und ausgeführt werden, z.B. sequentiell, parallel oder alternativ.

Von außen betrachtet stellt ATLANTIC ein Prozeßverwaltungssystem dar (siehe Abb. 2/2). Die verwalteten Prozesse sind aus Anwendungsspezifikationen bzw. -implementationen abgeleitete Anwender-, Kanal- und Timerprozesse. ATLANTIC bildet somit eine Schicht zwischen dem Nutzer, also der Anwendung, und dem Betriebssystem. Damit existieren zwei Schnittstellen, das Low-Level- und das High-Level-Interface, die an die jeweiligen Bedingungen angepaßt werden müssen. Das Low-Level-Interface beschreibt die Schnittstelle zwischen ATLANTIC und dem Betriebssystem. Die verwendete Hardware- und Softwareumgebung von ATLANTIC, wie Prozessor, Betriebssystem, Compiler und Anwendungssprache, wird näher spezifiziert, und die benötigten Werte werden definiert. Das High-Level-Interface definiert die Schnittstelle zwischen der verwendeten Spezifikations- bzw. Implementationssprache und ATLANTIC und beschreibt das für die Modellierung und Simulation der Prototypen notwendige Laufzeitsystem. Die entsprechenden Dateien werden als include-Files während der Compilierung in den Prototyp eingefügt.

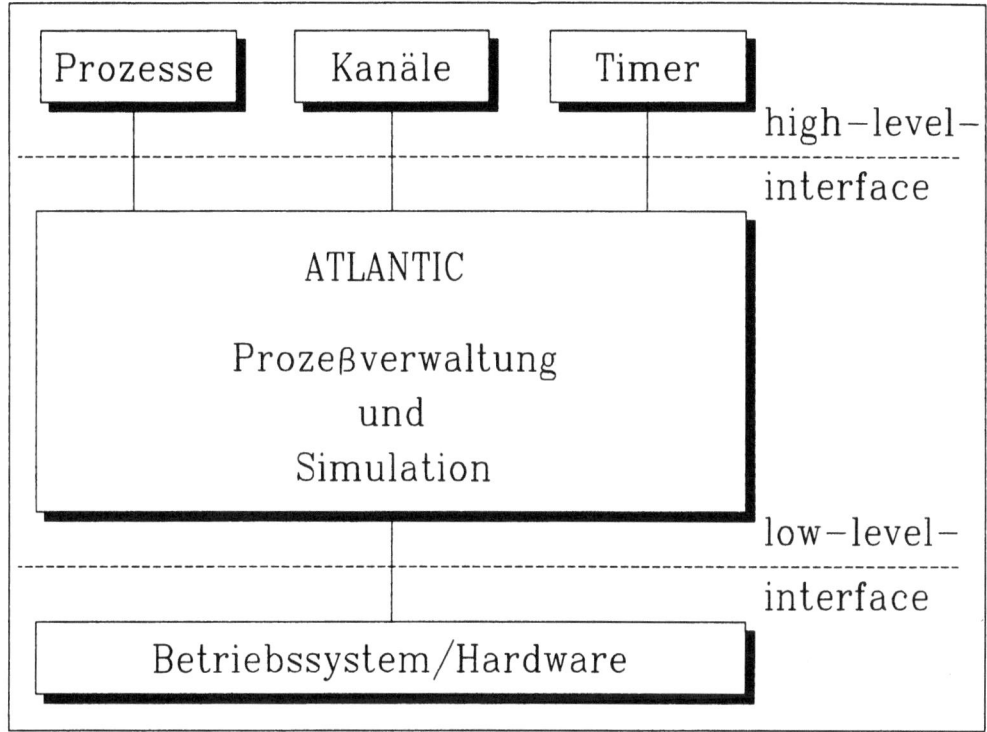

Abb. 2/2: Äußere Sicht auf ATLANTIC

ATLANTIC wird wie folgt genutzt. Zuerst werden alle Übersetzungseinheiten eines zu bildenden Software-Prototyps durch einen Compiler auf syntaktische Richtigkeit überprüft und in entsprechende ATLANTIC-C++-Quelltexte, also den eigentlichen Prototyp, transformiert. Dieser Prototyp besteht aus Prozeß-, Kanal- und Timertypen. Der so entstandene Quelltext wird durch einen C++-Compiler übersetzt. Dabei werden das Low-Level- und das High-Level-Interface als include-Dateien in die Übersetzung mit einbezogen. Durch einen Linker wird der entstandene Objektcode mit Hilfe von Laufzeitbibliotheken in eine abarbeitungsfähige Form gebracht. Nach dem Start von ATLANTIC werden im Dialog Informationen zum Simulationslauf abgefragt, z.B. Konfiguration, Parameter, Simulationsdauer, Umfang des Protokolls sowie zu beobachtende Modellgrößen. Wahlweise kann auch ein vorher erstelltes Testfile die gewünschten Informationen übergeben. Während der Simulation werden aus den Prozeß-, Kanal- und Timertypen die entsprechenden Instanzen erzeugt und als Prozesse verwaltet. ATLANTIC bietet die Möglichkeit, diese Instanzen zu beobachten und entsprechende Ausgabedateien zu generieren. Ausgewählte Simulationswerte können am Bildschirm verfolgt werden. Ein Neustart der Simulation ist jederzeit auch mit neuen Parametern möglich. Auf die Ergebnisse der Simulation werden wir in Abschnitt 4 noch ausführlicher eingehen.

3. Ablauf von Estelle-Spezifikationen unter ATLANTIC

Die Realisierung der Schnittstelle ATLANTIC - Estelle untergliedert sich in drei Teile (/Födisch 90/):

- die Implementation des Low-Level- und des High-Level-Interface,
- die Entwicklung eines Estelle-Compilers und
- die Modellierung eines Estelle-Laufzeitsystems.

Die Teilschritte sind eng miteinander verknüpft und wurden gleichzeitig bearbeitet. Im folgenden soll vor allem die Modellbildung behandelt werden. Die anderen beiden Schritte werden nur kurz dargestellt.

3.1. Interfaces

Im Low-Level-Interface wird neben der Hard- und Softwareumgebung auch die verwendete Spezifikationssprache, hier Estelle, angegeben. Das High-Level-Interface enthält Verwaltungsstrukturen und Funktionen, die zur Generierung von Protokoll-, Report-, Watch- und Trace-Files gebraucht werden. Auch Funktionen zur Fehlerbehandlung und speziell zur Estelle-Prototyp-Simulation sind hier deklariert.

3.2. Compiler

Der Estelle-Compiler transformiert eine Estelle-Spezifikation in die entsprechende Eingangsdarstellung von ATLANTIC. Der Spezifizierer hat die Möglichkeit, über als Estelle-Kommentare verkleidete Steueranweisungen die Arbeit des Compilers und auch die Abarbeitung unter ATLANTIC zu beeinflussen. Der Compiler wurde mit einem Compiler-Compiler-Tool implementiert. Realisiert sind z.B. eine bedingte Übersetzung, Dialogmöglichkeiten und die Indikation von Zeitsteuerungen.

3.3. Laufzeitsystem

Grundlage des Estelle-Laufzeitsystems ist die Modellierung der Estelle-Sprachelemente. Eine Estelle-Spezifikation beschreibt im allgemeinen die Struktur und das innere Verhalten eines Software-Systems. Die Aufgabe eines Estelle-Laufzeitsystems muß es daher sein, das vom Estelle-Standard vorgeschriebene Verhalten einer Spezifikation korrekt zu realisieren. Ausführliche Sprach- bzw. Verhaltensbeschreibungen sind in /ISO 9074/ und /ISO 5710/ enthalten.

Eine Estelle-Spezifikation besteht aus ineinander verschachtelten Moduln, die in einer Eltern-Kind Beziehung stehen und miteinander kommunizieren können. Die Modulstruktur kann sich dynamisch, im Rahmen der spezifizierten Möglichkeiten, ändern. Jeder Modul repräsentiert den Typ eines bestimmten erweiterten endlichen Zustandsautomaten, der aus drei Teilen besteht. Der Deklarationsteil enthält z.B. die Datenstrukturen und Kindermodulbeschreibungen eines Moduls. Der Initialisierungsteil beschreibt den Startzustand des Moduls und der Transitionsteil definiert die möglichen Zustandswechsel. Dieser Aufbau einer Estelle-Spezifikation erleichtert die Modellbildung. Jeder Modul kann bei der Modellbildung als ein Prozeßtyp angesehen werden. Die Verschachtelung der Module wird in den Prozeßtypen aufgelöst, bleibt aber über Verweise zum Elternprozeß und zu den Kindprozessen in den Prozeßtypen erhalten (siehe Abb. 3/1).

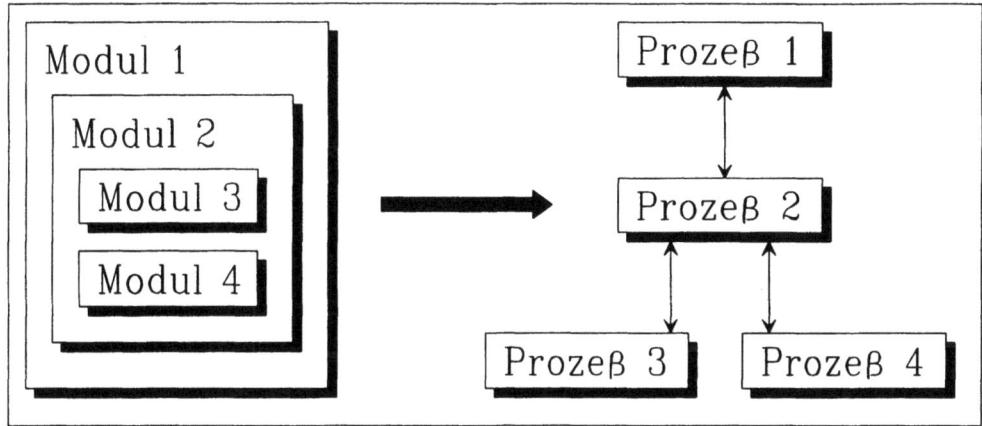

Abb. 3/1: Umsetzung verschachtelter Moduldeklarationen in Prozeßtypen

Eine Estelle-Spezifikation enthält nur Modulbeschreibungen, aus denen das Laufzeitsystem Instanzen, d.h. wirkliche Prozesse, generieren muß. Estelle benutzt dafür das Sprachelement *init*, mit dem angegeben wird, welche Modulvariable mit welchem Modultyp initialisiert werden soll. Unter ATLANTIC in der objektorientierten Programmiersprache C++ wird das unkompliziert über Konstruktorfunktionen realisiert. Zu jedem· spezifizierten Modultyp existiert eine eigene Konstruktorfunktion, die die Instanzen erzeugt und mit den spezifizierten Modulparametern aktualisiert. Bei der Generierung einer Modulinstanz erfolgt der Eintrag einer Instanzbeschreibung in die Ereigniszeitliste, die einerseits der Identifikation der Modulinstanz und andererseits der Verwaltung als Prozeß dient. Sie enthält auch Informationen über den Modul, die das Laufzeitsystem für seine Tätigkeit benötigt. Das sind z.B. die Angabe der Kindermodule und des Elternmoduls, die möglichen Zustände und "feuerbare" Transitionen des Prozesses. Die Vernichtung einer Modulinstanz, in Estelle *release*, wird über zu jedem Modultyp gehörende Destruktorfunktionen realisiert. Der entsprechende Modulprozeß wird dabei aus der Ereigniszeitliste entfernt und vernichtet. Den Anlauf und die Beendigung einer Estelle-Spezifikation unter ATLANTIC realisiert ein Monitor-Prozeß, der vom

Estelle-Compiler erzeugt und unter ATLANTIC gestartet wird. Er sorgt für die Generierung einer Instanz des äußersten Moduls einer Spezifikation.

Der Prozeßteil, das interne Verhalten eines jeden Modultyps, wird als Koroutine realisiert. Da mehrere Instanzen eines Modultyps die gleiche Prozeß-Koroutine nutzen, müssen die Daten jeder Modulinstanz gesondert aufbewahrt werden. Der Verweis auf die zugehörigen Daten ist in der entsprechenden Instanzbeschreibung enthalten. Die vom Compiler zur Verfügung gestellte Prozeß-Koroutine beinhaltet also nur den C-Code für den Initialisierungsteil und den Transitionsteil eines Modultyps. Nach dem Erzeugen eines Modulprozesses wird der entsprechende Initialisierungsteil abgearbeitet und damit der Prozeß in einen festgelegten Anfangszustand versetzt. Dieser Teil entspricht dem *initialize*-Ausdruck einer Estelle-Spezifikation. Nach der Abarbeitung geht der Prozeß in einen passiven Zustand über. Der Transitionsteil der Koroutine entspricht dem Transitionsteil eines Modultyps. Er enthält somit den eigentlichen Anweisungsteil. Für die Koroutine wurde das in /ISO 9074/ vorgeschriebene Verhalten in Bezug auf die Auswahl von "feuerbereiten" Transitionen in C nachgebildet. Steht ein Prozeß am Anfang der Ereigniszeitliste, so wird die zugehörige Koroutine aktiviert und alle "feuerbereiten" Transitionen werden mittels einer case-Anweisung (in C switch) bestimmt. Erhält dann der entsprechende Modulprozeß die Erlaubnis eine Transition auszuführen, wird aus den "feuerbereiten" Transitionen zufällig eine ausgewählt und "gefeuert". Durch Setzen von Optionen kann der Nutzer von ATLANTIC eingreifen und eine ihm angenehme Auswahl treffen. Anschließend wird der Prozeß passiv und neu in die Ereigniszeitliste einsortiert oder wartet auf ein noch nicht eingetretenes Ereignis.

Für die Realisierung des Estelle-Laufzeitsystems sind die Kommunikationsmechanismen zwischen den Modulinstanzen ebenfalls von Bedeutung. Eine in einer Estelle-Spezifikation beschriebene Verbindung besteht aus zwei Interaktionsendpunkten vom gleichen Kanaltyp. Eine von einem Endpunkt aus gesendete Interaktion (Nachricht) ist mit dem folgenden Zeittakt am Ziel in einer Warteschlange verfügbar.

Die Interaktionspunkte in ATLANTIC werden als Strukturen mit einem Verweis zur besitzenden Modulinstanz realisiert. Sie enthalten Warteschlangen für ankommende Interaktionen.

Beim Aufbau einer Verbindung in Estelle über
    connect IP1 to IP2;
müssen die beiden zu verbindenden Interaktionspunkte vom gleichen Kanaltyp sein und entgegengesetzte Rollen, z.B. Sender und Empfänger, besitzen. Im Laufzeitsystem wird durch *connect* ein Kanalprozeß mit Verweisen auf die zu verbindenden Interaktionspunkte generiert (siehe Abb. 3/2).

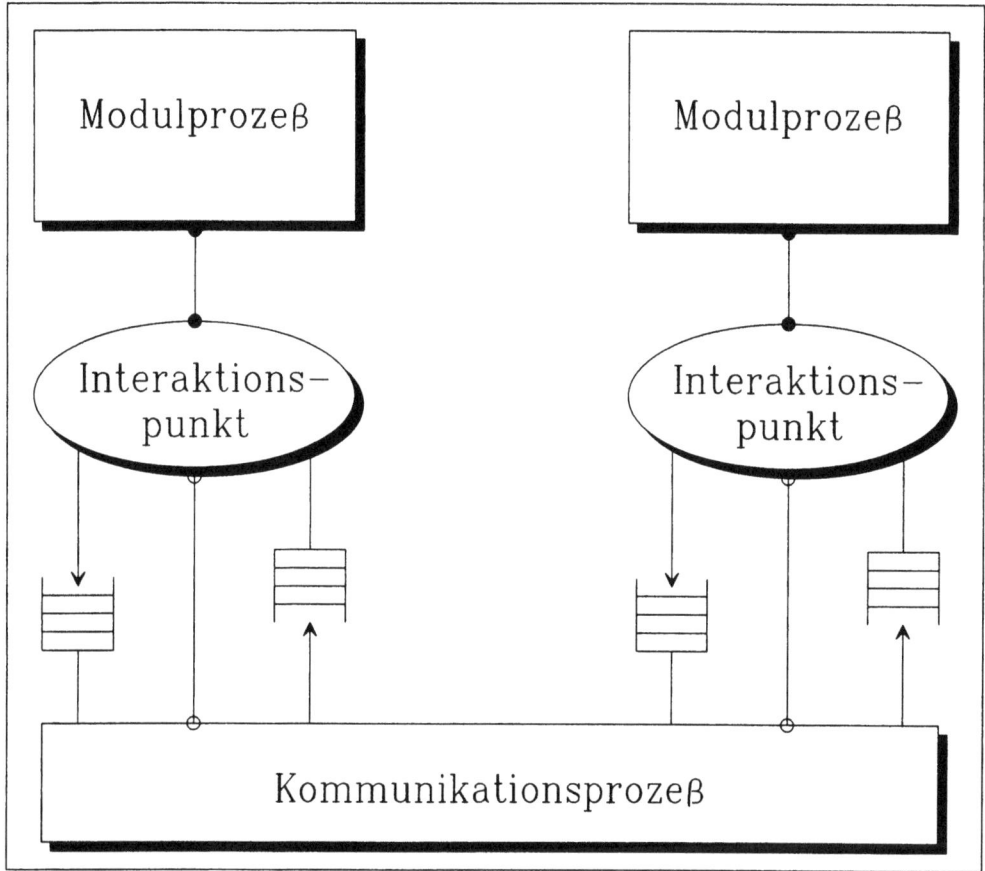

Abb. 3/2: Aufgebaute Kommunikationsverbindung mit Kanalprozeß

Die Auflösung bestehender Verbindungen erfolgt in Estelle durch die *disconnect*-Anweisung, im Laufzeitsystem durch eine Destruktorfunktion. Es reicht die Angabe eines Interaktionspunktes aus. Eine spezielle Möglichkeit von disconnect besteht in der Angabe einer Modulvariable. Ihre Angabe führt bei der Abarbeitung zur Auflösung aller Verbindungen der durch die Modulvariable identifizierten Modulinstanz.

Estelle bietet weiterhin die Möglichkeit, Verbindungen zu verlängern. Mittels der *attach*-Anweisung können zwei Interaktionspunkte des gleichen Kanaltyps und der gleichen Rollen miteinander verbunden werden. Bestehende Warteschlangen werden im neuen Verbindungsendpunkt zu einer gemeinsamen Warteschlange zusammengefaßt. Das Trennen einer solchen Verbindungsverlängerung geschieht mittels *detach* und der Angabe eines der beteiligten Interaktionspunkte. Dabei wird auch eine bestehende Warteschlange wieder getrennt. Die eben beschriebenen Abläufe sind im Laufzeitsystem über die Funktionen *attach* und *detach* verfügbar.

define I1_AB 1

struct IN_I1_AB
{
DATENTYP DATEN;
}

Beispiel 3/1:   Struktur einer Interaktion

Die Interaktionen im Laufzeitsystem sind ebenfalls Strukturen (siehe Beispiel 3/1). Für sie existieren Konstruktoren und Destruktoren, die nach Bedarf aktiviert werden. Mit der *output*- Funktion werden die Interaktionen an einen Interaktionspunkt übergeben und der entsprechende Kanalprozeß sorgt für die Übertragung auf der aufgebauten Verbindung zum Ziel-Interaktionspunkt, wo mit Beginn der nächsten Transition die Nachricht entnommen werden kann.

Wird eine Interaktion für eine Kommunikation gebraucht, muß der für diese Interaktion zuständige Konstruktor aufgerufen werden. Dieser Konstruktor prüft, ob bereits eine Interaktionsstruktur dieses Typs in einer speziellen Liste enthalten ist. In diesem Falle wird diese Struktur entnommen und andernfalls wird eine neue Interaktionsstruktur dieses Typs erzeugt. Es folgt die Aktualisierung der Struktur mit den aktuellen Interaktionsdaten.

Das Senden einer Interaktion wird in Estelle durch
   output IP.IAC(PARAMETER);
ausgelöst. IP gibt dabei den Interaktionspunkt an, über den die Interaktion *IAC* mit den aktuellen Werteparametern *PARAMETER* gesendet werden soll. Die *output* entsprechende Laufzeitsystem-Funktion ruft die Interaktionsgenerierungsfunktion auf und übergibt mit dem Ende der ausgeführten Transition die Interaktion dem Ausgabe-Interaktionspunkt. Vom Kanalprozeß wird diese Interaktion dem Ziel-Interaktionspunkt so übergeben, daß die Interaktion zum nächstfolgenden Zeitpunkt dort ohne Zeitverlust in der Warteschlange verfügbar ist. Dabei wird die gesamte Kette von Interaktionspunkten durchlaufen, die durch *connect* und *attach* in der entsprechenden Verbindung enthalten sind. Man kann also immer nur von Verbindungsendpunkt zu Verbindungsendpunkt senden.

Für das Testen von Echtzeitspezifikationen ist noch eine andere Version der Kommunikationsrealisierung im Gespräch. Hierbei werden die Interaktionspunkte als Prozesse realisiert. Das bietet die Möglichkeit, einen realen Zeitverbrauch für die Übertragung der Interaktionen zu realisieren, der ja bei Echtzeitproblemen nicht ohne Bedeutung ist.

Estelle bietet kein Sprachelement zum Empfang von Interaktionen an. Die Eingabe geschieht indirekt über die *when*-Klausel einer Transition.
   when IP.IAC(PARAMETER)

gilt dabei als zu erfüllende Bedingung der Transition. Die entsprechende Transition ist nur "feuerbereit", wenn die Interaktion *IAC* in der Warteschlange des Interaktionspunktes *IP* steht. Da mehrere Transitionen "feuerbar" sein können, muß genau eine ausgewählt werden. Erst wenn die ausgewählte Transition diejenige mit der *when*-Klausel ist, wird die Interaktion *IAC* aus der Warteschlange entfernt. Im Laufzeitsystem wird das durch die Funktion *input* realisiert.

Die Realisierung weiterer Estelle-Konstrukte erfolgt über Funktionen oder einfache Transformationen nach C, die allerdings ohne direkte Auswirkungen auf das hier beschriebene Laufzeitsystemverhalten unter ATLANTIC sind (siehe /Födisch 90/).

4. ATLANTIC-Ausgaben für Estelle-Prototypen

Als Ergebnis der Simulation eines Estelle-Prototyps werden 4 Filetypen generiert:
- Trace-File,
- Watch-File,
- Report-File und
- Protocol-File.

Das Trace-File protokolliert den gesamten Ablauf der Simulation (siehe Beispiel 4/1). In der 1. Spalte wird der auslösende Prozeß eingetragen. Die 3. Spalte enthält die ausgeführten Aktionen und die 2. das Ziel dieser Aktion. In Spalte 4 stehen, falls erforderlich, erläuternde Kommentare bzw. auch Zeitpunktangaben zum Verständnis des Trace-Ablaufs. Insgesamt können neben dem momentan aktiven Prozeß, über Optionen gesteuert, die folgenden Aktionen und ihr Ziel im Trace-File erscheinen:

- Erzeugen und Vernichten der einzelnen Instanzen,
- Auf- und Abbau von Verbindungen (d.h., Erzeugen und Vernichten von Kanalprozessen),
- Kommunikation über output und geteilte Variablen,
- Empfang von Interaktionen und andere Transitionsbedingungen,
- Modulzustände und
- Auswahlen über case- und if-Anweisungen.

Bei einer Auswertung kann demzufolge der Ablauf des gesamten Prototyps der Estelle-Spezifikation nachvollzogen werden.

```
------------------------------------------------------------------
                         TRACE - FILE
------------------------------------------------------------------
<MONITOR>   | *****         | *ESTELLE-MACHINE RUNS |*****
<MONITOR>   | SYSTEM        | CREATED               |
SYSTEM      | INSTANCE      | INIT                  |
SYSTEM      | CHILD1        | CREATED               |
CHILD1      | INSTANCE      | INIT                  |
CHILD1      | CHILD11       | CREATED               |
CHILD1      | START         | STATE                 |
SYSTEM      | CHILD2        | CREATED               |
CHILD2      | INSTANCE      | INIT                  |
CHILD2      | STATE0        | STATE                 | WAS SELECTED
SYSTEM      | IP1 TO IP2    | CONNECT               |
    :       |    :          |    :                  |    :
```

Beispiel 4/1: Trace-File

Das Watch-File dient der Beobachtung von Variablenwerten. Es wird über den Compiler mittels Optionen vom Spezifizierer gesteuert. Ist das Beobachten einer Variable, eines Records, einer Menge oder eines case-Wertes gefordert, wird bei jeder Änderung des Wertes desselben eine Ausgabefunktion erzeugt, die die Ausgabe von Name, Typ und Wert in das Watch-File realisiert. Die Intensität der Beobachtung wird über eine Watch-Variable gesteuert, die ganzzahlige Werte von 0, keine Beobachtung, bis n, maximaler Beobachtungsgrad, annehmen kann. Beispiel 4/2 zeigt einen Ausschnitt aus einem Watch-File.

```
------------------------------------------------------------------
                         WATCH - FILE
------------------------------------------------------------------
NAME            TYPE            VALUE
------------------------------------------------------------------
my-state        STRING          IDLE
varint          INTEGER         7
varint          INTEGER         6
signal          sigtyp          CONREQ
   :              :                :
```

Beispiel 4/2: Watch-File

Im Report-File sind statistische Angaben enthalten, die wahlweise ausgegeben werden können. Es besteht die Möglichkeit der Anlage von Zählern (tallies) zur Variablenbeobachtung, um aus den erhaltenen Werten Häufigkeitsverteilungen in Form von Histogrammen zu erzeugen. In Tabellenform erfolgt eine Ausgabe der einzelnen Warteschlangenlängen an den Interaktionspunkten wie Maximum-, Minimum- und mittlere Längen. Auch Statistiken zu bestimmten Moduloperationen (siehe Trace-File) und zu den Kanalprozessen werden tabellarisch erfaßt. Das Aussehen einer Reportausgabe ist in Beispiel 4/3 zu sehen. Hieraus ist z.B. zu erkennen, wie oft sich ein Kindermodul des Moduls SYSTEM in welchem Zustand befunden hat.

```
------------------------------------------------------------
                    REPORT - FILE
------------------------------------------------------------
 P R O C E S S     E N S E M B L E S

    ENSEMBLE:    SYSTEM

    TITLE        /IDLE    /STATE0  /STATE1 /  ...
    CHILD1         1        10       15
    CHILD2         :         :        :
      :            :         :        :

    ENSEMBLE:    CHILD1

    TITLE        ...
    CHILD11      ...
      :           :
```

Beispiel 4/3:   ATLANTIC-Reportausgabe

Das Protocol-File verdeutlicht den Ablauf innerhalb der Module sowie die Kommunikation zwischen ihnen. Ein Protocol-File enthält auch die beim Trace-File aufgeführten Konstrukte eines Moduls. Das Protokoll zielt hier auf die inneren Aktionen und Zustandswechsel der einzelnen Module mit der Angabe der Kommunikation nach innen (zu seinen Kindern) und nach außen (Eltern- und Geschwistermodule, Umgebung). Beispiel 4/4 zeigt eine solche Ausgabe.

Hier erkennt man, daß sich die Modulinstanz CHILD3 über die Transition trans1 in den Zustand START bewegt. Dabei werden über den Interaktionspunkt IP4 die Interaktion iac3 gesendet und die Kindermodulinstanzen CHILD31 und CHILD32 kreiert. Schließlich wird noch eine Kommunikationsverbindung zwischen dem eigenen Interaktionspunkt IP5 und dem Interaktionspunkt IP9 von CHILD31 aufgebaut. Der nächste Zustandswechsel erfolgt über die Transition trans4.

Mit dem Protocol-File hat man die Möglichkeit, den Ablauf innerhalb einer Modulinstanz und deren Kommunikation genau mitzuverfolgen. Für den Fall, daß die Aussagen in Bezug auf den Kommunikationsablauf nicht genügen, können alle input- und output-Operationen notiert und aus diesen Daten nach Abschluß der Simulation das gesamte Signalspiel zwischen den Modulen in eine günstige grafische Form gebracht werden.

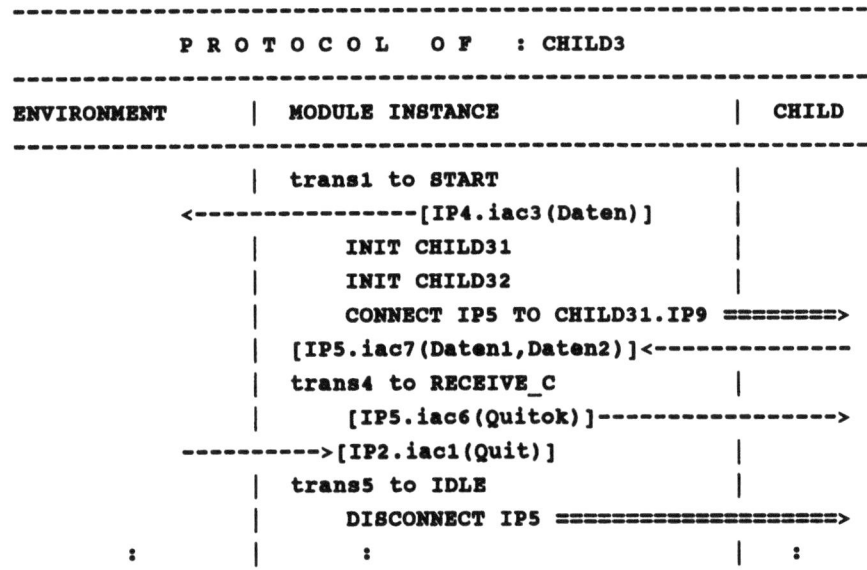

Beispiel 4/4: Estelle-Protocol-File

Die Generierung der einzelnen Ausgabedateien kann wahlweise, je nach Erfordernis, zu- oder abgeschaltet oder auf dem Bildschirm angezeigt werden. Durch aufgetretene Fehler und eine sich anschließende Auswertung der Ausgaben lassen sich Rückschlüsse auf die Qualität der simulierten Spezifikation und deren Verbesserung ziehen.

5. Schlußbemerkungen

Das Estelle-Tool zum Rapid Prototyping von Protokollspezifikationen ist eine Komponente eines Estelle-Arbeitsplatzes, der die rechnergestützte Entwicklung von Kommunikationsprotokollen von der Spezifikationen bis hin zur Implementation und Wartung ermöglichen soll. Er umfaßt einen Estelle-C-Compiler, eine Estelle-Laufzeitumgebung und weiterhin noch verschiedene Werkzeuge zum editieren und testen (/Födisch 91/). Damit ist zukünftig auch eine stärkere Modularisierung der Kommunikationssoftware möglich.

Der Einsatz des hier vorgestellten Estelle-Tools soll zum Testen und zur Leistungsmessung von Estelle-Spezifikationen erfolgen, bevor das spätere reale System verfügbar ist. Einfache Tests wurden mit der Estelle-Spezifikation der PROFIBUS-Norm (/Held 90/) durchgeführt. Komplexere Anwendungsmöglichkeiten ergeben sich bei der Betrachtung des gesamten ATLANTIC-Systems. Dazu gehört einmal die Möglichkeit verschiedene Spezifikationen eines Protokolls auf ihre Zusammenarbeit hin zu testen. Die Spezifikationen können dabei in verschiedenen formalen Spezifikationssprachen ausgeführt sein. Zum anderen kann eine Spezifikation eines Protokolls zusammen mit einer bereits implementierten Version als Partnerinstanz abgearbeitet werden, um so Aussagen über die Spezifikation und die Implementation zu erhalten.

Literaturverzeichnis

/Budko 91/   Budkowski, S.: ESTELLE DEVELOPMENT TOOLSET (EDT). Special Issue of Computer Networks and ISDN Systems on Tools and Formal Techniques for Protocol Engineering - to appear in 1991

/Chari 89/   Chari, V., Lenotre, J.-F., Lumbroso, L., Mariani, E.: An Estelle Simulator/Debugger Tool (edb). in Diaz, M. et al (ed), The Formal Description Technique Estelle, North-Holland, 1989, 381-396

/Dembi 89/   Dembinski, P., Budkowski, S.: Specification Language Estelle. in Diaz, M. et al (ed), The Formal Description Technique Estelle, North-Holland, 1989, 35-75

/Fischer 88/   Fischer, J.: ATLANTIS - a Software Simulator for Behavior Analysis of Protocol Specifications and their Target Implementations., System Analysis, Modelling and Simulation 5(1988)3

/Födisch 90/   Födisch, R.: Realisierung eines ESTELLE-Interface zum Simulationssystem ATLANTIC., Diplomarbeit, TU Magdeburg, Fakultät Informatik, 1990

/Foedisch 91/   Foedisch, R., Held, T., Koenig, H.: A Protocol Environment based on Estelle. angenommen für INDC 92, Helsinki, März 1992

/Held 90/   Held, T.: Spezifikation der PROFIBUS-FDL-Schicht in Estelle. Forschungsbericht TUM/IRB-002/90, TU Magdeburg, Fakultät für Informatik, 1990

/Hogrefe 89/   Hogrefe, D.: Estelle, LOTOS und SDL. Springer Verlag Berlin Heidelberg 1989

/ISO 5710/   ISO/IEC 5710:1991 Proposed Draft Amendment 1: Estelle Tutorial, 1991

/ISO 9074/   ISO/IEC 9074:1989 Estelle-A Formal Description Technique Based On An Extended State Transition Model, 1989

/König 90/   König, H.: Kommunikationsprotokolle. Akademie Verlag Berlin 1990

/König 91/   König, H.: Rechnergestützte Entwicklung von Kommunikationsprotokollen. in Löffler, H. (ed.): Kommunikations-Netzwerke und Offene Systeme (OSI): Stand und Entwicklungstrends. Online '91, Congress III, Hamburg, 1991, S. III.OS. 01-17

/Richard 89/   Richard, J.-L., Claes, T.: A Generator of C-Code for Estelle. in Diaz, M. et al (ed), The Formal Description Technique Estelle, North-Holland, 1989, 397-420

Formale Konzepte zur Lokalisierung von Funktionen in räumlich verteilten Systemen

Rainer Prinoth
GMD-Darmstadt
Rheinstr. 75
6100 Darmstadt

Einleitung

Von Beschreibungsmitteln erwartet man zunehmend mehr, daß sie es ermöglichen die in der Spezifikation von verteilten Systemen gewonnenen Einsichten auch für die Bereiche Implementation, Test und Wartung nutzbar zu machen. Diese Erwartung stellt an die Beschreibungsmittel zusätzliche Anforderungen. Das betrifft insbesondere die Formalisierung räumlicher und zeitlicher Aspekte für Implementation und Test. Dieser Beitrag zeigt an einfachen Beispielen auf, wie die Präzisierung räumlicher Zuordnungen von Funktionen die Aussagen über Beobachtbarkeit und Verlagerung von Entscheidungen (Unsicherheit über den Entscheidungsträger) beeinflußt.

1. Zwei Grundphänomene verteilter Systeme

Der Betrachtung zweier Grundphänomene räumlich verteilter Systeme, nämlich der **Lokalisierung von Entscheidungen** und der **Beobachtbarkeit** dienen die nachfolgenden Beispiele.

Beispiel 1.1: Eine **Person A** ergreift ein leeres Glas und schenkt anschließend Wasser oder Wein ein. Das Ergreifen des Glases sei als **Aktion a** modelliert, das Einschenken von Wasser durch Person A als **Aktion b** und das Einschenken von Wein durch Person A als **Aktion c**. Person A entscheidet sich vor oder nach dem Ergreifen des Glases, ob Wasser oder Wein eingeschenkt wird. **Abb. 1.1** modelliert beide Fälle in diesem Beispiel: die linke Zustandsmaschine (mit

Anfangszustand zl1) zeigt den Fall, in welchem A vor dem Ergreifen des Glases die Entscheidung fällt, ob Wasser oder Wein eingeschenkt werden soll, die rechte Maschine (mit Anfangszustand zr1) zeigt den Fall, in welchem A erst nach dem Ergreifen des Glases entscheidet, ob Wasser oder Wein eingefüllt wird.

In der Sprechweise der Darstellung in Abb. 1.1 fällt die Entscheidung im ersten Fall im Zustand zl1 der linken Maschine und im zweiten Fall im Zustand zr2 der rechten Maschine.

Beispiel 1.2: Eine **Person A** schreibt einen Brief und schickt ihn ab. Empfänger des Briefes sind die **Personen B** oder **C** (eine von beiden). Das Schreiben des Briefes durch Person A sei als **Aktion a** modelliert, der Empfang des Briefes durch Person B als **Aktion b** und der Empfang durch Person C als **Aktion c**. Abb. 1.1 modelliert auch für dieses Beispiel zwei Fälle: die linke Zustandsmaschine zeigt den Fall, in welchem die Person A vor dem Schreiben des Briefes die Entscheidung fällt, an wen dieser gerichtet ist, die rechte Maschine zeigt den Fall, in welchem diese Entscheidung erst nach dem Schreiben des Briefes fällt, wobei Unklarheit über den Entscheidungsträger herrscht.

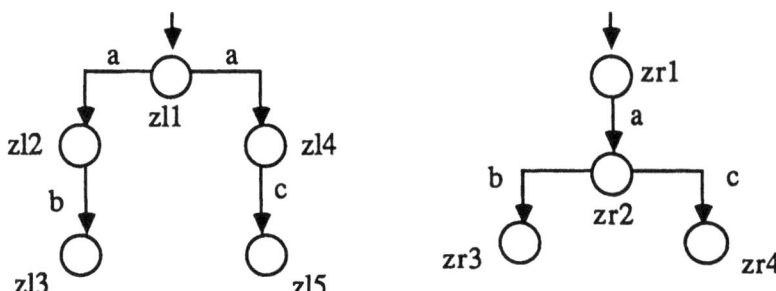

Abb. 1.1   zwei Modelle für die Aktionsfolgen ab und ac

Im ersten Beispiel ist nur eine Person in beiden Fällen mit der Entscheidung befaßt, im zweiten Beispiel sind drei Personen am gesamten Vorgang beteiligt und - bei größerer räumlicher Entfernung zwischen den Personen - auch noch ein Transportdienst. Der letzte Fall im zweiten Beispiel (rechte Maschine) unterscheidet sich von den drei anderen dadurch, daß der Zustand zr2, in dem die

Entscheidung fällt, nicht eindeutig einem Entscheidungsträger zugeordnet ist !

Beispiel 1.3: Die Aktionen der **Person A** im Beispiel 1.1 werden von einer **Person D** beobachtet. Im ersten Fall (linker Fall nach Abb. 1.1) des Beispiels beobachtet D, wie A ein Glas nimmt und vermutet zu diesem Zeitpunkt, daß A Wein einschenken wird bzw. vermutet, daß A Wasser einschenken wird (**Aktion a**); anschließend beobachtet D das Einschenken von Wasser (**Aktion b**) oder Wein (**Aktion c**).
Im zweiten Fall (rechter Fall nach Abb. 1.1) beobachtet D, wie A ein Glas nimmt (**Aktion a**); anschließend beobachtet D, daß A entweder Wasser (**Aktion b**) oder Wein (**Aktion c**) einschenkt.

Im Beispiel 1.3 wird angenommen, daß Aktion a nicht mit "a-Wasser" bzw. "a-Wein" benannt wird (siehe linkes Modell in Abb. 1.1). Aber selbst wenn dies der Fall wäre, könnte der Beobachter doch erst aus der nachfolgenden Aktion b bzw. c schließen, daß tatsächlich Wasser bzw. Wein eingeschenkt wurde.

Beispiel 1.4: Die Aktionen der **Person A** im Beispiel 1.2 werden von **Person B** beobachtet. Beobachter B wird als räumlich getrennt von den anderen Personen angenommen und kann deswegen nur erkennen, ob er einen Brief von A bekommen hat (**Aktion b**) oder (noch) nicht. Für B ist darüberhinaus nicht sichtbar, ob er Beobachter im Modell links oder rechts aus der Sichtweise von Abb. 1.1 ist.

In Beispiel 1.4 ist nach dem bisher Gesagten intuitiv klar, wie und was B beobachten kann; in Beispiel 1.3 ist das für den Beobachter D nicht klar: erst die Einbeziehung des Beobachters in das gesamte Modell und seine Lokalisierung zeigt das Verhältnis zwischen Beobachter und Beobachtetem auf.

Aus der obigen Diskussion wird klar, daß Beobachtbarkeit oder auch die Verlagerung von Entscheidungen in räumlich verteilten Systemen von den Annahmen über die Lokalisierung entsprechender Funktionen abhängen. Diese Annahmen dürfen nicht auf einer intuitiven Ebene verbleiben, wenn es in kritischen Fällen nicht zu Fehlinterpretationen von und nicht realisierbaren Annahmen über Eigenschaften verteilter Systeme kommen soll.

Benötigt werden *formale* Konzepte, die die räumliche Zuordnung von Konstrukten des Logikentwurfs verteilter Systeme zu implementations- und test-relevanten Konstrukten *formal* in Beziehung zu setzen gestatten.
Dieser Beitrag demonstriert den Einstieg in die Erweiterung von Beschreibungsmitteln um Konzepte, die räumliche Bezüge formal fassen. Diese Konzepte wurden in [P-82] und [P-83] eingeführt. Als Beschreibungsmittel, an dem diese Konzepte erläutert werden, dienen Produktnetze, eine vom Autor 1983 eingeführte Klasse beschrifteter Petrinetze.
Der Beitrag faßt einige Aspekte der GMD-Studie Nr. 192 [P-91] zusammen.

2. Netze und Partitionen

2.1 Unbeschriftete Netze

Ein Petrinetz N (auch Netz genannt) besteht aus zwei zueinander disjunkten Mengen S und T und einer Relation F zwischen diesen Mengen.

Ein **Netz** $N = (S,T,F)$ besteht aus:
- einer endlichen Menge S von **Stellen**
  (graphisch wird eine Stelle durch einen Kreis repräsentiert)
- einer endlichen Menge T von **Transitionen**
  (graphisch wird eine Transition durch ein Rechteck repräsentiert)
- einer **Flußrelation** $F \subset (S \times T) \cup (T \times S)$
  wobei die Elemente von F Kanten genannt werden
  (graphisch wird eine Kante durch einen gerichteten Pfeil zwischen den betreffenden Knoten repräsentiert)

Die Bedingung $\mathbb{F} \subset (S \times T) \cup (T \times S)$ bedeutet, daß
- Kanten von Stellen zu Transitionen führen (**Eingangskanten**) oder von Transitionen zu Stellen (**Ausgangskanten**); und daß
- höchstens eine Kante von einer Stelle (bzw. Transition) zu einer Transition (bzw. Stelle) führt.

Stellen, die über Eingangskanten mit einer Transition t verbunden sind, heißen **Eingangsstellen von t**. Diese Stellenmenge wird mit $^\bullet t$ bezeichnet:

$^\bullet t := \{ x \in S \mid (x,t) \in \mathbb{F} \}$.

Stellen, die über Ausgangskanten mit einer Transition t verbunden sind, heißen **Ausgangsstellen von t**. Diese Stellenmenge wird mit $t^\bullet$ bezeichnet:

$t^\bullet := \{ y \in S \mid (t,y) \in \mathbb{F} \}$.

In Netzen werden Zustände durch Markierungen ausgedrückt. Eine **Markierung** eines Netzes ist eine Abbildung

$M: S \rightarrow \mathbb{N}_0$, wobei $\mathbb{N}_0$ wie gewöhnlich die Menge der natürlichen Zahlen mit Einschluß der 0 bezeichnet.

Eine Markierung ordnet also jeder Stelle eines Netzes eine Zahl zu, die graphisch als entsprechende Anzahl von Marken dargestellt wird.

Das Verändern von Markierungen wird ausschließlich durch das Schalten von Transitionen des Netzes bewirkt.
Eine Transition t ist unter einer Markierung M **aktiviert**, wenn
$M(x) \geq 1$ für alle $x \in {^\bullet t}$ gilt.
Aktivierte Transitionen können schalten.
Wenn eine Transition, die unter der Markierung M aktiviert ist schaltet, wird M durch die **Nachfolgemarkierung M′** ersetzt. Der Schaltvorgang erniedrigt die Markenzahl jeder Eingangsstelle (die nicht Ausgangsstelle derselben Transition ist) um eins und erhöht die Markenzahl jeder Ausgangsstelle (die nicht Eingangsstelle derselben Transition ist) um eins. Alle anderen Stellen verändern ihre Markenzahl nicht:

$M'(x) := M(x)-1$ für $x \in {^\bullet t} \setminus t^\bullet$,

$M'(x) := M(x) + 1$ für $x \in t^\bullet \setminus {}^\bullet t$ und

$M'(x) := M(x)$ für alle anderen $x \in \mathbb{S}$.

2.2 Produktnetze

Die Lokalität und Berechenbarkeit der Schaltregel sicherzustellen, waren neben der Ausdrucksstärke die Triebfedern bei der Einführung der **Produktnetze** [EP-84], [BOP-89].

Der Übergang von unbeschrifteten zu beschrifteten Netzen vollzieht sich durch Einführen der Individualität für Marken - die jetzt auch **Objekte** heißen. Für Produktnetze wird jeder Stelle $s \in \mathbb{S}$ ein eigenständiger **Definitionsbereich** (Objektstruktur) $D_S = A_1 \times ... \times A_n$ zugewiesen. Nur Elemente des Definitionsbereichs treten als Objekte der betreffenden Stelle auf. n ist die **Dimension** der Stelle s.

Definitionsbereiche werden nach formal gefaßten Regeln aus entscheidbaren Mengen, berechenbaren Funktionen etc. konstruiert derart, daß D_S als Resultat der Anwendung der Regeln selbst eine entscheidbare Menge ist [BOP-89].

Jede Kante f trägt eine **Kantenanschrift K(f)**. Eine Transition t kann eine **Transitionsinschrift P(t)** enthalten, in der zusätzliche Bedingungen für das Schalten der Transition formuliert sind.

Abb. 2.1 zeigt ein Beispiel für ein Produktnetz zusammen mit einer Markierung.

Unbeschriftetes Netz:

$\mathbb{S} = \{s1, s2, s3, s4\}$ $\mathbb{T} = \{T1, T2\}$

$\mathbb{F} = \{(s2,T1), (s1,T2), (s4,T2), (T1,s1), (T1,s3), (T2,s4)\}$

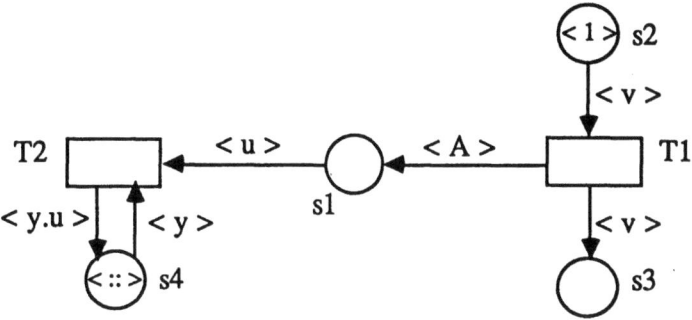

Abb. 2.1 markiertes Produktnetz vor dem Schalten...

Definitionsbereiche der Stellen:

$D_{s1} = D_{s4} = \{A\}^*$ $D_{s2} = D_{s3} = \mathbb{N}$,

Kantenanschriften:

K(s2,T1) = \<v\>   K(s1,T2) = \<u\>   K(s4,T2) = \<y\>
K(T1,s1) = \<A\>   K(T1,s3) = \<v\>   K(T2,s4) = \<y.u\>

u,v und y sind Variable, A ist eine Konstante. \<y.u\> bedeutet Konkatenation der für y und u stehenden Folgen.

Für die in Abb. 2.1 gegebene Anfangsmarkierung M0 gilt:

$M0_{s2}$ = \<1\> und $M0_{s4}$ = \<::\>, wobei mit :: das leere Wort bezeichnet ist.

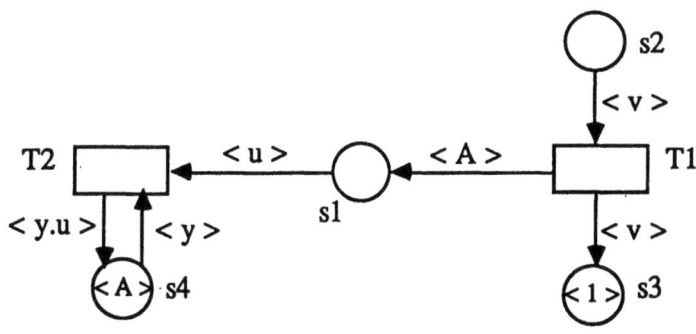

Abb. 2.2   ...und nach dem Schalten von T1 und T2

Das Schalten von Transitionen in Produktnetzen wird in Verbindung mit dem "Setzen" ("Interpretieren") von Variablen definiert. Solche Interpretationen sind dann jeweils für die Variablen, die zu einer Transition gehören, vorzunehmen [BOP-89].

Im vorliegenden Fall kann unter der Anfangsmarkierung die Transition T1 schalten. Für die hieraus resultierende Nachfolgemarkierung M1 gilt:

$M1_{s1} = <A>$, $M1_{s3} = <1>$ und $M1_{s4} = <::>$.

Unter M1 kann Transition T2 schalten. Für die hieraus resultierende Nachfolgemarkierung M2 gilt (siehe **Abb. 2.2**):

$M2_{s3} = <1>$ und $M2_{s4} = <A>$.

2.3 Konzepte zur Lokalisierung von Transitionen

Wie können räumliche Beziehungen im Modell explizit gemacht werden? Das kann dadurch geschehen, daß Aktionen jeweils eines Partners - also lokale Aktionen - durch die zugehörige Transitionsmenge gekennzeichnet werden (**Partition** der Transitionsmenge eines Netzes).

Eine **Partition** Π der Transitionsmenge T eines Produktnetzes P besteht aus Teilmengen $\Pi_i \subset T$, die Blöcke genannt werden und für die gilt:

- Jeder Block ist zu jedem anderen Block disjunkt, d.h. keine Transition kommt in mehr als einem Block vor.
- Die Blöcke überdecken T, d.h. jede Transition ist Element eines Blockes.

Auf Grund der gewählten Partition Π ergibt sich folgende Klassifizierung der Stellen des Produktnetzes P relativ zu einem Block Π_i:

- Menge der **blockinternen Stellen** Si-I: jede dieser Stellen hat Kanten ausschließlich von/zu Transitionen des Blocks Π_i.
- Menge der **Randstellen** Si-R: jede dieser Stellen hat Kanten sowohl

von/zu Transitionen des Blocks Π_i als auch von/zu Transitionen mindestens eines weiteren Blocks.

- Menge der **blockexternen Stellen Si-EX**: keine dieser Stellen hat Kanten von/zu Transitionen des Blocks Π_i.

Information als Markierung blockinterner Stellen ist beispielsweise für einen nicht zum Block gehörenden Beobachter nicht wahrnehmbar, anders als Information in Randstellen. Lediglich Letztere ist für den außenstehenden Beobachter sichtbar: auf Information in Randstellen können Transitionen "zugreifen", die Aktionen des Beobachters modellieren.

Ein durch die Transitionsmenge Π_i definiertes **Unterproduktnetz Pi** ist ein Produktnetz mit

- der Transitionsmenge Π_i
- der Stellenmenge Si, die aus allen bzgl. Π_i blockinternen- und Randstellen besteht
- der Flußrelation Fi, die aus der Einschränkung von F auf $(\Pi_i \times Si) \cup (Si \times \Pi_i)$ besteht.

Als Beispiel sei das Produktnetz P in Abb. 2.1 gewählt. Eine Partition Π der dortigen Transitionsmenge T sei wie folgt gewählt:
$\Pi_1 = \{T1\}$ und $\Pi_2 = \{T2\}$.

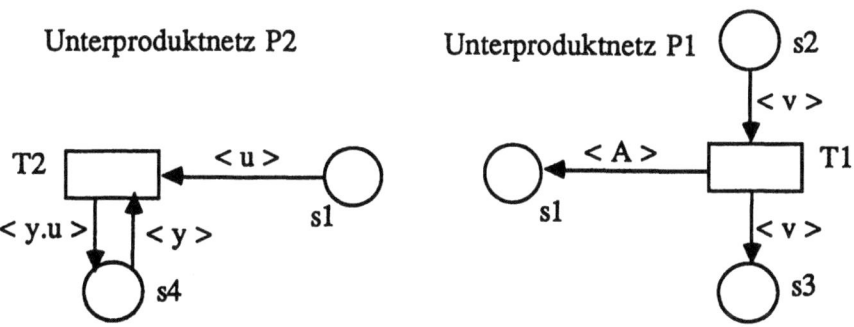

Abb. 2.3 aus der Partition Π abgeleitete Unterproduktnetze P1 und P2

Aus Sicht von Block Π_1 gilt: S1-I = {s2,s3}   S1-R = {s1}   und

S1-EX = {s4}.

Aus Sicht von Block Π_2 gilt: S2-I = {s4}   S2-R = {s1}   und

S2-EX = {s2,s3}

Die durch die obige Partition definierte Aufteilung des Produktnetzes P in die beiden Unterproduktnetze P1 und P2 ist der **Abb. 2.3** zu entnehmen.

3. Entscheidungen in räumlich verteilten Systemen

Nachfolgend werden die Beispiele 1.1 und 1.2 des Kap.1 vor dem Hintergrund der Formalisierung der räumlichen Zuordnung von Aktionen diskutiert. Hierzu werden die Modelle des Kap. 1 zunächst als Netze dargestellt und anschließend die Transitionsmenge unterschiedlichen Partitionierungen unterzogen, was Konsequenzen für die Stellen des Netzes hat: je nach Aufteilung der Transitionsmenge in Blöcke werden Stellen zu internen-, Rand- oder externen Stellen (siehe Kap. 2).

Abb. 3.1 zeigt eine Netzdarstellung der in Abb. 1.1 gegebenen Modelle.

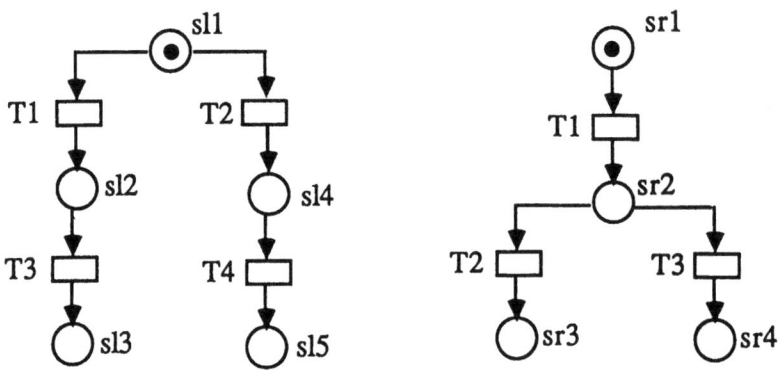

Abb. 3.1 Netzdarstellung der in Abb. 1.1 gegebenen Modelle

In Beispiel 1.1 werden sämtliche Aktionen - hier als Schalten entsprechender Transitionen dargestellt - von einer Person und an einem Ort durchgeführt. Dem entspricht sowohl im "linken" wie auch im "rechten" Fall eine jeweils nur aus einem Block bestehende Partition.

Aus der Wahl der Partition folgt, daß alle Stellen - Stellen sl1 bis sl5 im linken Fall und Stellen sr1 bis sr4 im rechten Fall - blockintern sind. Damit sind auch die Entscheidungen (Stelle sl1 im linken Fall bzw. Stelle sr2 im rechten Fall) intern.

Anders sind in Beispiel 1.2 die Aktionen räumlich zugeordnet: wenn die Aktionen jeder Person einem Block einer Partition zugeteilt werden, dann besteht die gewählte Partition in jedem der beiden Fälle aus drei Blöcken: Partition Π_{links} besteht aus den Blöcken Π_{links1} = {T1, T2}, Π_{links2} = {T3} und Π_{links3} = {T4}; für den rechten Fall gilt: Partition Π_{rechts} besteht aus den Blöcken $\Pi_{rechts1}$ = {T1}, $\Pi_{rechts2}$ = {T2} und $\Pi_{rechts3}$ = {T3}.

Aus der Partition Π_{links} folgt für die Stellen:
- sl1 ist interne Stelle des Blocks Π_{links1}, sl3 ist interne Stelle des Blocks Π_{links2} und sl5 ist interne Stelle des Blocks Π_{links3};
- sl2 ist Randstelle für die Blöcke Π_{links1} und Π_{links2}, sl4 ist Randstelle für die Blöcke Π_{links1} und Π_{links3}.

Die Stelle sl2 (zusammen mit den entsprechenden Markenbelegungen) definiert die Kommunikationserfordernisse zwischen den Personen A und B, Stelle sl4 die Kommunikationserfordernisse zwischen den Personen A und C [P-83], [P-91].
Die Entscheidung fällt intern in Stelle sl1. Sie wird sichtbar durch die Markenbelegung in den Randstellen sl2 und sl4 des zur Person A gehörenden Unterproduktnetzes.

Aus der Partition Π_{rechts} folgt für die Stellen:
- sr1 ist interne Stelle des Blocks $\Pi_{rechts1}$, sr3 ist interne Stelle des Blocks $\Pi_{rechts2}$ und sr4 ist interne Stelle des Blocks $\Pi_{rechts3}$;
- sr2 ist Randstelle für alle Blöcke !

Damit sind die Kommunikationserfordernisse zwischen den Unterproduktnetzen nicht zweiseitig und beschränken sich nicht auf den Transport von Information zwischen jeweils zwei Personen ! Vielmehr ist Randstelle sr2 auch die Stelle der Entscheidung zwischen den beiden Alternativen (T2 schaltet oder T3 schaltet): die Entscheidung fällt hier nicht intern, sondern - räumlich nicht zuordenbar - an den Rändern der Unterproduktnetze [P-91].

Entscheidungen der Personen sind nicht beliebig zwischen Systemen transferierbar. Randstellen sollten nur zweiseitige Kommunikation ausdrücken und nicht zu mehr als einer Person verzweigen (die Randstellen heißen dann Kommunikationsstellen, s.u.).
Eine Randstelle ist **Kommunikationsstelle**, wenn sie die nachfolgenden Eigenschaften K1 und K3 erfüllt:

K1: Eine Randstelle hat mindestens eine Eingangs- und eine Ausgangskante.

K3: Alle Ausgangskanten einer Randstelle koinzidieren mit Transitionen genau eines Blocks Π_i der Partition.

Bemerkung: Eigenschaft K2 macht Aussagen im Zusammenhang mit Verbots- und Abräumkanten von Produktnetzen [P-83], [P-91]. Da derartige Kanten in diesem Beitrag nicht vorkommen, wurde K2 weggelassen.

Eigenschaft K3 formuliert die Abwesenheit **globaler Konflikte**, wie sie in [P-82] und [P-83] eingeführt wurden. Ein globaler Konflikt bewirkt, daß eine

Transition, die eine Marke auf eine Randstelle legt, keinen Einfluß darauf hat, welches Unterproduktnetz diese Marke wegnimmt. Aus Sicht der "Implementation" einer derartigen Stelle bedeutet dies, daß der Absender nicht weiß, wer der Empfänger seiner Nachricht sein wird [CN-87], [P-91].

Abb. 3.2 veranschaulicht diesen Aspekt an einem Beispiel: in der linken Teilabbildung führen die Wege von A nach B bzw. nach C jeweils über ein der Zweierbeziehung eindeutig zugeordnetes "Fließband". A braucht also lediglich den Brief auf das fragliche Fließband zu legen und kann sicher sein, daß es beim gewünschten Empfänger ankommt.

Ganz anders verhält es sich mit der rechten Teilabbildung; hier besteht der Weg von A nach B oder C aus drei Fließbändern. A legt den Brief auf das bei ihm beginnende Fließband, jedoch wird der Brief am Ende dieses Fließbandes in nicht vorhersagbarer Weise auf das nach B oder nach C führende Fließband verzweigt.

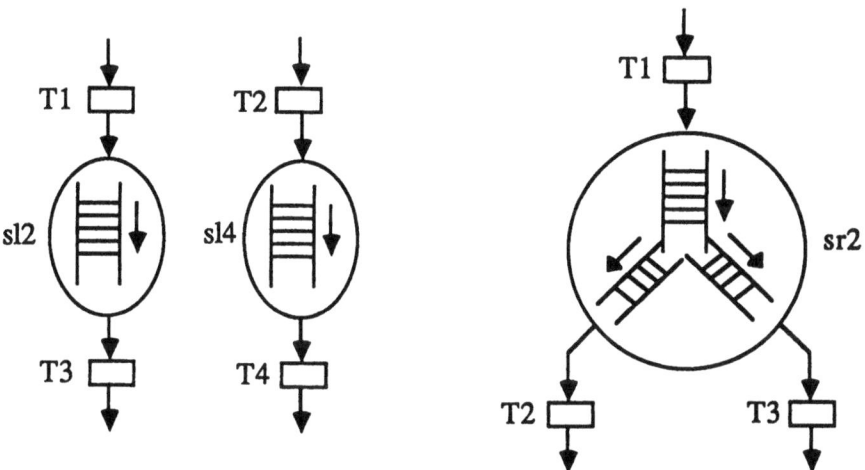

Abb. 3.2 Fließbänder als Beispiel zur Überbrückung der räumlichen Distanz zwischen A, B und C.

Wenn die Stellen sl2, sl4 und sr2 in den Abbildungen 3.1 und 3.2 als zum Postdienst gehörend gedeutet werden, dann modelliert die jeweils rechte Teilabbildung den Fall, in dem der Empfänger des Briefes von der Post und *nicht* vom Absender festgelegt wird. Diese Verlagerung der Entscheidung ist i.a. **jedoch nicht erwünscht.**

4. Beobachtbarkeit in räumlich verteilten Systemen

In diesem Abschnitt werden die beiden Fälle des Beispiels 1.3 vor dem Hintergrund der Präzisierung räumlicher Beziehungen diskutiert.

Es soll angenommen werden, daß in Produktnetzen nicht das Schalten einer Transition selbst beobachtet wird, sondern eine Markenkonstellation auf geeignet gewählten Stellen. Die Auswahl dieser Stellen schränkt beobachtbares Verhalten ein. Das gilt für beobachtete ebenso wie für beobachtende Systeme. Lediglich Markenkonstellationen an Randstellen sind beobachtbar (vgl. Abschnitt 2.3); das Innere beobachteter Systeme ist für den Beobachter nicht einzusehen !

Die Netzspezifikation des Verhaltens der im Beispiel 1.3 betrachteten Person A ist für beide Fälle der Abb. 3.1 zu entnehmen. Eine Spezifikation der beobachtenden Person D ist bislang nicht gegeben.

Wie wird die beobachtende Person D nun eingeführt ? Grob gesagt, wird D als Unterproduktnetz eingeführt als Ergänzung der Modellierung von A. Anschließend wird eine Partition gewählt, die alle Transitionen der beobachteten Person A einem Block der Partition zuordnet und alle neu hinzugekommenen Transitionen - die zur beobachtenden Person D gehören - einem anderen Block. Durch die Partition sind die Randstellen zwischen Beobachter und Beobachtetem eindeutig festgelegt.

Konkret im Beispiel 1.3 wird die beobachtende Person D durch eine Transition Tb und eine Stelle sb modelliert, außerdem wird eine Stelle s-fenster hinzugefügt, über die D etwas von A erfährt: Tb entnimmt dem "Beobachtungsfenster" (Stelle s-fenster) die vom Beobachteten A dorthin abgelegten Objekte und schreibt sie in der Reihenfolge der Abnahme als Folge in Stelle sb auf.

Genau die in Stelle sb vorliegende Information definiert das von Person D beobachtete Verhalten der Person A. Die in sb vorliegende Folge hängt ab von der durch A zur Verfügung gestellten Information und von der Art, wie D diese Information sammelt.

Stelle sb kann räumliche Verteilung zwischen Beobachter und Beobachtetem modellieren [P-91], was beispielsweise - in Analogie zu entsprechenden Stellen in Abb. 3.2 - durch ein Fließband illustriert werden kann. Eine Verfeinerung als Produktnetz für eine derartige räumliche Verteilung ist in Anhang A der GMD-Studie Nr. 192 [P-91] zu finden.

Abb. 4.1 zeigt eine Modellierung für den rechten Fall des Beispiels 1.3, **Abb. 4.2** eine Modellierung für den linken Fall.

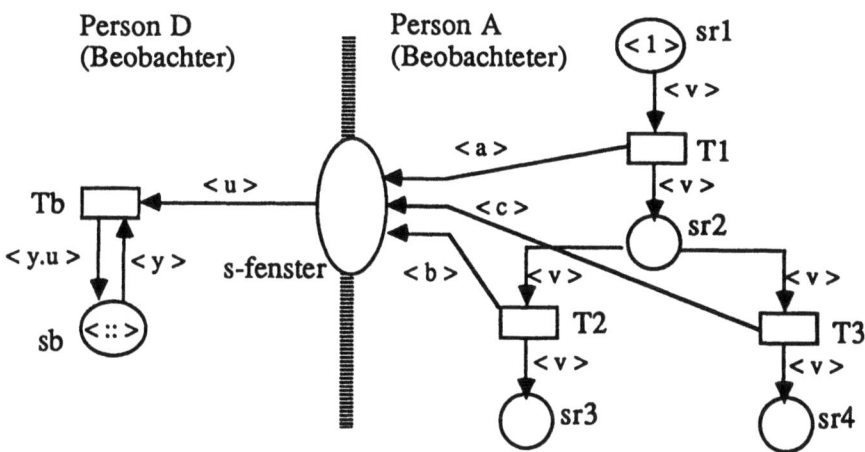

Abb. 4.1 Beobachtende Person D und beobachtete Person A als Produktnetz (rechter Fall)

Für das in Abb. 4.1 gegebene Produktnetz sind u.a.:
- S = {sr1,sr2,sr3,sr4,sb,s-fenster} und T = {T1,T2,T3,Tb}
- Definitionsbereiche: $D_{sr1} = D_{sr2} = D_{sr3} = D_{sr4} = N$ und

  $D_{sb} = D_{s\text{-fenster}} = \{a,b,c\}^*$
- v,u und y sind Variable, a,b und c Konstante; <y.u> bedeutet Konkatenation der für y und u stehenden Folgen;

Für die in Abb. 4.1 gegebene Anfangsmarkierung M0 gilt:

- $M0_{sr1} = <1>$ und $M0_{sb} = <::>$, wobei mit :: das leere Wort bezeichnet ist.

Als Partition Π_{rechts} der Transitionsmenge $\mathbb{T}$ wird die aus den Blöcken $\Pi_{rechts-A}$ und $\Pi_{rechts-D}$ bestehende gewählt mit:

- $\Pi_{rechts-A} = \{T1, T2, T3\}$ und $\Pi_{rechts-D} = \{Tb\}$.
- Das durch $\Pi_{rechts-A}$ definierte Unterproduktnetz P-rechts-A ist die beobachtete Person A und das durch $\Pi_{rechts-D}$ definierte Unterproduktnetz P-rechts-D ist die beobachtende Person D.
  Daraus ergibt sich s-fenster als einzige Randstelle sowohl für P-rechts-A als auch für P-rechts-D und damit als einzige Beobachtungsstelle. s-fenster erfüllt die Kriterien einer Kommunikationsstelle (Kriterien K1 und K3 in Kap. 3) und kann damit ohne weiteres räumlich verteilt werden [P-91].

Der gestreifte senkrechte Strich zwischen A und D gehört nicht zum Produktnetz; durch ihn wird illustriert, daß lediglich die Markenkonstellationen der Stelle s-fenster beobachtbar sind.

Auf eine dem rechten Fall entsprechende Beschreibung der Stellen, Definitionsbereiche, Partitionsblöcke usw. für den linken Fall wird hier verzichtet; sie folgt aus Abb. 4.2 in Analogie zur vorangegangenen Diskussion der Abb. 4.1.

Jetzt werden beide Modelle miteinander verglichen. Die in Abb. 4.1 und Abb. 4.2 in die Stelle s-fenster von Person A abgelegte Information modelliert aus Sicht der Person D die mit gleichen Labeln versehenen Aktionen in Abb. 1.1. Beispielsweise bewirkt das Schalten der Transition T1 die Ablage eines Objekts $<a>$ in Stelle s-fenster usw.
Nach Abb. 4.1 besteht die von A erzeugbare Information für Person D entweder

aus dem Objekt < a > gefolgt von < b > oder sie besteht aus dem Objekt < a > gefolgt von < c >. Diese Information wird von Person D reihenfolgeerhaltend

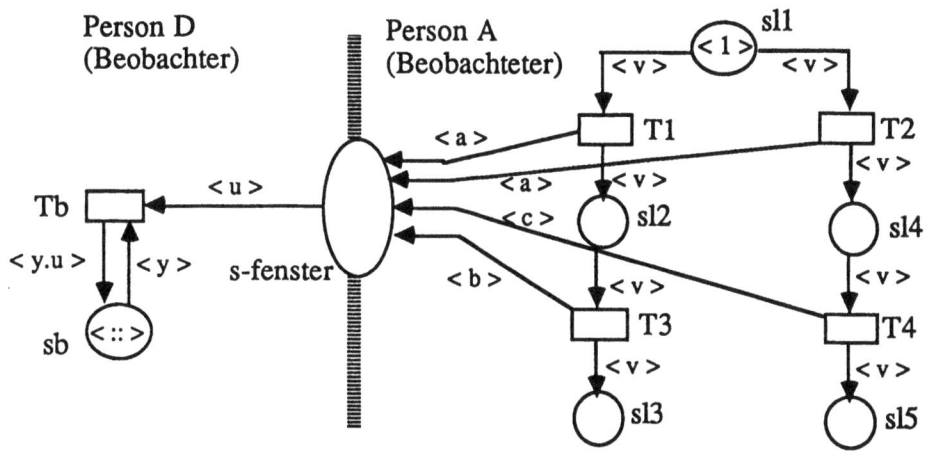

Abb. 4.2 Beobachtende Person D und beobachtete Person A als Produktnetz (linker Fall)

oder nicht reihenfolgeerhaltend in Stelle sb protokolliert [P-91]. Genau die gleiche Information wird von A im Modell der Abb. 4.2 erzeugt und in gleicher Weise von D protokolliert.

Damit sind aus Sicht der beobachtenden Person D die Modellierungen von A in den Abb. 4.1 und 4.2 nicht unterscheidbar (äquivalent).

Das bedeutet aber, daß der in der Literatur (z.B. [M-89]) oft gemachte Unterschied zwischen den in Abb. 1.1 links und rechts modellierten Abläufen unter der in diesem Beitrag angenommenen Präzisierung räumlicher Beziehungen nicht besteht.

Wie aus der hier sorgfältig geführten Diskussion geschlossen werden kann, müssen sich Objekte, die in Abb. 4.2 durch das Schalten der Transitionen T1 und T2 auf die Schnittstelle s-fenster gelegt werden unterscheiden, wenn der Beobachter diesen Unterschied auch wahrnehmen können soll.

Eine Möglichkeit dem Beobachter diesen Unterschied sichtbar zu machen besteht etwa darin, daß

Kante (T2,s-fenster) mit <ac> beschriftet wird und Kante (T1, s-fenster) mit <ab>, wobei a "Historie" ausdrückt (Aktion a hat stattgefunden) und c bzw. b ein Stück "Vorausschau" liefert (als nächstes ist Aktion c bzw. ist Aktion b möglich). Hierzu sind beliebige Varianten an Information denkbar, die der Beobachtete dem Beobachter zukommen lassen kann, z.B. reguläre Ausdrücke, die mögliche Sequenzen beschreiben. Allen diesen Möglichkeiten ist jedoch gemeinsam, daß Vorabanalysen notwendig werden. Derartige Information zu liefern ist bei realen Systemen unüblich und wegen der hohen Komplexität derartiger Systeme auch praktisch kaum möglich.

In vielen Fällen wird es sich bei Beobachter und Beobachtetem um autonome Systeme handeln, die zudem nicht direkt miteinander umgehen, sondern über ein - Zeit verbrauchendes - Kommunikationsmedium Information austauschen. Der Beobachter wird stets damit rechnen müssen, daß der Beobachtete nicht wunschgemäß bzw. überhaupt nicht (re)agiert. In jedem Fall erfährt er nur etwas über die "Vergangenheit" des Beobachteten; bei Briefpost können es mehrere Tage sein, die verstreichen bis der Beobachter die Post erhält, die der Beobachtete an ihn gerichtet hat, in Kommunikationsnetzen kann es mehrere Sekunden dauern, bis eine vom Beobachteten produzierte Nachricht den Beobachter erreicht.

5. Abschließende Bemerkungen

Die Formalisierung der räumlichen Beziehungen zeigte, daß
- die Verlagerung von Entscheidungen an Ränder räumlich verteilter Systeme zur Unsicherheit über den Entscheidungsträger führen kann,
- ein Beobachter interne Unterschiede des Beobachteten (z.B. zwischen dem linken und dem rechten Modell der Abb. 1.1) nicht wahrnehmen kann.

Äquivalenzbegriffe, die einen Vergleich verschiedener Systeme ermöglichen, sollten vor dem Hintergrund der Präzisierung räumlicher Beziehungen von Beobachter und Beobachtetem überprüft werden.

Die hier an Beispielen erarbeiteten Aussagen sind von Bedeutung überall dort, wo Modelle in mehrere Teile zerlegt werden und diese Teile einzelnen lokalen Systemen zugeordnet werden. Derartige Modelle können aus Produktnetzen, aus Zeichenketten usw. bestehen. In Produktnetzmodellen werden Zerlegungen in einzelne Teile über die Einführung geeigneter Partitionen der Transitionsmenge formal faßbar sein; Modelle aus Zeichenketten lassen sich durch eine Partitionierung zugehöriger Alphabete geeignet zerlegen [P-82]. Manipulationen von Zeichenketten können dort kritisch werden, wo die Wirkung der Manipulation Teilketten betrifft, die auf mehr als ein räumlich verteiltes System abgebildet werden.

Literatur

[P-82] R. Prinoth; An Algorithm to Construct Distributed Systems from State-Machines; Protocol Specification, Testing, and Verification, North-Holland; IFIP, 1982

[P-83] R. Prinoth; Modularisierung von Stellen/Transitionsnetzen; Arbeitspapiere der GMD 26, 1983

[EP-84] H. Eckert, R. Prinoth; Produktnetze - Definition eines PROSIT- Beschreibungsmittels; Arbeitspapiere der GMD 92, 1984

[CN-87] H. J. Burkhardt, H. Eckert, R. Prinoth, E. Raubold; A model of cooperation and its specification with nets; Concurrency and Nets, Springer, 1987

[BOP-89] H. J. Burkhardt, P. Ochsenschläger, R. Prinoth; Product Nets A Formal Description Technique for Cooperating Systems; GMD-Studien 165, 1989

[M-89] R. Milner; Communication and Concurrency; Prentice Hall, 1989

[P-91] R. Prinoth; Beschreibungsmittel und Konzepte zur Realisierung verteilter Systeme - Überlegungen an Hand von Produktnetzen; GMD-Studien 192, 1991

A Simple Toy Example of a Distributed System: On the Design of a Connecting Switch

Thomas F. Gritzner

Institut für Informatik
Technische Universität München
Postfach 20 24 20, Arcisstr. 21
W–8000 München 2 (Germany)

Abstract

In this paper the development life-cycle of a design method for distributed systems is explained in close connection with the example of the connecting switch. A connecting switch is a system where stations may get connected, may then send actions, and may get disconnected; it exhibits the behaviour of a very simple protocol. A development life-cycle commonly includes four phases: (1) requirement specification; (2) design specification; (3) abstract program design; (4) concrete program design. This requires collecting various description formalisms into one design method. Each transition of one phase to a more concrete one is guided by a design decision; each verification of the corresponding transition result remains as proof obligation. By the case study of a connecting switch the application of the design method will be demonstrated by treating this example within all four phases in order to make the design method more transparent.

1. Introduction

A distributed system consists of a family of components that are able on the one hand to work independently (*concurrency*) and on the other hand to communicate with each other in order to exchange data or to synchronize for avoiding conflicts (*cooperation*). A development method for distributed systems provides a framework for organizing the development of a distributed system. Systematic development of distributed systems is important, because a distributed system with more than one active parallel components is hard to test appropriately.

A suitable development method should provide an appropriate collection of development steps descending several levels of abstraction guided by design decisions. Usually, one may list the following levels integrated in a development life-cycle starting from the informal level as the most abstract one:

0. informal problem description
1. requirement specification
2. design specification
3. abstract program (aiming at applicative programming languages)
4. concrete program (aiming at procedural programming languages for efficient implementations)

Different levels of abstraction may require various description formalisms to be integrated into one design method. The formalisms of the approach to be presented are listed as follows:

1. trace specifications (predicates over traces)
2. functional specifications (stream processing functions)
3. implementation (dealing with abstract & concrete programs)

Trace specifications aim at the description of the required behaviour of a distributed system. *Functional specifications* allow to specify networks of agents representing the family of components of the concerning distributed system. The *implementation* step includes two notational programming languages: an applicative one for writing abstract programs, and a procedural one for writing concrete programs.

A design method should also support the derivation of a less abstract specification from a more abstract one. This comprises exactly the transition step from one level of abstraction to the next to be also described formally. These transition steps are two-way: The downwards direction is guided by a design decision, the upwards direction is the verification of the derived more concrete specification against the more abstract one. The latter one requires a collection of proof techniques as formal representation. Closely related to transitions are *refinements* of a specification within one level of abstraction. In fact, any design decision leads to a refinement of a specification in the usual sense.

When building a family of specifications along the development life-cycle one observes particular (pairs of) aspects as follows:

Safety/Liveness: The behaviour of a distributed system may be captured by a set of histories. Mathematically, histories are represented by finite or infinite sequences of actions, while observations correspond to prefixes of histories. A **safety** property is a predicate that holds for a history iff none of the finite observations violates the condition, which rules out unwanted reactions of the system. A **liveness** property is a predicate that requires that each finite observation can be continued to a liveness correct history. These two aspects can be combined together by demanding for liveness that only *safety* correct finite observations need to be continued to a liveness and also *safety* correct history.

- *System/Environment:* The following three pairs of aspects are closely connected. The first concerns the fundamental decomposition of a distributed system in the whole into two large components. One is called the **system** in the sense of a *main task*, i.e. a "system" includes all components to be implemented. The other is called the **environment**, which works rather like an operating system, because it has to prepare some services for the "system".

- *Open/Closed System View:* A distributed system consists of a collection of components that interact. While the "environment" part feeds actions to the "system" part, the "system" part gives back the amount of reactions to the "environment". Thus, a distributed system in the whole may be regarded as a *closed* circulation; this is what is called the **closed system view**. The **open system view** rests on the consideration of the specification only of certain selected components of the concerning distributed system; the treatment of the specification of the "system" part only appears as an important special case.

Rely/Guarantee (Assumption/Commitment): The open system view, as a generalisation of the division into system and environment, leads to a division of the distributed system into the considered and into the unconsidered part. Sometimes the specification of the considered part may be done without respect to the unconsidered, i.e. the rest may behave arbitrarily. But, as e.g. in the case of the circulation thought, in the most cases the unconsidered part has to follow some restrictions: the considered part **relies** on the right behaviour of the unconsidered one, and only then it **guarantees** the right reaction. This concept is comparable to *Hoare triples* $\{p\}STEP\{q\}$, which says "if condition p is satisfied, then the program step $STEP$ leads to condition q after its execution".

By our simple example of a connecting switch we want to present a large cross-section of the design method meeting every level of abstraction and we will also give formalizations and examples of the just discussed aspects of distributed system design. This case study is intended to be not only a kind of cross section of a "state of the art" of the chosen approach, but also a feed-back, i.e. this case study shall show how the approach can be refined for a certain class of special cases.

This paper is organized as follows:

The following section contains introduction of basic notions and basic notations. In order to deal with definitions more uniformly, we propose a notational framework similar to algebraic specification techniques.

Then, in section 3 we first introduce our example of a fairly simple protocol: the connecting switch. We also treat the techniques of building trace specifications: pure trace logic, macro techniques for a more state-oriented view, transition systems as a directly state-oriented concept, TRANSACT for specifying systems that execute a fixed recurrent sequence of actions more conveniently. The different forms of trace specifications of our example are verified one against another. Furthermore, the aspects of modifiability of trace specifications with respect to manipulations of the action set are dealt with: action refinement, action enrichment, notion of persistent trace specifications, and the specification with REACTION, which is a technique to build specifications with actions pressed into a more uniform form.

Mechanisms to step forward from a requirement specification to a design specification is explained in section 4. First, the technique of component-oriented trace specification is explained, which allows to form component trace specifications according to the selected separation of components from the distributed systems. Then, the notion of a differentially strategic system is introduced; differentially strategic systems give insight to the causality structure of a component trace specification in order to enable the construction of an agent specification from the component specification. The last part of the section is dedicated to the proper application of component-orientation to our example: a general design decision comes out from the system/environment view as described above; the component trace specifications of "system" and "environment" for the connecting switch are given and proved correct.

Section 5 deals with two forms of functional specifications. The first one is that of partial order processing specifications, which is intended to express and exploit the causality of inputs or of outputs. The other is the well-known one of stream processing functions. But, the continuity requirement of stream processing functions is strong enough to establish the immediate transition from design specification to the first implementation level of abstract programs.

The two phases of implementation are described in section 6: abstract programs written in functional style, concrete programs written in procedural style. Finally, the transition from abstract program to concrete program is sketched by giving examples for transformation rules.

2. Basic Structures

Streams and Operations on Streams

Let a set $\mathcal{A}$ be given, then $\mathcal{A}^*$ will denote the set of all finite sequences over $\mathcal{A}$, and $\mathcal{A}^\infty$ denotes the set of all infinite sequences over $\mathcal{A}$. The set of all **streams** over $\mathcal{A}$ is given by $\mathcal{A}^\omega := \mathcal{A}^* \cup \mathcal{A}^\infty$.

On streams we use the following operations:

ε	denotes the empty sequence,
$\langle a_1, a_2, \ldots \rangle$	denotes for $a_i \in \mathcal{A}$ the sequence made of $a_1, a_2, \ldots$,
$s \& t$	denotes the sequence yielded by the concatenation of $s, t \in \mathcal{A}^\omega$,
ft s	denotes the *first* (left-most) element of the sequence $s \in \mathcal{A}^\omega \setminus \{\varepsilon\}$,
rt s	denotes the *rest* of the sequence, i.e. the sequence omitting the first element if any (i.e. rt $\varepsilon := \varepsilon$),
lt s	denotes the *last* (right-most) element of the finite sequence $s \in \mathcal{A}^* \setminus \{\varepsilon\}$,
ld s	denotes the *lead* of the sequence, i.e. the sequence omitting the last element if any (i.e. ld $s := s \iff s = \varepsilon \vee s \in \mathcal{A}^\infty$),
$\#s$	denotes the length of $s \in \mathcal{A}^\omega$, i.e. the *number* of its elements,

$S©s$ denotes for $S \subseteq A$ and $s \in A^\omega$ the "subsequence" of s consisting of elements of S only (*filtering*),

$S \S s$ abbreviates $\#(S©s)$,

x in s abbreviates the condition $\{x\} \S s > 0$.

Moreover, on streams we define the **prefix ordering** $\sqsubseteq$ by:

$$\forall s, t \in A^\omega: s \sqsubseteq t :\iff \exists s' \in A^\omega: s\&s' = t.$$

A function $f: A^\omega \to B^\omega$ is called a **stream processing function** iff f is **prefix continuous**, i.e. continuous wrt. $\sqsubseteq$. Let $(A^\omega \to B^\omega)$ denote the set of all stream processing functions from A^ω into B^ω.

Additional Notations

We make use of the following notations which occur usually in connection with functional programming languages:

a **where** b := <<expression a, where condition b is satisfied>>;

$(b\ ?\ e_1 : e_2)$:= $\begin{cases} e_1, & \text{if condition } b \text{ holds,} \\ e_2, & \text{if condition } b \text{ does not hold} \end{cases}$;

$\phi[x_0 \leftarrow v_0]$:= ϕ' **where** $\forall x: \phi'(x) = (x = x_0\ ?\ v_0 : \phi(x))$.

$(b\ ?\ e_1 : e_2)$ sometimes serves also as abbreviation of the formula $(b \Rightarrow e_1) \land (\neg b \Rightarrow e_2)$. We also use different ways of denoting function application: $\phi(x), \phi[x], \phi\{x\}$, as usual, or $\phi.x$, or even ϕ_x.

A Notational Framework for Introductions

If any definition within a specification has to be given, we use algebraic specification style so that each such definition looks like a part of a complete algebraic specification. For this we use the following notations:

[used] sort *Sort*

marks the introduction of a sort *Sort*; *Sort* is a set which contains exactly one undefined element usually denoted by $\bot_{Sort}$. The **used** attribute means that *Sort* has no further particular requirements to meet; usually, *Sort* is assumed to be isomorphic to the basic sort *Nat* of natural numbers by a function also denoted by *Sort* (e.g. $Sort(\bot_{Nat}) = \bot_{Sort}$).

fct *constant*: *Sort*

introduces a constant *constant*; *constant* is an element of *Sort*; *Sort* may also be a composed sort like $Sort_1 \times Sort_2$, but not a functional sort.

fct *function*: *functionality*

introduces a strict continuous function *function* with functionality *functionality*; *functionality* has not only the common form $Sort_1[\times \ldots \times Sort_k] \to Sort_0$, but also special forms to indicate the notation of the application of *function*.

fun *operation*: *functionality*

introduces a function *operation* with functionality *functionality*; *operation* need neither satisfy any strictness nor any continuity condition except particularly marked: e.g. **Sort** says that *operation* must be strict and continuous wrt. this argument place.

rel *predicate*: *domain*

introduces a predicate *predicate* over the set *domain*; *predicate* may be used as a set, namely as subset of *domain*, but predicate application is usually denoted by $predicate[d_1, \ldots, d_k]$ instead of $(d_1, \ldots, d_k) \in predicate$.

such that *formula*
   indicates that the objects mentioned in the formula *formula* have to satisfy *formula*, i.e. it establishes a mobile notation for the satisfaction of axioms as "laws".

Predefined objects: Sorts: Nat (natural numbers $0, 1, \ldots$), $\overline{Nat}$ (closed line of natural numbers $0, 1, \ldots, \infty$ linearly ordered), $Bool$ (boolean values uu, tt, ff where $uu = \perp_{Bool}$), $Pot(Sort)$ (power set of $Sort \setminus \{\perp_{Sort}\}$) etc.; predicates: δ_{Sort} (definedness predicate concerning $Sort$), etc.

As an example we treat the specification of streams over a sort S:

> **used sort** S;
> **sort** S^ω;
> **fct** ε: S^ω;
> **fct** ft: $\mid S^\omega \to S$;
> **fct** rt: $\mid S^\omega \to S^\omega$;
> **fun** &: $S \stackrel{\downarrow}{\times} S^\omega \to S^\omega$ /* & is an overloaded operation symbol, here = append */
> **such that** $\forall x: S, s: S^\omega \left[\begin{array}{l} \neg \delta_S[\text{ft } \varepsilon] \wedge \text{rt } \varepsilon = \varepsilon \\ \wedge \text{ ft } x\&s = x \wedge (\delta_S[x] ? \text{ rt } x\&s = s : x\&s = \varepsilon) \end{array} \right]$;

3. Introduction to the Example and Trace Specifications

In this section we introduce the example of the connecting switch. In order to try a more formal introduction we also treat the variants of trace specifications of this system.

3.1 A Very Simple Protocol: The Connecting Switch

The concerning example in this form, including the first variant of a trace specification of it, is taken from [Broy 88, 3.1.1–3.1.2]. The informal description reads as follows:

> "A connecting switch is a system where stations may get connected, may then send messages and may get disconnected."

We present now a more formal treatment by giving a first requirement specification of it, i.e. a first variant of trace specification. By this we explain the notion of a trace specification and demonstrate by which means such a specification can be composed.

3.1.1 Fundamental Remarks to Trace Specifications

Traces and Trace Specifications: Trace specifications are intended to describe the behaviour of a distributed system in an abstract manner such that one needs not think about placing the components. For this, a behaviour is expressed by a complete observation of the history of the *actions* executed by the system. Such a history is made under the assumption of a sequential observer, i.e. parallel sequences of executed actions are sequentialised. Instead of a history we speak of a *trace*. Technically we represent traces as streams: **traces** are *streams of actions*[1]. Now we can also treat the notion of a trace specification formally: a **trace specification** is a *predicate* over *traces*, i.e. the behaviour of a distributed system is described by a set of traces, namely the truth set of the predicate of the concerning trace specification.

---
[1] Therefore, traces in our sense are *finite* or *infinite sequences*.

Writing Trace Specifications: When composing a trace specification one has to follow at least the following lines:

- specify a set of actions, say Act; thence, the set of all possible traces is Act^ω;
- now specify a predicate over traces, say TS by

  rel $TS: Act^\omega$ **such that** $\forall s: Act^\omega \bigl[TS[s] \iff formula\bigr]$,

  where $formula$, the kernel of the trace specification, is written in first-order predicative style[2].

Actions: Actions are *atomic execution steps*. A general *execution step* is carried out by one of the components of the distributed system and causes a transformation of the system's state. *Atomic* means that in the sequential observer's view an action is an observation unit, because an action is regarded as "uninterruptable" or "undivisible". For example, let $Act := \{a, b\}$, i.e. we have two different actions a and b. The trace set describing the parallel composition a∥b of the two actions is simply equal to $\{\langle a, b\rangle, \langle b, a\rangle\}$. Now assume a and b being not atomic program steps, then their parallel composition yields more possibilities of traces, since it is unknown in which atomic step the control changes the active process.

Structuring Actions by Action Generating Functions: $Act := \{a, b\}$ contains unstructured and uninterpreted actions. But, in general, we have often objects the actions refer to. At this, algebraic specification techniques may be used to get more structured action sets. More precisely, into a (new) signature we collect some sort symbols expressing the action set, say Act, the object sets, and other things together with some function symbols of the functionality $Obj\_Sort_1 \times \ldots \times Obj\_Sort_k \to Act$ representing **action generating functions**. Now, take the *term algebra* belonging to this signature as the desired set of actions.

Auxiliary Concepts for Conveniently Writing Trace Specifications: First, we give the definition of the *actual* predicate, which concerns the consideration of an observed actual situation and which is therefore of good use for specifying safety conditions of all possible trace specifications:

rel $actual: Act^\omega \times Act \times Act^\omega$

such that $\forall p: Act^\omega, a: Act, s: Act^\omega \bigl[actual[p, a, s] \iff p \in Act^* \land p \& \langle a\rangle \sqsubseteq s\bigr]$.

Intuitively, $actual[p, a, s]$ holds iff p is a *finite* observed "past" and a is to be the considered actual action of trace s.

Now, we want to introduce two auxiliary concepts concerning action generating functions and the action set. Assume that we have introduced an action generating function act by

fct $act: OS_1 \times \ldots \times OS_k \to Act$

which we abbreviate by **Act fct** $act: OS_1 \times \ldots \times OS_k$.

The word symbol **Act fct** says that:
(1) the sort symbol for the set of all possible actions is called Act;
(2) act has Act for its result type;
(3) act is variously overloaded by the following definition scheme:

fun act **overload:** $OS_1 \times \ldots \times OS_i \to Pot(Act)$ {**forall** $0 \leq i \leq k$}

such that $\forall x: OS_1 \times \ldots \bigl[act(x) = \{a: Act \mid \exists y: OS_{i+1} \times \ldots \times OS_k\, [a = act(x, y)]\}\bigr]$.

---

[2]The underlying logic is called *trace logic*, because, above all, operations on traces are allowed as special operations.

In addition to this, we extend the domain of ©, and thence also of §, so that it accepts elements of sorts of the form $OS_1 \times \ldots \times OS_i$ belonging to one or more action generating functions, say to

    Act fct act_l: $OS_1 \times \ldots \times OS_i \times RestSort_l$,

where l ranges between 1 and a fixed number, on its first argument place. Then, © has the following meaning:

    fun © overload: $OS_1 \times \ldots \times OS_i \times Act^\omega \to Act^\omega$

    such that $\forall x : OS_1 \times \ldots \times OS_i \left[x \text{ © } s = \left(\bigcup_l act_l(x) \right) \text{ © } s \right]$,

and the same for §.

3.1.2 The First Trace Specification of the Connecting Switch: Pure Trace Logic

Now, we develop the first formal treatment of the connecting switch following along the lines of composing trace specifications.

Action Set Specification: Objects that actions will refer to: There are two not further specified sorts of objects:

    used sort *Stations* /* sort of possible stations */;

    used sort *Messages* /* sort of messages to be sent */.

Action generating functions: According to the description, we have three appropriate functions:

    Act fct *conn*: *Stations* × *Stations* /* connect to another station */;

    Act fct *send*: *Stations* × *Messages* /* send a message to the (unique) partner */;

    Act fct *disc*: *Stations* /* give up the (unique) communication link */.

Trace Specification: A formal requirement specification for the connecting switch is now given by the predicate *con_switch* over Act^ω:

    rel *con_switch*: Act^ω
        such that
$$\forall s : Act^\omega \Big[con\_switch[s] \iff \forall t : Stations \begin{bmatrix} (L1)\ conn(t) \S s = disc(t) \S s\ \wedge \\ \forall p : Act^\omega, a : Act \Big[actual[p, a, s] \implies \\ (S1)\ \big((a \in conn(t) \implies conn(t) \S p = disc(t) \S p)\ \wedge \\ (S2)\ (a \in send(t) \implies conn(t) \S p > disc(t) \S p)\ \wedge \\ (S3)\ (a = disc(t) \implies conn(t) \S p > disc(t) \S p)\big) \Big] \end{bmatrix} \Big]$$

Explanation: The *con_switch* predicate basically specifies the following properties:

(L1) A station may only be and is altogether as often disconnected as being connected; this says, whenever a connection has been established, this connection has to be given up once in the future.

And in every intermediate state of the system the following conditions must hold:

(S1) A station may only be allowed to get connected, if it has been as often disconnected as connected.

(S2) A station may only be allowed to send a message, if it has been more often connected than disconnected.

(S3) A station may only be allowed to get disconnected, if, again, it has been more often connected than disconnected.

As one can see, this specification, which is taken directly from [Broy 88, 3.1.2] (with slight changes), is not a really ad-hoc variant of a trace specification of a connecting switch. It is presented here in order to show how a trace specification can be stated in "pure" trace logic. Later variants of trace specifications will show several techniques for obtaining a trace specification more intuitively and more systematically.

3.1.3 Safety and Liveness Conditions at Trace Specifications

As above mentioned, for a specification of a distributed system one can distinguish safety and liveness conditions for the behaviour of the system. In fact, the division of a collection of requirements contained by a system specification according to these two categories is complete in the sense that every specification can be obtained by stating safety and liveness requirements separately and then composing them [Dederichs Weber 90].

For trace specifications the notions of safety and of liveness may be formalized as follows:

- A predicate S over traces of Act^ω is called a **safety predicate** iff
$$\forall t: Act^\omega \big[S[t] \iff \forall t': Act^* \,[\, t' \sqsubseteq t \implies S[t'] \,] \big].$$
- A predicate L over Act^ω is called a **liveness predicate** iff
$$\forall t': Act^* \big[\exists t: Act^\omega \,[\, t' \sqsubseteq t \land L[t] \,] \big].$$
- (*Combination of both aspects:*) Let S be a *safety predicate* over Act^ω. A predicate L_S over Act^ω is called **liveness predicate for** S iff
$$\forall t': Act^* \big[S[t'] \implies \exists t: Act^\omega \,[\, t' \sqsubseteq t \land L_S[t] \land S[t] \,] \big].$$

We consider from the first trace specification only the conditions (S1) and (L1). (S1) reads in full detail as follows:

rel $S1$: Act^ω such that $\forall s: Act^\omega \big[S1[s] \iff$

$$\forall t: Stations,\, p: Act^\omega,\, a: Act \left[\begin{array}{l} actual[p, a, s] \land a \in conn(t) \\ \implies conn(t) \,\S\, p = disc(t) \,\S\, p \end{array} \right] \big];$$

and is, therefore, obviously recognizable as a safety condition. Whereas (L1) is the following predicate:

rel $L1$: Act^ω such that $\forall s: Act^\omega \big[L1[s] \iff \forall t: Stations\, [conn(t) \,\S\, s = disc(t) \,\S\, s] \big]$,

which may easily be identified as a mere liveness condition: add to an observation the missing *conn* or *disc* actions to complete it. In order to prove that (L1) is a liveness condition wrt. the conjunction of (S1), (S2), and (S3), which is assumed to be represented by a predicate over Act^ω called *safe_con_switch*, one has to show that

$(\star)\quad \forall s: Act^\omega, t: Stations \big[safe\_con\_switch[s] \implies conn(t) \S s \not< disc(t) \S s \big]$

holds; then, one knows that one has only to add an appropriate sequence of *disc* actions to complete a safe observation to a safe trace.

3.2 State-Oriented Trace Specifications

As we have seen, a trace specification describes the operational behaviour of a distributed system by (sequential) histories of the actions executed by the system. However, an action, and so does any execution step, causes a transformation of one system's state to another. Therefore, it seems to be equivalent to use state-oriented concepts for requirement specifications. As we will show in this subsection, working with states can be used to obtain trace specifications more easily.

3.2.1 Applying Macro Techniques

For the state-oriented concept to be explained now, a state is conceived as a property of traces, because a trace as a sequence of actions is an execution step transforming one of the potential initial states to the current state. Here, we use the realisation of states rather as another auxiliary concept like *actual* and so on.

Thus we introduce the "states" belonging to the connecting switch in a special way: we use appropriate auxiliary predicate definitions as the following:

rel $state\_one$: $Stations \times Act^\omega$;
rel $state\_two$: $Stations \times Act^\omega$

such that $\forall t: Stations, p: Act^\omega \begin{bmatrix} state\_one[t, p] \iff conn(t)\S p = disc(t)\S p \land \\ state\_two[t, p] \iff conn(t)\S p > disc(t)\S p \end{bmatrix}$.

The aimed reformulation of the trace specification of *con_switch* is not difficult and is left up to the reader. But we give some remarks on particular circumstances:

1. *Initial state* is *state_one*, because $\forall t: Stations[state\_one[t, \varepsilon]]$ holds.

2. "Final" state is *state_one*, as it may be read from the liveness condition (L1).

3. *conn* actions may only be performed iff the assumed protocol machine is in *state_one*.

4. *state_two* is reached by *conn* actions and enables the performing of either a *send* or a *disc* action, but it may be left by a *disc* action only.

These remarks may be depicted as shown in FIGURE 1.

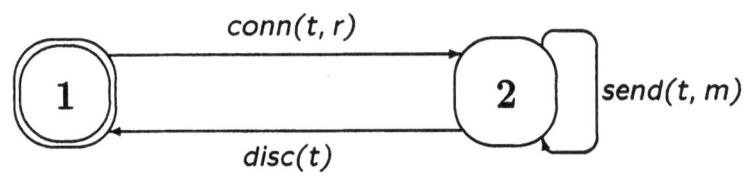

Figure 1: Transition System for each Station t of the Connecting Switch

There are two concluding remarks remaining: (1) we could have chosen some more placative names for the predicates representing the states, but the present selection is done intentionally with respect to the second technique described next and in order to stress the macro character of these predicates; (2) another example how one can write a trace specification of this style is shown in [Dederichs 90], where the well known problem of the dining philosophers is treated.

3.2.2 Using Transition Systems

The foregoing state-oriented concept does not introduce the set of states directly. In contrary to this, we explain now a state-oriented concept which uses such a set of states explicitly: the notion of a *transition system*. Such a transition system gets a trace as its input, and works on it by changing the state for each executed action starting from one of the possible initial states.

More precisely, we introduce a *sort* of states as concrete objects in the following manner: while actions are built up by action generating functions, thence we may speak of being *synthesized*, the set of states is assumed to be sufficiently large and each state can be "read" by predicates having them as parameters, hence we may speak of being *analysed* and of *state reading predicates*.

Now we explain the necessary components of a **transition system**:

1. a sort of actions: **sort** *Act*;
   together with some action generating functions;

2. a sort of states: **sort** *States*;
   together with some state reading predicates: **rel** *stat*: ...× *States* × ...;

3. the transition relation: **rel** $\longrightarrow$: *States* × \ *Act* \ × *States*;

4. a subset of states containing the initial states, say *Init*;

5. finally, a subset of states containing the liveness states, say *Final*.

As usual, one may introduce the extension $\Longrightarrow$ of $\longrightarrow$ from actions to finite traces:

rel $\Longrightarrow$: *States* × \ *Act** \ × *States*
such that
$$\forall s: Act^*, \sigma_0, \sigma_1: States \left[\sigma_0 \stackrel{s}{\Longrightarrow} \sigma_1 \iff \begin{pmatrix} (s = \varepsilon \land \sigma_0 = \sigma_1) \lor \\ \exists a: Act \, [s = \langle a \rangle \land \sigma_0 \stackrel{a}{\longrightarrow} \sigma_1] \lor \\ \exists \sigma': States \, [\sigma_0 \stackrel{\tilde{s}}{\Longrightarrow} \sigma' \land \sigma' \stackrel{\tilde{s}}{\Longrightarrow} \sigma_1] \end{pmatrix} \right]$$

Once the transition system is completely specified, a trace specification is constructed as a predicate of *accepted* traces:

Case 1: a *finite* trace $s \in Act^*$ is called **accepted** iff
$\exists \sigma_0, \sigma_1: States \left[\sigma_0 \in Init \land \sigma_0 \stackrel{s}{\Longrightarrow} \sigma_1 \land \sigma_1 \in Final \right]$ holds;

Case 2: an *infinite* trace $s \in Act^\sim$ is called **accepted** iff
$\forall s': Act^* \left[s' \sqsubseteq s \implies \exists \tilde{s}: Act^* \, [s' \sqsubseteq \tilde{s} \land \tilde{s} \sqsubseteq s \land \tilde{s} \text{ is accepted}] \right]$ holds.

We let *Acc_Traces* denote the set of all accepted traces of the specified transition system.

Now we want to apply this technique to the example of the connecting witch. What remains to do is specifying the components *States*, *Init*, $\longrightarrow$, and *Final*.

State reading predicate: There is only one state reading predicate. It collects all currently connected stations:

    rel *is_conn*: *States* × *Stations*
    such that /* *States* **is sufficiently large** */
    $\forall T: Pot(Stations) \left[\exists \sigma: States \, [\forall t: Stations \, [is\_conn[\sigma, t] \iff t \in T]] \right]$,

where *is_conn*[σ, t] says station t to be connected in state σ.

Initial configuration; Liveness Condition: The following condition must hold in the unique initial state *init*, i.e. *Init* = {*init*}:

    fct *init*: *States* such that $\forall t: Stations \left[\neg is\_conn[init, t] \right]$.

init is also the only liveness state, i.e. *Final* = {*init*}.

State transition: The transition relation $\longrightarrow$ is specified as follows:

$$\text{rel} \longrightarrow: States \times \setminus Act \setminus \times States \text{ such that}$$
$$\forall a: Act, \sigma_0, \sigma_1: States \Big[\sigma_0 \xrightarrow{a} \sigma_1 \iff$$
$$\exists t: Station\, [a \in conn(t) \wedge \neg is\_conn[\sigma_0, t] \wedge is\_conn[\sigma_1, t]] \vee$$
$$\exists t: Station\, [a \in send(t) \wedge is\_conn[\sigma_0, t] \wedge is\_conn[\sigma_1, t]] \vee$$
$$\exists t: Station\, [a = disc(t) \wedge is\_conn[\sigma_0, t] \wedge \neg is\_conn[\sigma_1, t]] \Big].$$

A kind of transition graph, or its projection to each station, may be found in FIGURE 1.

Trace Specification: As explained above, the trace specification is constructed from Acc_Traces as follows:

$$\text{rel } Acc\_con\_switch:\ Act^\omega$$
$$\text{such that } \forall s: Act^\omega \Big[Acc\_con\_switch[s] \iff s \in Acc\_Traces \Big],$$

where $s \in Acc\_Traces$ stands for the specification of the conditions of accepted traces mentioned above:
$(s \in Act^* \ ?\ init \xrightarrow{s} init : \forall p: Act^* \big[p \sqsubseteq s \implies \exists \tilde{s}: Act^* \,[p \sqsubseteq \tilde{s} \wedge \tilde{s} \sqsubseteq s \wedge init \xrightarrow{\tilde{s}} init]\big]\,)$.

Comparing the two trace specifications con_switch and Acc_con_switch, as for *correctness* aspects, one wants to prove that, at least, one of them is a refinement of the other. But, actually, they are equivalent, i.e. they have the same truth set.

Theorem 1 con_switch and Acc_con_switch are equivalent predicates, i.e.
$\forall s: Act^\omega \big[con\_switch[s] \iff Acc\_con\_switch[s]\big]$ holds.

Proof: "$\Longrightarrow$": First we prove the following

Lemma: Assume that $s \in Act^*$ and $safe\_con\_switch[s]$ are satisfied. Then
$$\forall t: Stations\, [conn(t)\S s = disc(t)\S s \iff \forall \sigma: States\, [(init \xrightarrow{s} \sigma) \implies \neg is\_conn[\sigma, t]]]$$
holds.

Proof of lemma: Let $t \in Stations$ and $s \in Act^*$ be given.

For convenience, we abbreviate the condition of the left hand side, $conn(t)\S s = disc(t)\S s$, by $L1[t, s]$ and the condition of the right hand side, $\forall \sigma: States\,[(init \xrightarrow{s} \sigma) \implies \neg is\_conn[\sigma, t]]$, by $\Lambda 1[t, s]$.

The case of $s = \varepsilon$ is clear. Assume that s is non-empty and the claimed condition is satisfied (*induction hypothesis*).

Let $a \in Act$. *Case 1:* Assume $a \in conn(t)$. If $L1[t, s]$ holds, neither $L1[t, s\&(a)]$ holds, nor does $\Lambda 1[t, s\&(a)]$. But $\neg L1[t, s]$ turns to $\neg con\_switch[s\&(a)]$ because of (S1). *Case 2:* If $a \in send(t)$, obviously $L1[t, s\&(a)]$ and $\Lambda 1[t, s\&(a)]$ both hold (if $L1[t, s]$) or fail (if $\neg L1[t, s]$). *Case 3:* If $a = disc(t)$, $L1[t, s]$ is ruled out by (*) or by (S3). But if $\neg L1[t, s]$, then $\Lambda 1[t, s\&(a)]$ holds. $L1[t, s\&(a)]$ may be shown by (S1): a $conn(t)$ action, say c, is only allowed for some past $p \in Act^*$ iff $L1[t, p]$, but then $\neg L1[t, p\&(c)]$ holds, thence, s is only allowed to contain one $conn(t)$ action more than the number of $disc(t)$ actions. $\diamond$

We have also shown by the lemma that:
$$(s \in Act^* \wedge con\_switch[s]) \implies \forall \sigma: States\,[(init \xrightarrow{s} \sigma) \implies \sigma = init].$$

From the obvious property $s \in Act^* \wedge safe\_con\_switch[s] \implies \exists \tilde{s}: Act^*\,[s \sqsubseteq \tilde{s} \wedge con\_switch[\tilde{s}]]$ we can conclude that it remains to show that $(s \in Act^* \wedge safe\_con\_switch[s]) \implies \exists \sigma: States\,[init \xrightarrow{s} \sigma]$:

Case 1: $s = \varepsilon$ directly implies $init \xrightarrow{s} init$. *Case 2:* Let a be an action, and assume the claimed condition being satisfied for $s \in Act^*$ (*induction hypothesis*). Further, assume $safe\_con\_switch[s\&(a)]$, and let $t \in Stations$. *Case 2.1:* If $a \in conn(t)$, by (S1) $\neg L1[t, s]$ is ruled out. Let $\sigma \in States$ be such that $init \xrightarrow{s} \sigma$ holds; σ exists because of the induction hypothesis. By the lemma and $L1[t, s]$ we get $\neg is\_conn[\sigma, t]$. The specification of $\longrightarrow$ and the assumption of $States$ being sufficiently large now gives us a state σ' such that $\sigma \xrightarrow{a} \sigma'$ (and $is\_conn[\sigma', t]$). It follows that $init \xrightarrow{s\&(a)} \sigma'$. *Cases 2.2, 2.3:* may analogously be established. $\diamond$"$\Longrightarrow$"

"$\Longleftarrow$": While assuming $s \in Act^*$ and also $init \xrightarrow{s} init$ instead of $safe\_con\_switch[s]$, we reprove the lemma above.

Reproof of lemma: Let $t \in Stations$. The case of $s = \varepsilon$ is trivial. Assume that $s \neq \varepsilon$ and the claimed condition holds for s (*induction hypothesis*).

Let $a \in Act$. *Case 1:* Assume $a \in conn(t)$. If $L1[t, s]$ holds, then $L1[t, s\&(a)]$ and $\Lambda 1[t, s\&(a)]$ both fail to hold. But $\neg L1[t, s]$ is tackled by the specification of $\longrightarrow$, because by induction hypothesis this says
$$\exists \sigma: States\,[(init \xrightarrow{s} \sigma) \wedge is\_conn[\sigma, t]],$$

which is equivalent to $\forall \ldots$, for $\longrightarrow$ is deterministically specified, and, thence, implies $\neg \exists \sigma': States\, [\, init \stackrel{s\&(a)}{\Longrightarrow} \sigma'\,]$.

Now it is clear, how the reproof of the lemma can be completed: instead of using the safety conditions (S1) to (S3) one takes the corresponding subformulas of the specification of $\longrightarrow$. $\Diamond$

From the lemma we can derive $Acc\_con\_switch[s] \implies L1[s]$. It remains to deal with the safety parts: $\exists \sigma : States\, [\, init \stackrel{s}{\Longrightarrow} \sigma\,] \implies safe\_con\_switch[s]$.

Case 1: If $s = \varepsilon$ holds, $init \stackrel{s}{\Longrightarrow} init$ and $safe\_con\_switch[s]$ both obviously are satisfied. *Case 2:* Let a be any action, and assume that there is a state σ' such that $init \stackrel{s\&(a)}{\Longrightarrow} \sigma'$. But then we have also a state σ such that $init \stackrel{s}{\Longrightarrow} \sigma$ and $\sigma \stackrel{a}{\longrightarrow} \sigma'$. Assume that $safe\_con\_switch[s]$ already holds (*induction hypothesis*). *Case 2.1:* Assume $a \in conn(t)$. Then, we have $\neg is\_conn[\sigma, t]$, and by the lemma, which may be (re)proved by assuming $\exists \sigma : States\,[\, init \stackrel{s}{\Longrightarrow} \sigma\,]$ instead of $Acc\_con\_switch[s]$, it follows that $L1[t, s]$ holds, which, in turn, implies (S1) being satisfied for $s\&(a)$. But $a \in conn(t)$ is not concerned by (S2) and (S3) and, therefore, $s\&(a)$ is proved safe. *Cases 2.2, 2.3:* may be shown analogously. $\Diamond$ "$\Longleftarrow$" □

The proof of our first theorem shows, how safety and liveness are expressed by the two description techniques. Further, it exposes a proof technique which is merely a classical Noetherian induction, because $\sqsubseteq$ well-orders Act^* (and even Act^ω). This technique is used throughout all proofs concerning safety properties. But, in this example, it has also turned out that finding a proposition on some kind of states, like the ones of the macro technique, may lead to an easy proof of liveness: the lemma contains $L1[t, s]$ which is obtainable from $L1[s]$ by omitting the universal quantification over stations. In general, proofs of liveness properties are harder to achieve; they require an analysis of the underlying formulae of the liveness conditions belonging to the concerning trace specification.

3.3 TRANSACT – A Notational Framework for Trace Specifications of Transactional Systems

Now we discuss a technique which deals with a class of distributed systems that may be characterised by the property of repeatedly carrying out certain sequences of actions called **transactions** and that we therefore call **transactional**. This class of distributed systems comprises the behaviours of all protocols, because protocols describe such appropriate sequences of actions, which have to be executed by the participating communication partners.

The technique to be explained offers a notational framework for formulating special predicates over traces called $TRANSACT$. $TRANSACT$ allows to specify transactions as sequences of actions by some kind of *regular* composed language over the set of actions as the alphabet.

A special aspect of the composition of a sequence of actions rests on the fact that each transaction may have one unique owner. In the example of the connecting switch, we have the simple case of any action belonging to the station found in its (first) argument place. Therefore, the complete "transaction" of connecting, then sending, and finally disconnecting has one station as its owner. Such a special situation is called the **owner principle**.

A $TRANSACT$ specification takes this principle also into account; it consists, therefore, of an owner part expressed by the sort of potential owners and of the transaction part specifying the sequence of actions to be carried out. Now we explain this in detail:

[*owner* :: *transaction*] is the usual $TRANSACT$ statement; it specifies the requirement of following the sequence of actions *transaction*, which is owned by some element of the sort *owner*.

owner is, as we already mentioned, the sort of potential owners, rather its name.

transaction specifies a sequence of actions; it is composed as follows:

- atom: *action*;
- sequential composition: *transaction.transaction*;
- nondeterministic choice: *transaction/transaction*;

- "Kleenian star": *transaction*;
- bracketing: (transaction).

action is of the form $actionname\bigl[-arg2\bigl[-arg3\ldots\bigr]\bigr]$, where $arg2$ etc. are names of sorts.

We only have given the general syntactical concerns of $TRANSACT$; the semantical aspects will be considered when dealing with our example.

Now, the intended $TRANSACT$ specification of the connecting switch simply reads as follows:

    rel con_switch_∞: Act^ω
    such that [Stations :: conn-Stations.*send-Messages*.disc],

which is equivalent to the following (extended) trace logic specification:

$$\ldots\textbf{such that } \forall s: Act^\omega \Bigl[con\_switch_\infty[s] \iff$$

$$\forall t: Stations \left[\exists \begin{cases} k: \overline{Nat}, \\ r: \overline{Nat} \to Stations, \\ m: \overline{Nat} \to Messages, \\ n: \overline{Nat} \to Nat \end{cases} \right\} \left[\begin{array}{l} t\copyright s = \\ \&_{i=1}^{k}[\,\langle conn(t,r_i)\rangle\& \\ \langle send(t,m_i)\rangle^{n_i}\& \\ \langle disc(t)\rangle\,] \end{array} \right] \right],$$

where $t\copyright s$ is one of the auxiliary concepts, $\&_{i=1}^{k}$ has the usual meaning, in particular, for $k=\infty$ it leads a least upper bound expression, and $\langle send(t,m_i)\rangle^{n_i}$ is the n_i-length sequence consisting of $send(t,m_i)$ actions only.

As for the transition system specification, we want to compare the TRANSACT specification with the pure trace logic specification con_switch: it turns out that con_switch is equivalent to con_switch_∞.

Theorem 2 con_switch_∞ is equivalent to con_switch, i.e.
$\forall s: Act^\omega \bigl[con\_switch_\infty[s] \iff con\_switch[s]\bigr]$ holds.

Proof: "$\Longrightarrow$": Let $s \in Act^\omega$ be such that $con\_switch_\infty[s]$. L1[s] is obvious, as is $L1[t,s]$ (see proof of Th.1) for each station t.

For convenience, we abbreviate the condition on the right hand side in the trace logic specification of con_switch_∞ by $con\_switch_\infty[t,s]$ for $t \in Stations, s \in Act^\omega$.

Fix a station t. *(S1)*: Let $p \in Act^*$ and $a \in conn(t)$ be such that $p\&a \sqsubseteq s$, i.e. $actual[p,a,s]$ holds. From the special form of s we may conclude that $con\_switch_\infty[t,p]$ because $conn(t)$ actions follow only and exclusively $disc(t)$ actions in $t\copyright s$. But then $L1[t,p]$ holds and (S1) is satisfied. *(S2)*: Let $p \in Act^*$ and $a \in send(t)$ be such that $p\&a \sqsubseteq s$. Then, we get a $p' \in Act^*$ such that $p' \sqsubseteq p$, $con\_switch_\infty[t,p']$, and $p = p'\&(\langle conn(t,r)\rangle\&\langle send(t,m)\rangle^n)$ for some appropriate r, m, n. Thence, we have
$$conn(t)\S p - disc(t)\S p = 1 > 0$$
and (S2) is satisfied. *(S3)*: may be shown analogously to (S2). $\diamond$ "$\Longrightarrow$"

"$\Longleftarrow$": Let t be any station. Let $s \in Act^\omega$ be such that $safe\_con\_switch[s]$ and $L1[t,s]$ (see proof of Th.1) are satisfied. The case of $s = \varepsilon$ is clear, thus assume that $s \neq \varepsilon$. W.l.o.g. assume $s = t\copyright s$. Let p, a be such that $actual[p,a,s]$ and $con\_switch_\infty[t,p]$ are satisfied. *(induction hypothesis)*

We show that there exists $\tilde{p}$ such that $p\&\langle a\rangle \sqsubseteq \tilde{p}$, $\tilde{p} \sqsubseteq s$, and $con\_switch_\infty[t,\tilde{p}]$.

By $con\_switch_\infty[t,p]$ it follows that $L1[t,p]$. But, by (S2), (S3), and $L1[t,p]$ we immediately get $a \in conn(t)$. Furthermore, by $L1[t,p]$ and $L1[t,s]$ we get some b such that $actual[p\&\langle a\rangle, b, s]$. But by (S1) $b \notin conn(t)$ holds. If $b = disc(t)$ we immediately have $\tilde{p} := p\&\langle a, b\rangle$.

But if $b \in send(t)$, there exists some c such that $p\&\langle a, b\rangle\&c \sqsubseteq s$, $conn(t)\copyright c = \varepsilon$, and It $c = disc(t)$ by $L1[t,s]$. By (S1) and (S3) ld c must be a (maybe empty) sequence of send actions. But, then, put $\tilde{p} := p\&\langle a,b\rangle\&c$.

The last claim shows $con\_switch_\infty[t,s]$, and we are done. $\diamond$ "$\Longleftarrow$"   □

We have shown, how liveness and safety conditions may be composed within a compact notation by $TRANSACT$: **safety:** any admissible trace must follow the prescribed transaction when restricted to each station, **liveness:** any begun transaction must be completed.

$TRANSACT$ specifications may not only be used to abbreviate the specifying formula as described above, but also to define predicates over traces itself. This allows to use $TRANSACT$ statements freely in a trace logic specification in order to combine transaction requirements with trace logic constraints. Taking these remarks into account, for our example, the following specification is obtained, where the $TRANSACT$ specification appears now as a separate predicate:

> rel con_switch_∞: Act^ω
> such that $\forall s : Act^\omega \big[con\_switch_\infty[s] \iff$
>   [Stations :: conn-Stations.*send-Messages*.disc]$[s] \big]$.

3.4 Modifiability of Trace Specifications

In this subsection we want to explain some techniques or aspects that concern modifications of an existing trace specification to get another more refined or supplemented trace specification; we only want to consider modifications aiming at the set of all possible actions. It is also explained how the proof obligation of verifying the resulted trace specification against the older one may be solved.

3.4.1 Action Refinement

Action refinement says that the actions of the old specification are to be more detailed. In a development process it is convenient to give a trace specification based on a set of abstract actions first and, then, to use appliances to get a correct trace specification containing lesser abstract actions.

An action refinement is characterised by a (one-one) mapping from the old action set into some execution units consisting of elements of the new action set. Actually, this mapping is in the simplest case from Act_{old} into $Pot(Act^*_{new})$, or, more elaborately, if some reference to the past is needed, from $Act^*_{old} \times Act_{old}$ into $Pot(Act^*_{new})$.

Technically, we may proceed as follows: let ρ be an action refinement, i.e. a mapping $Act^*_{old} \times Act_{old} \to Pot(Act^*_{new})$ respectively; we extend ρ to a mapping $\bar\rho$ by

> fun $\bar\rho: Act^\omega_{old} \to Pot(Act^\omega_{new})$
> such that
> $\bar\rho(\varepsilon) = \{\varepsilon\} \wedge$
> $\forall s: Act^\omega_{old} \big[s \neq \varepsilon \implies \forall s': Act^\omega_{new} \big[s' \in \bar\rho(s) \iff$
> $\forall p: Act^*_{old}, a: Act_{old}, q: Act^\omega_{old} [s = p\&\langle a\rangle\&q \iff$
> $\exists p', a', q': Act^\omega_{new} [p' \in \bar\rho(p) \wedge a' \in \rho(p, a) \wedge q' \in \bar\rho(q) \wedge s' = p'\&a'\&q']]\big]\big]$;

let $T_{old} \subseteq Act^\omega_{old}$ be the trace set of the old specification; then, we are able to form $T_{new} \subseteq Act^\omega_{new}$, the trace set of the new specification, by putting

> $T_{new} := \{s': Act^\omega_{new} \mid \exists s: Act^\omega_{old} \big[s \in T_{old} \wedge s' \in \bar\rho(s)\big]\}$.

Thus, an action refinement in our sense is fully described by the specification of the components of the following quintuple:

> $(Act_{old}, T_{old}, Act_{new}, \rho)$.

The first two components are part of the presupposed trace specification; Act_{new} is specified with the same method as for Act_{old} (term algebra over some set of action generating functions).

Here, we want to give now a notational framework for specifying the proper component ρ of an action refinement:

*pred*1 **refines** *pred*2 **by** *refinement* says that the predicate *pred*1 over Act^ω_{new} is obtained from the predicate *pred*2 over Act^ω_{old} by an action refinement specified by *refinement*. Thus, *pred*1 is a reference to Act_{new} and *refinement* concerns ρ.

refinement ::= '[{'*refmapentry*{','*refmapentry*}*'}]' specifies ρ; each string *refmapentry* belongs to the specification of ρ.

refmapentry ::= *old_act_term* ↦ *block* says how to refine an action of the form *old_act_term* into execution sequences specified by *block*.

old_act_term is simply a term over Act_{old}. It may contain free variables together with type information (may be omitted if the context is very clear), which, in turn, may used in *block*, the right side of a *refmapentry* specification.

block ::= [*statement*] determines a set of finite sequences over Act_{new}, which is needed for the specification of any image of any action under ρ.

statement is the kernel of our notational framework; it allows to build a set of (finite) execution sequences by the aid of programming-language-like combinators; there are the following possibilities:

- atom: *new_act_term*, a term over Act_{new};

- empty command: **skip**;

- sequential composition: *statement* ';' *statement*;

- conditional choice: **if** *condition* **then** *statement* **else** *statement* **fi**,
  where *condition* is a (trace logic) formula, which contains references to the past by the word symbol **past**;

- binary nondeterministic choice: *statement* **or** *statement*;

- arbitrary nondeterministic choice: **select** *new_act_term* **where** *condition*;

- very special implicit conditional choice: *new_act_term* **if needed**, see later.

We enrich now our notational framework by adding some object-based concepts. Now we assume for trace specifications that all objects, which actions refer to, have a state of existence: an action may only occur in a trace, if all objects found as its arguments are in the state "existent", otherwise these objects have to be *created*. But, if all action generating functions concerning some sort are defined in the common way, we assume that the system creates all objects of this sort *before* executing any action; namely, in order to bring the object-based concepts into the foreground we use two new word symbols **creat** and **delet** in the place of **Act fct**:

    creat *Act_fct_name*: *sort_name*
                          /* *Act_fct_name* is a create action for objects of *sort_name* */;

    delet *Act_fct_name*: *sort_name*
                          /* *Act_fct_name* is a delete action for objects of *sort_name* */.

Whenever **creat** and **delet** is used, each trace specification implicitly contains safety requirements saying that any action referring to any sort concerned by any **creat** or **delet** line may only occur iff the corresponding objects are created by a create action before, but not deleted by a delete action.

Now, we are able to apply the concept of an action refinement to the connecting switch. The example specification reads as follows:

/* Act_{new} (here: Act') */
sort $Links = Stations \times Stations$;
creat $create: Stations \to Act'$;
delet $del: Stations \to Act'$;
creat $link: Links \to Act'$;
delet $unlink: Links \to Act'$;
Act fct send overload: $Stations \times Messages \to Act'$
         /* $Messages$ is a sort carrying always existent objects! */;
/* Refinement */
rel $refd\_con\_switch$: Act'^ω **such that**
  $refd\_con\_switch$ **refines** con_switch **by**

$$\left\{ \begin{bmatrix} \begin{pmatrix} conn(t,r) & \mapsto [create(t) \text{ if needed}; create(r) \text{ if needed}; link(t,r)], \\ send(t,m) & \mapsto [send(t,m)], \\ disc(t) & \mapsto [\text{select } unlink(t,r) \\ & \quad \text{where } (\,conn(t)\copyright\textbf{past} \neq \varepsilon \\ & \quad\quad ? \text{ lt } (conn(t)\copyright\textbf{past}) = conn(t,r) \\ & \quad\quad : r \in Stations \setminus \{t\} \,) \\ & \quad ; \\ & \quad [\textbf{skip or } del(t)]] \end{pmatrix} \end{bmatrix} \right\}$$

From this specification, one immediately reads that

$$\begin{aligned} \rho(p, send(t,m)) &= \{\langle send(t,m)\rangle\} \\ \rho(p, disc(t)) &= (\ conn(t)\copyright p \neq \varepsilon \\ &\quad ?\ \{\langle unlink(t,r)\rangle, \langle unlink(t,r), del(t)\rangle | \text{lt}(conn(t)\copyright p) = conn(t,r)\} \\ &\quad :\ \{\langle unlink(t,r)\rangle, \langle unlink(t,r), del(t)\rangle | r \in Stations \setminus \{t\}\}\ ) \end{aligned}$$

As an exception, $\rho(p, conn(t,r))$ is not so simple reconstructable; the **if needed** construct adds a safety requirement to the new system specification that refers not to the past of the $conn$ action, but to the past of the translation itself: e.g. "$create(t)$ **if needed**" says that, iff the past of the translation does not support the state of existence for t, we have to put $create(t)$ onto its corresponding place, otherwise the construct stands for **skip**.

Now, the proof obligation of verifying $refd\_con\_switch$ against con_switch does not become visible, because it appears as a mere validation problem of the refinement itself. In the general case, one goes the other way round: let two trace logic specifications be given; so as to verify one specification against the other, there is the hard possibility of searching for a "suitable" refinement mapping ρ; remember ρ not only influencing the action sets but also the safety and liveness conditions.

3.4.2 Action Enrichment

Another modification option for the action set is the addition of actions. This immediately entails the addition of suitable conditions to the old specifications. On the one hand, the extension of the action set is a good test for formal description techniques about the flexibility of the underlying notational framework.

On the other hand, one gets the problem of which conditions being in some sense suitable for the new actions in connection with the older ones. As the technique of **action enrichment** is related to hierarchization of algebraic data types, we use the term **persistency** in the same situation: let two trace predicates p_1 and p_2 be given together with their action sets Act_1 and Act_2, where $Act_1 \subseteq Act_2$; p_2 is called **persistent** wrt. p_1 iff (a) for every trace t_2 of p_2 $Act_1 © t_2$ is a trace of p_1 and (b) for every trace t_1 of p_1 there is a trace t_2 of p_2 such that $t_1 = Act_1 © t_2$.

In our example, we want to extend the action set, which contains connecting, sending, and disconnecting actions so far, by a receive action. But, for simplicity, the receive action is not owned by the receiver, but by the sender. Moreover, we want to let any receiving action only occur as a consequence of a foregoing sending action (*safety*), but also vice versa, i.e. we expect for each sending action a corresponding receiving action (*liveness*). More detailed, the enriched specification reads as follows:

> **sort** Act_r **extends** Act;
> **Act fct** $recv: Stations \rightarrow Act_r$;
> **rel** $con\_switch_r: Act_r^\omega$ **such that**
> con_switch_r **extends** con_switch **by**
> $$s: Act_r^\omega \left[\forall t: Stations \left[\begin{array}{l} (L2)\ send(t) \S s = recv(t) \S s\ \wedge \\ \forall p: Act_r^\omega, a: Act_r \left[actual[p, a, s] \Longrightarrow \right. \\ (S4)\ (a = recv(t) \Rightarrow send(t) \S p > recv(t) \S p) \end{array} \right] \right],$$

where we have renounced to introduce the required extensions of our notational framework formally.

Now, the proof of persistency appears as the corresponding proof obligation with respect to verification:

Theorem 3 con_switch_r is persistent wrt. con_switch.

Proof: Let $s \in Act_r^\omega$. Ad (a): Assume $con\_switch_r[s]$, and put $s_0 := Act © s$. $L1[s]$ holds, therefore, also $L1[s_0]$, because $conn(t) © s = conn(t) © s_0$ and $disc(t) © s = disc(t) © s_0$. For the same reason, from $safe\_con\_switch[s]$ we may conclude that $safe\_con\_switch[s_0]$ holds. Thus, $con\_switch[s_0]$ is satisfied.
Ad (b): Assume $con\_switch[s]$. We construct $s_1 \in Act_r^\omega$ from s as follows:
$$s_1 = \varrho(s) \text{ where}$$
$$\varrho(\varepsilon) = \varepsilon \wedge \varrho(\langle a \rangle \& \hat{s}) = (\ a \in send\ ?\ \langle a, recv(t) \rangle \text{ where } a \in send(t)) : \langle a \rangle\ ) \& \varrho(\hat{s}).$$
$s = Act © s_1$ is obvious; $L2[s_1]$ clearly holds by the definition of ϱ; by a simple induction with reference to ϱ one also shows $S4[s_1]$: if $actual[p_1, recv(t), s_1]$ holds, so does $\exists a: Act_r\ [a \in send(t) \wedge p_1 = p'_1 \& \langle a \rangle]$ such that $send(t) \S p'_1 = recv(t) \S p'_1$. Thence, we have shown $con\_switch_r[s_1]$. □

This kind of proof obligation and the notion of persistency, respectively, are important to the design method, because a persistent action enrichment allows to proceed with the development steps for the old specification independently forward, because the causality connections within the specified traces are not abandoned by the enrichment.

3.4.3 REACTION – Towards a More Natural Modelling

The given pure trace logic specification con_switch of the connecting switch is not the natural derivation from the informal requirements, as we have already admitted. As shown, one way is to develop a $TRANSACT$ specification, possibly combined with trace logic; another is to specify a transition system.

But to make the view of a connecting switch more transparent, one goes back to the notion of object in the fundamental version: (a) objects have a state each, its *knowledge*, and (b) objects interact with another objects by *sending messages*. For the latter point, we may reorganize *Act* such that there is one uniform action generating function called *message* containing sender, receiver, and the message content.

This reorganization may also be used as a way to develop a design specification rather than a requirement specification. Once, all actions are transformed into the uniform representation by message actions, one has selected the causality structure of the system: objects may turn into agents.

In order to reflect the division of a distributed system into "system" and "environment" a little bit, we introduce a new object *environment*, to which the overall control of the system is assigned.

An important point of this view is the direct usage of the messages of *Messages* within the communications, i.e. the proper messages are sent by a station (within a send command) and are delivered via *environment* to the target station.

REACTION is a notational framework for descriptions of the interaction of the objects of a system. We do not go into detail, while the specification of our example with REACTION is listed in FIGURE 2.

From the REACTION specification, one obtains a trace specification by the following steps:

(a) add the object *environment* by introducing a new sort *Envs* and a constant definition, form **sort** $Comps = Stations + Envs$.

(b) reorganize *Messages* by *adding* messages of the form
$hello(t)$, $ok(t, \hat{m})$, $says\_hello(t)$, $error(t, \hat{m})$, $send(m)$, bye, and $says\_bye(t)$
where $\hat{m} \in \{hello(t), send(m), bye \mid t, m\}$, each m is an old message.

(c) reorganize *Act* by using **Act fct** *message*: $Comps \times Messages \times Comps$ alone.

(d) use the transition system technique: **send**'s are executed actions, and **knows**'s determine state transitions; the **final** and **initial** blocks have their usual meaning.

The model behind REACTION leads to a more realistic specification of a connecting switch also for another reason: within traces we also allow erroneous situations, which end up with *environment* sending an *error* message to the ordering station. This is because every **block** is intended to be executed by nondeterministic choice. But, in turn, this even requires an appropriate error handling.

Once, the trace specification, say *con_switch_reaction*, is derived from the REACTION specification, we are able to show by identifying the *ok* messages with the corresponding actions of the old set that the result is a *persistent* action enrichment, but we omit the proof here.

system CON_SWITCH_REACTION:
  objects $environment$(ENVIRONMENT), r, t(STATION), m(MESSAGE);
  global
    $environment$ knows $said\_hello$(STATION), $said\_hello$(STATION, STATION)
  endglobal;
  initial
    $environment$ knows $\forall t[\neg said\_hello(t)] \land \forall t, r[\neg said\_hello(t, r)]$
  endinitial;
  final
    $environment$ knows $\forall t[\neg said\_hello(t)] \land \forall t, r[\neg said\_hello(t, r)]$
  endfinal;
  block $stations\ may\ get\ connected$:
    $\forall t, r[t$ sends $hello(r)$ to $environment$;
    if $\neg said\_hello(t)$
    then
      $environment$ sends $ok(t, hello(r))$ to t,
        sends $says\_hello(t)$ to r,
        knows $said\_hello(t) \land said\_hello(t, r)$
    else $environment$ sends $error(t, hello(r))$ to t endif]
  endblock;
  block $may\ then\ send\ messages$:
    $\forall t, m[t$ sends $send(m)$ to $environment$;
    if $said\_hello(t)$
    then
      $environment$ sends $ok(t, send(m))$ to t,
        sends m to r where $said\_hello(t, r)$ /* here, m is delivered directly */
    else $environment$ sends $error(t, send(m))$ to t endif]
  endblock;
  block $and,\ finally,\ may\ get\ disconnected$:
    $\forall t[t$ sends bye to $environment$;
    if $said\_hello(t)$
    then
      $environment$ sends $ok(t, bye)$ to t,
        sends $says\_bye(t)$ to r where $said\_hello(t, r)$
        knows $\neg said\_hello(t) \land \neg said\_hello(t, r)$ where $said\_hello(t, r)$
    else $environment$ sends $error(t, bye)$ to t endif]
  endblock
endsystem

Figure 2: A REACTION Specification for the Connecting Switch

4. From Trace Specifications to Design Specifications

Trace specifications are used for describing the behaviour of a distributed system on an abstract level. Towards an implementation, one does not consider the system in the whole, but one begins to distinguish some *components*. Basically, components themselves may be regarded as systems, to which a trace specification may be assigned. Technically, one partitions the set of actions and formulate requirements on traces appropriately restricted, as we will show in 4.1. Actually, a component is an entity that has its own causality structure of its own executed actions.

Besides the potentially arbitrary divisibility into components, one may follow the following principle: one divides the distributed system into the "system" component and the "environment". These two components interact within a closed circulation, as depicted in FIGURE 3. How this interaction

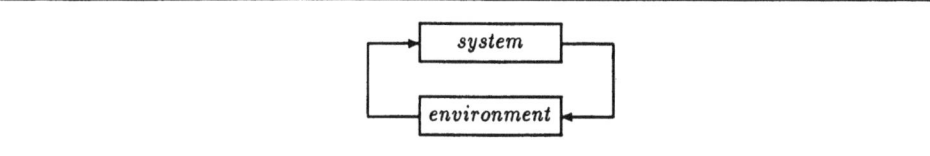

Figure 3: "System" and "Environment"

may be formally described, more with respect to a design specification, is shown in 4.2. But, in addition, this separation method requires the distinction of the input actions and the output actions of the "system" part in order to get a causality structure that may be later interpreted by mapping, as for e.g. by a stream processing function.

Finally, after we have prepared the technical appliances, in 4.3 we explain how the division into "system" part and "environment" part is made for a connecting switch. The described division will be the one that is taken into account for the aimed development process from the level of requirement specification down to an implementation.

4.1 Component-Oriented Specifications

A component-oriented specification deals with the formal division of a trace specification of a distributed system in the whole into single trace specifications of the aimed set of components.

We do not go more into detail and refer to [Broy et al. 91] for a better introduction to this method, but, at least, we describe here the fundamental setting applicable to simple cases: *Step 1:* divide Act into appropriate parts Act_1, Act_2, ... corresponding to the set of components (note that we do not require that the decomposition is obliged to generate disjoint sets of actions); *Step 2:* specify each component by a trace predicate yielding C_1, C_2, ...; *Step 3:* this is the verification step, prove that
$$\forall s: Act \left[C_1[Act_1 \textcircled{c} s] \wedge C_2[Act_2 \textcircled{c} s] \wedge \ldots \implies C[s] \right],$$
where C is the trace specification belonging to the whole system.

Now, we discuss two intuitive possibilities of decomposing the connecting switch. The first will go by (bidirected) links, the other by stations. But, to achieve this we have to assume *Stations* to be finite, otherwise we obtain an infinite verification claim because the conjunction becomes infinite. This shows a drawback of the method, which, but, does not play a great rôle to practice.

Decomposition by Links: The connecting switch is regarded as a family of bidirected links $con\_switch_{\{t,r\}}$:

> **begin** $\{\text{forall } t, r \in Stations \wedge t \neq r\}$
> **sort** $Act_{\{t,r\}}$ such that $Act_{\{t,r\}} = \{a : Act \mid a \in \begin{Bmatrix} conn(t,r), send(t), disc(t), \\ conn(r,t), send(r), disc(r) \end{Bmatrix}\}$;
> **rel** $con\_switch_{\{t,r\}}: Act^\omega_{\{t,r\}}$ such that
> $\forall s : Act^\omega_{\{t,r\}} \Big[con\_switch_{\{t,r\}}[s] \iff con\_switch[Act_{\{t,r\}} \copyright s] \Big]$
> **end.**

The proof obligation is immediately satisfied, the conjunction even is equivalent to the specification of the whole connecting switch itself.

Decomposition by Stations: Most naturally, one decomposes the connecting switch into the (*finite*) set of stations. The resulting specifications are determined by the following scheme:

> **begin** $\{\text{forall } t \in Stations\}$
> **sort** Act_t such that $Act_t = \{conn(t,r) : Act \mid r \in Stations\} \cup \{send(t), disc(t)\}$;
> **rel** $con\_switch_t: Act^\omega_t$ such that
> $\forall s : Act^\omega_t \Big[con\_switch_t[s] \iff con\_switch[Act_t \copyright s] \Big]$
> **end.**

It is the same case like at the decomposition by links. Only for convenience, we have used con_switch itself and filtering $\copyright$ in the specifications of the components. Moreover, con_switch is clearly a *persistent* action enrichment wrt. con_switch_t for each single station t.

4.2 Differential Strategies

The decomposition of a distributed system into "system" and "environment" parts is of good use for developing a design specification with functional agents. "System" and "environment" interact like depicted in FIGURE 3. This view leads to a splitting of the action set in either two parts: the system's input actions In and the system's output actions Out such that $In \cup Out = Act$ and $In \cap Out = \emptyset$.

"System" and "environment" may be viewed as *game players* when they cooperate; both may apply so-called *strategies* [Broy Dederichs Dendorfer Weber 91, §2.2] to perform their interaction. But, in the case of transactional systems, the game view can be described by a *differentially* strategic system, whose kernel is intended to formalize reactions by mapping changes of the input trace to changes of the output trace for each of the players.

More formally, a **differentially strategic system** is a quadruple $\partial \Sigma = (\mathcal{I}, \partial \mathcal{C}, \partial \mathcal{E}, \mathcal{F})$ of sets such that:

$\mathcal{I} \subseteq In^*$ is the set of **initial strategies** to be applied by the environment, which acts always as the first player.

$\partial \mathcal{C} \subseteq In \rightarrow Out^*$, $\partial \mathcal{E} \subseteq Out \rightarrow In^*$ are the sets of **differential strategies** to be applied by the game players: $\partial \mathcal{C}$ concerns "system", while $\partial \mathcal{E}$ concerns "environment".

$\mathcal{F} \subseteq Out^*$ is the set of **final situations** concerning the output of the "system" component: a trace may be finished up, if the output "last" made by "system" is element of $\mathcal{F}$.

Once, a differentially strategic system is defined, one may introduce the set of traces generated by it in order to install a device for comparing differentially strategic specifications to trace specifications (cf. similar definitions in [Broy Dederichs Dendorfer Weber 91, §2.2]):

$traces(\partial\Sigma) := traces_\varepsilon(\partial\Sigma) \cup traces_+(\partial\Sigma) \cup traces_\infty(\partial\Sigma);$

$traces_\infty(\partial\Sigma) :=$
$\left\{ \bigsqcup \{t_n : Act^\omega \mid n \in Nat\} \mid \forall n : Nat \left[t_n \sqsubseteq t_{n+1} \wedge t_n \in ptraces_\infty(\partial\Sigma, n) \right] \right\};$

$ptraces_\infty(\partial\Sigma, 0) := \mathcal{I} \setminus \{\varepsilon\};$

$ptraces_\infty(\partial\Sigma, 2n) :=$
$\{s\&e \mid s \in ptraces_\infty(\partial\Sigma, 2n-1) \wedge \exists \partial E : \partial \mathcal{E} \, [\, e = \partial E(\mathrm{lt}\,(Out©s))\,]\}$

$ptraces_\infty(\partial\Sigma, 2n+1) :=$
$\{s\&c \mid s \in ptraces_\infty(\partial\Sigma, 2n) \wedge \exists \partial C : \partial \mathcal{C} \, [\, c = \partial C(\mathrm{lt}\,(In©s))\,]\}$

$traces_+(\partial\Sigma) := \bigcup \{ptraces_+(\partial\Sigma, n) \mid n \in Nat\};$

$ptraces_+(\partial\Sigma, n) :=$
$\{s\&c \mid s \in ptraces_\infty(\partial\Sigma, 2n) \wedge \exists \partial C : \partial \mathcal{C} \, [\, c = \partial C(\mathrm{lt}\,(In©s))\,] \wedge c \in \mathcal{F}\};$

$traces_\varepsilon(\partial\Sigma) := (\, \varepsilon \in \mathcal{I} \,?\, \{\varepsilon\} : \emptyset \,)$

Now, the differentially strategic specification of the connecting switch is a differentially strategic system $\partial\Sigma_{con\_switch}$ such that:

$In := conn;\ Out := send \cup disc;$

$\mathcal{I} := \{\langle a \rangle \mid a \in In\} \cup \{\varepsilon\};$

$\partial \mathcal{C} := \{\partial C : In \to Out^* \mid \exists n : Nat,\, m : Messages \, [\partial C = \partial C_{n,m}]\};$

$\partial C_{n,m}(conn(t, r)) := \langle send(t, m)\rangle^n \& \langle disc(t)\rangle;$

$\partial \mathcal{E} := \{\partial E : Out \to In^* \mid \partial E(disc(t)) \in conn\};$

$\mathcal{F} := \{\langle send(t, m)\rangle^n \& \langle disc(t)\rangle \mid m, n, r, t\}.$

Theorem 4 $\partial\Sigma_{con\_switch}$ satisfies con_switch, i.e.
$$\forall s : Act^\omega \left[s \in traces(\partial\Sigma_{con\_switch}) \Longrightarrow con\_switch[s] \right]$$
holds.

Proof: For convenience, we write $\partial\Sigma$ for $\partial\Sigma_{con\_switch}$.
(1) $ptraces_\infty(\partial\Sigma, n)$ contains safe traces only for all $n \in Nat$: $n = 0$: $ptraces_\infty(\partial\Sigma, 0)$ contains only traces with a single conn action. $n > 0$: Assume $ptraces_\infty(\partial\Sigma, n)$ containing only safe traces (induction hypothesis). Case 1: $n = 2k$: $ptraces_\infty(\partial\Sigma, n+1)$ contains all traces $s\&c$ such that $s \in ptraces_\infty(\partial\Sigma, n)$ and there is a differential strategy $\partial C \in \partial \mathcal{C}$ such that $c = \partial C(conn(t, r))$ where $t, r \in Stations$ are appropriate. We get a $n \in Nat$ and a $m \in Messages$ such that $c = \langle send(t, m)\rangle^n \& \langle disc(t)\rangle$. Thence, each $s\&c$ is again safe. Moreover, each $s\&c$ is also live. Case 2: $n = 2k - 1$: may be shown analogously: the differential strategy sees only the disc action last executed. ◇"(1)"
(2) $ptraces_+(\partial\Sigma, n)$ satisfies con_switch for all $n \in Nat$: within the proof of (1) we have already shown in case 1 of the induction conclusion that $ptraces_\infty(\partial\Sigma, 2n + 1)$ contains only safe and live traces, but by the definition of $\mathcal{F}$ the latter set is equal to the former; this proves (2). ◇"(2)"
(3) $traces_+(\partial\Sigma_{con\_switch})$ satisfies con_switch: immediately by (2). ◇"(3)"
(4) the traces in $traces_\infty(\partial\Sigma)$ satisfy con_switch: each one is also a least upper bound of a prefix-monotonic trace sequence (t_n) such that $t_n \in ptraces_\infty(\partial\Sigma, 2n + 1)$, but every such trace is safe and also satisfies $\forall t : Stations \, [conn(t)\S t_n = disc(t)\S t_n]$. ◇"(4)"
Because con_switch$[\varepsilon]$ is true, we are done. □

What we have not shown in the last theorem is apparently clear: the differentially strategic specification catches only the normal forms of the traces performed by the connecting switch with respect to the partition of Act into In and Out. It can be shown, as done in 5.2, that a differentially strategic system may be used in order to obtain a functional specification by ordinary continuous stream processing functions.

To make the scope of this technique larger, one may attach states to the differential strategies, e.g. $\partial C \subseteq States \times In \to States \times Out^*$. In the end, this specifies nothing else than a step relation of the form $\sigma \xrightarrow{i|o} \sigma'$ where σ, σ' are states, $i \in In$ is the seen input, and $o \in Out^*$ is the performed reaction. Then, such a system of "state-oriented differential strategies" would lead to a corresponding state-oriented functional specification, which is not discussed in this work.

4.3 "System" and "Environment" Parts of a Connecting Switch

In 4.1 we have discussed some examples of distributions of the components that does not have any overlapping actions. But, if one considers the "immediate" division of a distributed system into "system" and "environment" components, it turns out that all actions are shared by both of them. The reason for this is already displayed in FIGURE 3; namely, "system" and "environment" interact within a closed circulation, i.e. the domain of output actions of the one is exactly the domain of input actions of the other.

The component-oriented specification according to the separation of "system" and "environment" may be achieved by the following proceeding (cf. [Dederichs 90, §3]):

1. divide Act into In and Out where In contains the input actions of "system", while Out contains the output actions of "system"; as mentioned, this determines the corresponding sets for "environment";

2. according to the selected splitting of Act, divide the safety and liveness conditions into two corresponding groups each;

3. now group the safety and liveness conditions, which should be renamed e.g. by prefixing $In\_$ or $Out\_$ appropriately, together in several, but suitable ways to obtain the two component specifications; the following very rough variants are recommendable:
   (a) $s' \sqsubseteq s \Longrightarrow (I\_safe[s'] \Longrightarrow O\_safe[s'])$ or (a') $O\_safe[s]$,
   (b) $I\_safe[s] \land I\_live[s] \Longrightarrow O\_live[s]$.

The treatment of the connecting switch along this proceeding reads as follows:
(A) rewritten safety and liveness conditions:

    rel $In\_S1: Act^\omega$ such that $In\_S1 = S1$;

    rel $Out\_S2: Act^\omega$ such that
$$Out\_S2[s] \iff (actual[p,a,s] \land a \in send(t) \Longrightarrow conn(t)\S p > disc(t)\S p);$$
    rel $Out\_S3: Act^\omega$ such that
$$Out\_S3[s] \iff (actual[p,a,s] \land a = disc(t) \Longrightarrow conn(t)\S p > disc(t)\S p);$$
    rel $Out\_L1: Act^\omega$ such that
$$Out\_L1[s] \iff (p \sqsubseteq s \land conn(t)\S p > disc(t)\S p \Longrightarrow$$
$$\exists p'\,[p \sqsubseteq p' \land p' \sqsubseteq s \land conn(t)\S p' = disc(t)\S p'])$$

(B) component specifications:

    rel $cs\_sys: Act^\omega$ such that
$$cs\_sys[s] \iff Out\_S2[s] \land Out\_S3[s] \land (In\_S1[s] \Longrightarrow Out\_L1[s]);$$
    rel $cs\_env: Act^\omega$ such that $cs\_env[s] \iff In\_S1[s]$.

One easily verifies that cs_sys and cs_env together form a suitable component-oriented specification for con_switch, i.e.
$$\forall s\,[\,cs\_sys[s] \land cs\_env[s] \Longrightarrow con\_switch[s]\,],$$
even that "$\iff$" holds.

Now, we are able to prove the correctness of the differential strategic system against the specification of the "system" part:

Theorem 5 $\partial \Sigma_{con\_switch}$ is a causality structure for cs_sys, i.e.

$$\forall s: Act^\omega \begin{bmatrix} s \in traces(\partial \Sigma_{con\_switch}) \Longrightarrow cs\_sys[s] \wedge \\ cs\_sys[s] \Longrightarrow \forall i: In, o: Out, p, q: Act^\omega \, [\, s = p\&\langle o, i\rangle\&q \Longrightarrow cs\_sys[p\&\langle i, o\rangle\&q] \,] \end{bmatrix}$$

holds.

Proof: The first part of the condition is clear from:

$$s \in traces(\partial \Sigma_{con\_switch}) \overset{Th.4}{\Longrightarrow} con\_switch[s] \Longrightarrow cs\_sys[s].$$

The second part first deserves some explanation: as we model asynchronous communication, the output may be postponed due to some delay; therefore, for all such modified traces, one is obliged to show that they can be carried out by the component.

Let $s \in Act^\omega$ be such that $cs\_sys[s]$, and let p, i, o, q be such that $s = p\&\langle o, i\rangle\&q$. Then, put $\tilde{s} := p\&\langle i, o\rangle\&q$. We have to show: $cs\_sys[\tilde{s}]$.

Assume $o \in send(t) \cup disc(t)$. If $i \notin conn(t)$, we are done, because $p\&\langle i\rangle$ satisfies the safety conditions for station t, as p does. Assume $i \in conn(t)$. $p\&\langle i, o\rangle$ is safe, i.e. it satisfies Out_S2 and Out_S3, and so does $\tilde{s}$. For $p\&\langle i, o\rangle$ does not satisfy In_S1 (either p itself does not satisfy In_S1, or $p\&\langle o\rangle$ and, thus, p satisfy In_S1, but then $p\&\langle i\rangle$ surely does not), it immediately follows that $\neg In\_S1[\tilde{s}]$. Thence, we have $In\_S1[\tilde{s}] \Longrightarrow Out\_L1[\tilde{s}]$, and $cs\_sys[\tilde{s}]$ is completely established. □

5. Functional Specifications

Design specifications consider the concerned distributed systems as a family of components, where the components may be more or less detailed parts of the system. Components of a distributed system exhibits their behaviour by taking input and forming output as a corresponding reaction, i.e. there is assumed a functional relation between the performed input actions and the performed output actions.

The formal description technique of functional specifications expresses this by describing components of a distributed system by *communicating agents*, which are connected by *directed asynchronous channels*. A communicating agent may have an arbitrary, but finite number of input and output ports depending on a decision of the designer.

As a basic technique, communicating agents may be modelled by a predicate over stream processing functions. Stream processing functions take the input stream(s) and produce from it the output stream(s) continuously. When the predicate does not specify the stream processing function uniquely, this means that the agent is intended to exhibit a nondeterministic behaviour.

Unfortunately, the continuity requirement of stream processing functions may lead to an inconsistent specification, i.e. the predicate has an empty truth set: if two actions occur on two different channels, one never can reconstruct the causal ordering of these two actions, even if this is desired.

Now, in this section we first describe the technique of using functional specifications with partial order processing functions, which tackle the drawback of ordinary stream processing functions by the ability of expressing the causality of actions explicitly.

The continuity requirement of stream processing functions works for our example as a design decision that provides the transition from design specification to abstract program. This is described in 5.2, where also the technique of using stream processing functions is explained.

5.1 Partial Order Processing Functions

So far we have modelled concurrency by the principle of interleaving: $a\|b$ is rather expressed by the set of traces $\{\langle a,b\rangle, \langle b,a\rangle\}$. This view may be also communicated itself to the streams of inputs and outputs belonging to agents described by stream processing functions. A serious drawback of the view has been already mentioned: if we distribute the actions onto more than one channel, the causal ordering of these is abandoned and unreconstructable.

If one wants to reflect the causal ordering of executed actions more directly, one uses a technique supporting explicit concurrency modelling. For this, we describe the usage of *partial orders* instead of the merely sequential ordering of traces: one introduces a sort of events and provide it with an ordering relation expressing the causality of occurrences; furthermore, we label the events with the actions to be executed in their right place. Now, if we use continuous functions mapping such partial orders to each other, we represent the causality of incoming or of outgoing actions of an agents directly, and we do not need more than one input and one output port each for communications between *partial order processing* agents.

Now, the definition of the sort of **labelled partial orders** (lpo's) over Act reads as follows:

used sort $Events$;
sort Act^p;
fun $\mathcal{E}: Act^p \to Pot(Events)$;
fun $\alpha: \downarrow Act^p \downarrow \times Events \to Act$
 such that $\forall p: Act^p, e: Events\,[\,\delta_{Act}[\alpha_p(e)] \iff e \in \mathcal{E}(p)\,]$;
rel $\prec: Events \times \downarrow Act^p \downarrow \times Events$
 such that $\forall p: Act^p, e, e': Events$ $\begin{bmatrix}(\delta_{Bool}[e \prec_p e'] \iff \{e,e'\} \subseteq \mathcal{E}(p)) \wedge \\ (e \prec_p e' \implies \begin{pmatrix} e \neq e' \wedge e' \not\prec_p e \wedge \\ \forall e''[e \not\prec_p e'' \vee e'' \not\prec_p e']\end{pmatrix})\end{bmatrix}$.

We introduce the causality relation $\leq_p$ based on the direct causality $\prec_p$:

rel $\leq$ **overload**: $Events \times \downarrow Act^p \downarrow \times Events$
 such that $\forall p: Act^p$ $\begin{bmatrix}\leq_p = \prec_p^* \cup \prec_p^\infty \wedge \\ \forall e: Events\,[\,e \in \mathcal{E}(p) \implies \exists e_0\,[\,e_0 \leq_p e \wedge minimal[e_0, p]\,]\,]\end{bmatrix}$;
rel $minimal: Events \times Act^p$ such that
 $\forall e: Events, p: Act^p\,\bigl[\,minimal[e,p] \iff$
  $e \in \mathcal{E}(p) \wedge \forall e'\,[\,e' \in \mathcal{E}(p) \wedge e' \leq_p e \implies e = e'\,]\,\bigr]$.

The second condition in the specification of $\leq_p$, which also influences the one of $\prec_p$, is called the **axiom of bounded causes** or the **axiom of the process start**, because it prescribes that a process cannot be started from infinity, whereas a process is able to be continued arbitrarily; this is also called **left-boundedness** of processes.

On lpo's there analogously exists a prefix ordering:

rel $\sqsubseteq$ **overload**: $Act^p \stackrel{\downarrow}{\times} Act^p$
 such that $\forall p, p': Act^p\,\bigl[\,p \sqsubseteq p' \iff$
  $p = p'|_{\mathcal{E}(p)} \wedge \forall e, e'\,[\,e \in \mathcal{E}(p) \wedge e' \in \mathcal{E}(p') \wedge e' \leq_{p'} e \implies e' \in \mathcal{E}(p)\,]\,\bigr]$;
fun $|: Act^p \times \downarrow Pot(Events) \downarrow \to Act^p$
 such that $\forall p: Act^p, E: Pot(Events)$ $\begin{bmatrix}(1) & \forall e\,[\,e \in \mathcal{E}(p|_E) \iff e \in \mathcal{E}(p) \cap E\,] \wedge \\ (2) & \forall e\,[\,e \in E \implies \alpha_{p|_E}(e) = \alpha_p(e)\,] \wedge \\ (3) & \forall e, e'\,[\,\{e,e'\} \subseteq E \\ & \implies (e \prec_{p|_E} e' \iff e \prec_p e')\,]\end{bmatrix}$.

Like for streams, we are able to construct an *actual* predicate:

rel $actual: Act^p \times Act \times Act^p$ **such that**
$$\forall p, q: Act^p, a: Act \left[actual[p, a, q] \iff \right.$$
$$\left. |\mathcal{E}(p)| < \infty \wedge p \sqsubseteq q \wedge \exists e [minimal[e, q|_{Events \setminus \mathcal{E}(p)}] \wedge \alpha_p(e) = a] \right].$$

Finally, the set of all **partial order processing functions** is denoted by $(\mathcal{A}^p \to \mathcal{B}^p)$.

With these concepts, we are able to specify the connecting switch by:

rel $popf\_stations: (In^p \to Out^p)$ **such that**
$$\forall \phi: (In^p \to Out^p) \left[popf\_stations[\phi] \iff \right.$$
$$\forall p: Act^p \left[\exists E: Events \to Pot(Events) \left[\right. \right.$$
$$\forall e\left[c \in \mathcal{E}(p) \implies \begin{pmatrix} E(e) \subseteq \mathcal{E}(\phi(p)) \wedge \\ \exists \partial C : \partial C \left[|E(e)| = \#\partial C(\alpha_p(e)) \right] \wedge \\ \exists e_d \begin{bmatrix} e_d \in E(e) \wedge \\ \forall e_E \left[e_E \in E(e) \implies e_E \preceq_{\phi(p)} e_d \right] \wedge \\ (\alpha_{\phi(p)}(E(e) \setminus \{e_d\})) \subseteq send(t) \wedge \\ \alpha_{\phi(p)}(e_d) = disc(t)) \\ \text{where } \alpha_p(e) \in conn(t) \end{bmatrix} \end{pmatrix} \right] \wedge$$
$$\forall e, e' \left[\{e, e'\} \subseteq \mathcal{E}(p) \wedge e \neq e' \implies E(e) \cap E(e') = \emptyset \right] \wedge$$
$$\bigcup \{ E(e) \mid e \in \mathcal{E}(p) \} = \mathcal{E}(\phi(p)) \wedge$$
$$\forall e_1, e_2 \left[\{e_1, e_2\} \subseteq \mathcal{E}(\phi(p)) \implies \right.$$
$$(e_1 \prec_{\phi(p)} e_2 \iff \Xi(e_1) \preceq_p \Xi(e_2))$$
$$\left. \left. \left. \text{where } \forall e [e \in \mathcal{E}(\phi(p)) \implies e \in E(\Xi(e))]) \right] \right]_{\exists E} \right]_{\forall p} \right].$$

The specification says the following: every lpo p containing only $conn$ actions is moved by a ϕ such that $popf\_stations[\phi]$ to a lpo $\phi(p)$, where the events of p are replaced by "pieces" of partial orders representing the reaction of the agent. Such a reaction of the agent is determined by the set ∂C of differential strategies of the "system" part: on any $conn(t)$ action it does an (finite) amount of $send(t)$ actions closing with a $disc(t)$ action. As expressed in the conditions for e_d it is not necessary to execute all $send(t)$ actions sequentially, but they may be executed in an arbitrary order; only the $disc(t)$ action is fixed, it marks the end of each reaction (cf. FIGURE 4).

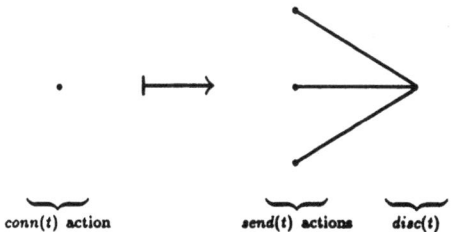

Figure 4: Example Reaction of *popf_stations*

Unlike for the other techniques, we do not treat the topic of correctness of the given specifications here, except the informal argument that the aimed behaviour is completely determined by that set ∂C of differential strategies belonging to the differentially strategic system $\partial \Sigma_{con\_switch}$, which have been proved to be a causality structure for the specification of the "system" part in Th.5, and only the connection between the differentially strategic system and the partial order processing function constructed from it has to be established, while this is in our case easy to do: there is an "epimorphism" between the $\phi(p)$'s and the p's; namely, the one "compressing" $E(e)$ to e.

5.2 The Continuity Requirement of Stream Processing Functions as a Design Decision

Starting from a requirement specification, a design specification is achieved by the selection of components that the distributed system shall be divided into. Components may be regarded as communicating agents, which process their input and give their output as reaction. Whereas partial order processing functions reflect the causal ordering of the input and output, stream processing functions provide input and output channels with sequentially flowing messages.

More formally, each function in $(A^\omega \rightarrow B^\omega)$ is called a **stream processing function**; a **stream processing agent** is the extension of this notion to functions processing tuples of streams, where the underlying ordering of the stream tuples is the canonically extended prefix ordering.

In our example, the situation is like depicted in FIGURE 3: two stream processing functions *system* and *environment* are to be specified; these are combined together by a fixed point equation representing the closed circulation.

We only concentrate on the specification of *system*; we regard *environment* as to be implemented by someone else following the *open system view*.

An agent specification consists of a predicate over stream processing functions; thus, an agent is rather represented by a *set* of stream processing functions. This is because one wants to allow the concerning agent to exhibit a nondeterministic or, equivalently, an underspecified behaviour.

One method of building agent specifications is to define the concerning predicate recursively, i.e. it is obtained as the weakest solution of a predicate equation, which is commonly specified in the following form:

rel $P: (A^\omega \rightarrow B^\omega)$ such that $\forall f \left[P[f] \iff formula[P, f] \right]$,

where $formula[P, f]$ is mostly represented as conjunction of formulae of the form

$\exists g \left[P[g] \wedge \forall i, x \left[f(i\&x) = reaction(i)\&g(x) \right] \right]$

and of the form $f(\varepsilon) = initial\_reaction$.

The specification of *system* reads now as follows:

rel $spf\_stations: (In^\omega \rightarrow Out^\omega)$
such that $\forall f \left[spf\_stations[f] \iff \right.$
$f(\varepsilon) = \varepsilon \wedge$
$\left. \exists g, \partial C \left[spf\_stations[g] \wedge \forall a, x \left[f(\langle a\rangle\&x) = \partial C(a)\&g(x) \right] \right] \right]$,

where ∂C is the set of differential strategies to be applied by "system" belonging to $\partial \Sigma_{con\_switch}$.

The first proof obligation for functional specifications with stream processing functions does not concern any verification question, but we have to prove that the specification is able to be satisfied by at least one continuous stream processing function.

Proposition 6 *spf_stations* is <u>consistent</u>, i.e. $\exists f \left[spf\_stations[f] \right]$ holds.

Proof: Take that f such that f maps each $conn(t)$ action to $disc(t)$ within the processed stream. f is clearly continuous and satisfies *spf_stations*. □

In order to verify an agent specification against a trace specification, one has to show that the traces which can be generated by the agent meet the corresponding requirement specification. The set of the (normal form) traces generated by an agent is defined as follows:

$$traces(P) := \{t : Act^\omega \mid \exists f \, [\, P[f] \wedge t \in traces(f) \,]\};$$

$$traces(f) := \left\{ t : Act^\omega \,\middle|\, \forall t' \begin{bmatrix} Out©t = f(In©t) \wedge \\ (t' \sqsubseteq t \Longrightarrow Out©t' \sqsubseteq f(In©t')) \wedge \\ \forall i \, [\, actual[t', i, t] \Longrightarrow Out©t' = f(In©t') \,] \end{bmatrix} \right\}$$

The notion of a causality structure communicates itself from differentially strategic systems to agent specifications. Moreover, the correctness proof is achieved, when the agent specification forms a corresponding causality structure, because agents are connected by *asynchronous* channels, thus, output delay is possible.

But, because the differentially strategic system is directly involved in the construction of $spf\_stations$, it seems to be more advantageous to prove that all (normal form) traces generated by $spf\_stations$ are also in $traces(\partial \Sigma_{con\_switch})$. For, if we have proved this, $spf\_stations$ obviously forms a causality structure for the cs_sys predicate, which is the component trace specification of $system$.

Theorem 7 $spf\_stations$ forms a causality structure for cs_sys; therefore, $spf\_stations$ <u>satisfies</u> cs_sys.
   Proof: Fix a f such that $spf\_stations[f]$. What we have to show is:
$$traces(f) \subseteq traces(\partial \Sigma_{con\_switch}).$$
Let t be a trace of $traces(f)$. Assume $t \neq \varepsilon$, for ε is already in $traces(\partial \Sigma_{con\_switch})$. But then, we get a $t' \in Act^\omega$ and $i \in In$ such that $t = \langle i \rangle \& t'$, since $f(\varepsilon) = \varepsilon$.
   Case 1: $Out©t' = t'$: it immediately follows that
$$f(\langle i \rangle) = f(In©t) = Out©t = Out©t' = t'.$$
Therefore, there is a differential strategy $\partial C \in \partial C$ such that $t = \langle i \rangle \& \partial C(i) \in traces_+(\partial \Sigma_{con\_switch})$.
   Case 2: There are $o \in Out^*$ and $j \in In$ such that $t = \langle i \rangle \& o \& \langle j \rangle \& t''$: first put $s := \langle i \rangle \& o$. From the definition of $traces(f)$ we conclude
$$f(\langle i \rangle) = f(In©s) = Out©s = o.$$
Similarly to case 1, we get a ∂C such that $t = \langle i \rangle \& \partial C(i) \& \langle j \rangle \& t''$. But, we have also a g such that $spf\_stations[g]$ and $\partial C(i) \& (Out©t'') = f(\langle i \rangle \& [In©(\langle j \rangle \& t'')]) = \partial C(i) \& g[In©(\langle j \rangle \& t'')]$.
It is not too difficult to show $\langle j \rangle \& t'' \in traces(g)$. By some kind of induction hypothesis, we assume that $traces(g)$ is a subset of $traces(\partial \Sigma_{con\_switch})$. Then, one may conclude that t is, finally, also a trace of $\partial \Sigma_{con\_switch}$. □

It is an interesting exercise to prove that all traces of the differentially strategic system are traces of the agent specification. The usage of differentially strategic systems for the construction of agent specification is a methodly advantageous device for the transition from trace specifications as requirement specifications to functional specifications as design specifications, for the verification of a differentially strategic system may be achieved more easily and immediately yields the correctness of the corresponding functional specification.

The continuity requirement of stream processing function is an incisive design decision: the stations agent behaves in some kind of monotonous matter; each *conn* action of a station t get its reaction not interspersed by actions concerning other stations than t itself. Indeed, the way to obtain the agent specification has been all the same as for partial ordering processing functions, where, by the way, their continuity requirements do not influence the causality of input and output each.

On the other hand, as described in the next section, the form of $spf\_stations$ is already appropriate to give the corresponding definition within an abstract program. Therefore, we view the continuity requirement for our example also as a design decision leading to the level of implementation.

6. Implementations of a Connecting Switch Simulator

The implementation phase is divided into two parts: the level of an *abstract* program and the level of a *concrete* program. Each part is provided with an appropriate language interface towards a more algorithmic language.

In the case of abstract programs one wants to implement the concerning distributed system by an applicative program. The usage of a functional language as a direct continuation of the level of a functional specification seems to be obvious. The transition step from design specifications to abstract implementations may be carried out by a correctness proof that compares the denotational semantics of the abstract program with the set of stream processing functions implied by the design specification.

Concrete programs are written in a more procedural-styled language. Procedural or imperative languages are a closer interface to current machine architectures. Once the distributed system to be implemented is given by an abstract program, one may apply *transformational programming*-style techniques: the transition step from abstract to concrete implementations requires the development of a set of semantically sound transformation rules to obtain correct programs. The appearing proof obligation is then solved by showing that the concrete program is resulted by a correct application of the transformation rules to the abstract program.

So as to write implementations, we make use of two languages called AL+ (*applicative language*) and PL+ (*procedural language*) which are slight modifications of Frank Dederichs' developments, AL and PL, [Dederichs 91].

AL+ provides concepts of applicative languages concentrating on the processing of streams:

- nondeterminism (mainly restricted on reactions);
- composition of functions;
- function applications;
- conditional choice expressions by an **if – then – else – fi** construct;
- stream equations as commands, stream expressions as instructions
- specification of functional modules:

    - **programs** as main units specifying each that stream outcoming from the closed system,
    - **agents** as nondeterministic functions processing streams,
    - **functions** as auxiliary nondeterministic functions

More precisely, single functional modules processing streams, i.e. agents and programs, are represented by couples of stream equations of the form *stream = expression*, guarded by declarations of e.g. subagents and so on. Thus, AL+ supports the hierarchical structuring of networks of agents.

Agent definitions usually correspond to predicates of stream processing functions that are specified recursively by the method described above except the consideration of initial reactions, which is repeated here for convenience:

$\quad$ **rel** $P: (A^\omega \to B^\omega)$ **such that** $\forall f \left[P[f] \iff formula[P, f] \right]$,

where $formula[P, f]$ is mostly represented as conjunction of formulae of the rough form

$\quad \exists g \left[P[g] \land \forall i, x \left[f(i\&x) = reaction(i)\&g(x) \right] \right]$.

AL+ is a typed language; each defined name of any sort has its boldface-type equivalent. The composed types are specially treated: stream types are written as **chan Sort**, product types are replaced by their component types and a special tuple processor as $(component_1, component_2)$, sequence types appropriate to express reactions are written as **sequ Sort**.

We do not go into detail of the formal definitions of AL and its denotational semantics. Now the abstract program of a connecting switch simulator reads as presented in FIGURE 5, where the *stations* agent is (almost) the direct translation of *spf_stations*.

```
program con_switch_AL+ = → chan Out c :
  agent stations = chan In e → chan Out c :
    funct C = In a → sequ Out :
      select ⟨disc(t)⟩ where a ∈ conn(t) tceles
      or ⟨select ⟨send(t, m)⟩ where a ∈ conn(t) ∧ m ∈ Messages tceles&C(a)⟩
    endfunct
    c = C(ft e)&stations(rt e)
  endagent
  agent environment = chan Out c → chan In e :
    funct E = Out b → sequ In :
      if b ∈ disc
      then select ⟨conn(t, r)⟩ where t, r ∈ Stations tceles or empty
      else empty fi
    endfunct
    agent init = chan In i → chan In o :
      o = (select ⟨conn(t, r)⟩ where t, r ∈ Stations tceles or empty)&i
    endagent
    e = init(E(ft c)&environment(rt c))
  endagent
  c = stations(environment(c))
endprogram
```

Figure 5: An AL+ Program for a Connecting Switch Simulator

The initial reaction is treated by a function appending the outcome of the initial reaction to the whole output of *environment*. This is because of the view of streams being a sequential or, better, a FIFO ordered storage of messages: initial reactions are to be the first elements of every output stream.

In addition to AL+, PL+ comprises some of the usual concepts of procedural or imperative languages:

- variables;
- assignments;
- loops;
- procedures (mainly functional ones);
- streams turn to channels used in order to establish directed one to one communication;
- send (*channel?var*) and receive (*channel!expr*) commands.

PL+ distinguishes two kinds of agent declarations:

- so-called *sequential* or *imperative* agents, whose body is the usual pair consisting of variable declarations and of imperative statements;
- *parallel*, *hierarchical*, or *equational* agents, whose body, like for AL+ agent declarations, consists of a couple of agent calls of the same form as AL+ stream equations.

I.e. the parallel combinator is replaced by an applicative-style syntax:

$t = f(s), s = g(r)$ is synonymous to something like [**call** $f(s\#t)$||**call** $g(r\#s)$].

As for AL+, we do not treat the syntax or the denotational semantics here more formally. The concrete program for a connecting switch simulator may be found in FIGURE 6.

```
program con_switch_PL+ = → chan Out c :
  chan In e;
  agent stations = chan In e → chan Out c :
    var In a; var Bool finished;
    loop
      e?a; finished := ff;
      do
        [c!select disc(t) where a ∈ conn(t) tceles; finished := tt]
        or c!select send(t, m) where a ∈ conn(t) ∧ m ∈ Messages tceles
      until finished od
    pool
  endagent
  agent environment = chan Out c → chan In e :
    var Out b;
    e!select ⟨conn(t, r)⟩ where t, r ∈ Stations tceles or stop;
    loop
      c?b;
      if b ∈ disc
      then e!select conn(t, r) where t, r ∈ Stations tceles or skip fi
    pool
  endagent
  c = stations(e),
  e = environment(c)
endprogram
```

Figure 6: A PL+ Program for a Connecting Switch Simulator

The verification obligation of the concrete program against the abstract one, as already mentioned, is achieved by showing that the correct application of some set of transformation rules leads from the abstract one to the concrete one.

The transformation rules mainly considered translate AL+ agents into semantically equivalent or, at least, refining PL+ agents. Proving the soundness of the transformation rules, therefore, requires the satisfaction proof of two predicates over stream processing functions.

We do not treat the latter topic here either and refer you to [Dederichs 91] for more detailed information, but we give some of the transformation rules decisive to our example:

$$
\begin{aligned}
&\textbf{agent } f = \textbf{chan u } i \to \textbf{chan v } o: \\
&\quad o = F[\text{ft } i] \& f(\text{rt } i) \\
&\textbf{endagent}
\end{aligned}
$$

──────────────────────────────── recursion-to-iteration

$$
\begin{aligned}
&\textbf{agent } f = \textbf{chan u } i \to \textbf{chan v } o: \\
&\quad \textbf{var u } x; \\
&\quad \textbf{loop} \\
&\qquad i?x;\ o!F[x]_1;\ \ldots;\ o!F[x]_{\#F[x]} \\
&\quad \textbf{pool} \\
&\textbf{endagent}
\end{aligned}
$$

$$
\begin{aligned}
&\textbf{agent } f = \textbf{chan v } i \to \textbf{chan v } o: \\
&\quad o = E \& i \\
&\textbf{endagent}
\end{aligned}
$$

──────────────────────────────── initial-reaction

$$
\begin{aligned}
&\textbf{agent } f = \textbf{chan v } i \to \textbf{chan v } o: \\
&\quad \textbf{var v } x; \\
&\quad o!E_1;\ \ldots;\ o!E_{\#E}; \\
&\quad \textbf{loop } i?x;\ o!x\ \textbf{pool} \\
&\textbf{endagent}
\end{aligned}
$$

7. Concluding Remarks

The example of a very simple protocol has been subjected to a development process complete upto the implementation level inclusive. We have also dealt with the proof obligations, which mainly concern verification problems, i.e. correctness questions.

It has turned out that the example of the connecting switch is mainly appropriate to elucidate the aspects of the trace specification technique. But, even if one has liked to develop a more realistic implementation of a connecting switch rather than of a connecting switch simulator, the example as treated here has served for making the several description techniques and their positions in the development life-cycle more transparent.

Acknowledgements

I wish to thank Manfred Broy for his kind encouragements and for his important correction remarks on draft versions. Furthermore, I am grateful to Frank Dederichs for his stimulating contrariness; he has made the importance of the Devil's Advocate for scientific discussion visible again; it should be noted that, besides that, his contributions [Dederichs 90] and [Dederichs 91] have incisively inspired this work. But also the contributions of my colleagues to [Broy et al. 91] have been of great help. I still have to mention the useful comments of the two anonymous referees..

A part of this work was presented in a talk with the same title at the *Second COMPASS Subgroup Meeting on "Concurrency and Object-Orientation"*, held in Braunschweig, on June 27-28, 1991. *COMPASS* (= A *COMP*rehensive Algebraic approach to *S*ystem *S*pecification and development) is the ESPRIT Basic Research Action No. 3264.

Finally, this work was partially supported by the *Sonderforschungsbereich 342 "Werkzeuge und Methoden für die Nutzung paralleler Architekturen"* ("Tools and methods for the exploitation of parallel architectures").

References

[Broy 88] Broy, M., *Towards a design methodology for distributed systems*, in: Broy, M.(ed.), *Constructive Methods in Computing Science*, Springer NATO ASI Series F **55** (1989), 311-364

[Broy Dederichs Dendorfer Weber 91]
Broy, M., F. Dederichs, C. Dendorfer, R. Weber, *Characterizing the behaviour of reactive systems by trace sets*, Technische Universität München, Technical Report, SFB-Bericht Nr. 342/2/91 A (1991)

[Broy et al. 91] Broy, M., F. Dederichs, C. Dendorfer, M. Fuchs, T. F. Gritzner, R. Weber, *The design of distributed systems — an introduction, in preparation*

[Dederichs 90] Dederichs, F., *System and environment: the philosophers revisited*, Technische Universität München, Technical Report TUM-I9040 (1990)

[Dederichs 91] Dederichs, F., *A Transformational Calculus for Efficient Implementations of Distributed Systems*, in preparation as doctoral dissertation

[Dederichs Weber 90]
Dederichs, F., R. Weber, *Safety and liveness from a methodological point of view*, in: Information Processing Letters **36:1** (1990), 25-30

Testfallgenerierung aus Petri-Netzen –
Probleme, Konzepte, Systeme

Bernd Baumgarten
GMD
Rheinstr. 75
W-6100 Darmstadt

Zusammenfassung: Die automatische bzw. rechnergestützte Generierung von Testfällen aus formalen Protokollbeschreibungen ist eine aussichtsreiche Alternative zur fehleranfälligen manuellen Erstellung umfangreicher Testsuites. Die formalen Beschreibungen der PROSIT-Gruppe basieren auf Produktnetzen, einer speziell entwickelten Klasse höherer Petri-Netze. Die standardmäßigen Funktionen der Produktnetzmaschine, eines integrierten Systementwurfs- und -analysewerkzeuges, wurden nunmehr um Testgenerierungsprogramme erweitert, die die Produktion von Testsuites nach den Richtlinien der OSI-Testmethodik-Norm unterstützen. In diesem Papier werden die Konzepte, deren Verwirklichung und die verbleibenden Aufgaben dieses Ansatzes erläutert. Außerdem werden Restprobleme identifiziert, die auf Schwächen der Protokoll- oder Testmethodik-Normen zurückgehen, und Lösungsvorschläge gemacht.

1. Einführung

Im Bereich Konstruktionsmethodik des Instituts für Systemtechnik der GMD, der sog. PROSIT-Gruppe (PROtokollSpezifikation, -Implementierung und -Test), wird seit geraumer Zeit an Methoden und Werkzeugen zur Modellierung und Analyse verteilter Systeme gearbeitet. Die bisher verfolgten PROSIT-Aktivitäten [PROSIT 88] umfassen u.a.

- die formale Definition des Beschreibungsmittels Produktnetze, Untersuchungen über deren "reduzierte" Erreichbarkeitsanalyse, sowie die Implementierung eines dedizierten Arbeitsplatzes, der "Produktnetzmaschine",

- Untersuchungen über Zusammenhänge zwischen Anforderungen und sie erfüllenden Netzen sowie über Verhaltenszusammenhänge zwischen Netzen untereinander, bis hin zu Korrektheitsbeweisen

- die Modellierung und Analyse konkreter Dienste und Protokolle (Alternating Bit, Mutual Exclusion, Vermittlung, Transport, CCR, Recovery Line, ISDN, FTAM, ...).

In diesem Rahmen wurden auch Konzepte zur Generierung von Testfällen für Protokollimplementationen aus einem Netzmodell des betreffenden Protokolls entwickelt [BurEckGie 86, BauGiePla 89/90]. Mit ihrer Implementierung wurde eine Lücke im Instrumentarium des Protocol Engineering mittels Petri-Netzen geschlossen.

Inhalt der vorliegenden Arbeit ist eine Bestandsaufnahme der Testgenerierung mit Netzen, gegliedert nach den Konzepten (Abschnitt 4), den gegenwärtig implementierten Werkzeugen (Abschnitt 5) und den verbleibenden Aufgaben und Problemen (Abschnitt 6). Vorbereitend werden in Abschnitt 2 die OSI-Architektur und das Beschreibungsmittel Netze sowie in Abschnitt 3 die OSI-Testmethodik kurz skizziert.

2. Protokollbeschreibung mit Netzen

2.1. OSI-Protokolle

Kommunikationsprotokolle sollen die typischen kommunikations- und verteilungsbedingten Störfaktoren bei der Kooperation verteilter Systeme vermeiden, beheben oder zumindest deren Schaden minimieren. Sie erfüllen Aufgaben wie Verbindungsauf- und -abbau, Redundanzaddition und darauf basierende Fehlererkennung bei der Datenübertragung, Zerlegen und Zusammensetzen von Datenblöcken bzw. Datenströmen, Flußkontrolle, Synchronisation, usw. Die Gesamtheit aller Protokollfunktionen wird vernünftigerweise nicht in einem monolithischen Block erbracht, sondern gemäß einer modularen Architektur.

Die derzeit wohl wichtigste Protokollarchitektur, das *OSI Basic Reference Model* [ISO 7498,CCITT X.200,Stö 89], realisiert das Prinzip aufeinander aufsetzender virtueller Maschinen in einem verteilten Kontext (Abbildung 1). Auf beiden Seiten einer zweiseitigen Kommunikationsbeziehung sind Schichten von 1 bis n realisiert. Ein Kommunikationsdienst der Schicht n wird von den zwei Partnerinstanzen dieser Schicht erbracht. Diese tauschen n-PDUs (protocol data units) gemäß dem n-Protokoll unter Benutzung des (n-1)-Dienstes aus, indem sie diese PDUs in (n-1)-SPs (service primitives, in der Testmethodik: "abstract SP", ASP) verpacken. Für den (n-1)-Dienst gilt eine Schicht tiefer das gleiche. Nur das physikalische Medium wird als globales System betrachtet.

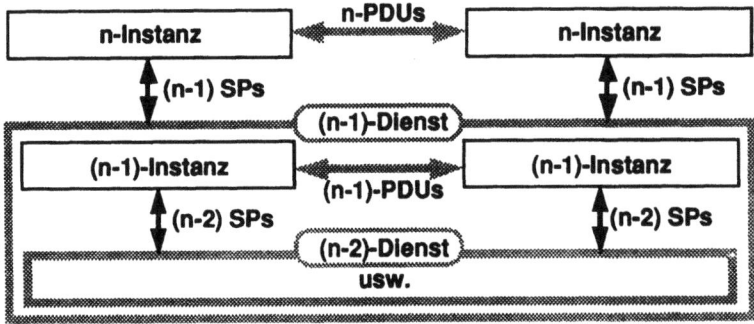

Abbildung 1: Prinzip des OSI-Referenzmodells (vereinfacht)

Implementiert sein müssen die beteiligten lokalen Systeme nicht unbedingt gemäß der OSI-Architektur. Sie müssen sich, um mit anderen Systemen zusammenarbeitsfähig zu sein, lediglich nach außen so verhalten, als ob sie diesem Aufbau folgten. Ausschlaggebend ist letztlich die Korrektheit des Bitstromes, den sie auf die Leitung geben (Black-Box-Prinzip).

Neben dem "Normalverhalten" enthalten die einzelnen OSI-Protokolle auch jeweils Vorschriften, wie mit einer Partnerinstanz umzugehen ist, die sich nicht protokollgemäß verhält. Es gibt also genormte *implizite Fehlermodelle*, die aber nicht unbedingt vollständig oder einheitlich sind. Sie bereiten i.a. bei der formalen Modellierung einiges Kopfzerbrechen, und die entsprechenden Tests beanspruchen einen beträchtlichen Teil des gesamten Testaufwandes.

2.2. Petri-Netze, Produktnetze

Petri-Netze sind ein formales Beschreibungsmittel zur Modellierung und Analyse diskreter, insbesondere verteilter Systeme [Bau 90, Rei 86, Sta 90]. Sie bieten die Möglichkeit, ausschließlich die problemimmanente Kausalstruktur zu beschreiben. Der Formalismus liefert daraufhin automatisch Nebenläufigkeit und Nichtdeterminismus mit, die daher nicht mehr explizit angegeben werden müssen. Die anschauliche graphische Darstellungsweise begünstigt ihren praktischen Einsatz bei Entwurf und Dokumentation.

Der Begriff Petri-Netze umfaßt leere Netze, mit anonymen Marken versehene sogenannte Stellen-Transitions-Netze sowie auch die höheren Netze, deren Marken mathematische Objekte sind, mit denen gerechnet werden kann. Dazu kommen zahlreiche Spezialformen und Varianten, insbesondere mit expliziten Zeitangaben und/oder Wahrscheinlichkeitsverteilungen.

Ein *Netz* ist ein gerichteter Graph mit zwei Typen von Knoten: den Stellen (meist als Kreise dargestellt) und den Transitionen (meist Rechtecke). Entlang der Kanten (Pfeile) wechseln sich Stellen und Transitionen strikt ab. Höhere Netze enthalten Kantenanschriften und Transitionsinschriften, die die Markendynamik durch qualitative und quantitative Bedingungen an die verbrauchten und erzeugten Marken steuern. Eine *Markierung* kann man sich vereinfacht als einen virtuellen Gesamtzustand des Systems vorstellen (das tatsächlich aber zu keinem Zeitpunkt einen beobachtbaren Gesamtzustand haben muß). Sie besteht aus einer Verteilung mathematischer Objekte auf die Stellen, so daß jeder Stelle eine Multimenge von Marken des stellenspezifischen Typs zugeordnet ist. Im Netz der Abbildung 2 haben drei Stellen den Typ N×N und eine (die linke) den Typ N.

Eine Transition kann unter einer gegebenen Markierung *schalten*, wenn entsprechend den erwähnten An- und Inschriften hinreichend viele geeignete Marken auf ihren Eingangsstellen vorhanden sind: In Abbildung 2 erfordert das Schalten von t drei Marken mit der innerhalb des Rechteckes angegeben Bedingung an deren Komponenten. Die Schaltung transformiert die Markierung wie abgebildet, indem sie die benötigten Marken wegnimmt und stattdessen (wiederum gemäß den An- und Inschriften) andere Marken auf den Ausgangsstellen der Transition ablegt. Die verbrauchten und erzeugten Marken entsprechen der in Klammern angegebenen Bindung der Variablen rund um die Transition.

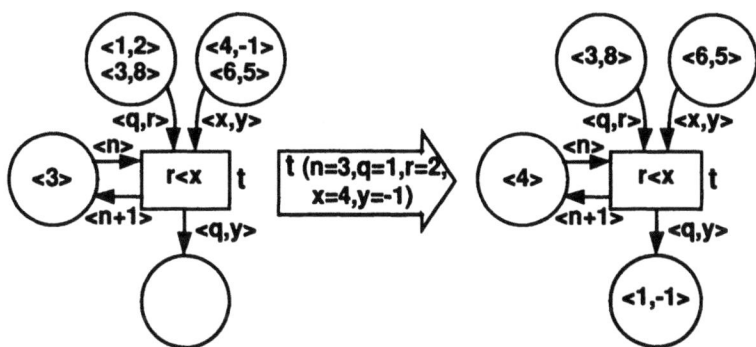

Abbildung 2: Eine Schaltung in einem höheren Netz

Eine *Erreichbarkeitsanalyse* (EA) bestimmt den *Erreichbarkeitsgraphen*. Dessen Knoten sind alle aus der Anfangsmarkierung durch schrittweises Schalten erreichbaren Markierungen, und seine Kanten entsprechen den jeweils durchgeführten Schaltungen. In Abbildung 2 wäre alternativ die Schaltung t(n=3,q=1,r=2,x=6,y=5) möglich.

Für Petri-Netze (vor allem für Stellen-Transitions-Netze) gibt es außer der meist recht umfangreichen EA noch weitere Analysemethoden, z.B. mittels strukturell-dynamischer Zusammenhänge, durch die linear-algebraische Berechnung von Invarianten oder durch netzverkleinernde Reduktionstechniken [Bau 90, Sta 90].

Eine spezielle Klasse höherer Netze sind die *Produktnetze* [BurOch Pri 89], die sich durch folgende Eigenschaften auszeichnen:

- Konstruktive Definition aller Datenmengen (alle entscheidbar) und Operationen (alle berechenbar) in einem Vorspann – daher Berechenbarkeit aller Schaltungen;
- Einbeziehung von Verbots- und Abräumkanten, die sowohl anschaulich als auch implementierbar sind und in gewöhnlichen höheren Netzen nur durch komplexe Hilfskonstruktionen ausgedrückt werden können. Verbotskanten laufen von einer Stelle zu einer Transition. Sie bestimmen keine verbrauchten Marken, sondern hindern die Transition am Schalten, wenn eine passende Markenmultimenge auf der ("Verbots-") Stelle vorhanden ist. Auch Abräumkanten laufen von einer Stelle zu einer Transition. Sie entnehmen nicht wie bei gewöhnlichen Kanten eine Markenmultimenge festgelegter Größe, sondern alle ggf. vorhandenen Marken des angegeben Musters.

Die *Produktnetzmaschine* ist ein integrierter Werkzeugsatz für Entwurf und Analyse von Systemen mittels Produktnetzen. Sie ist in Common Lisp implementiert und läuft auf Symbolics-Systemen wie auch auf Apple- und SUN-Rechnern mit Ivory-Karten [Och 91].

2.3. Formale Protokollbeschreibung

In der PROSIT-Gruppe wurde eine Methodik der Modellierung von verteilten Systemen und insbesondere OSI-Protokollen mittels Produktnetzen entwickelt, die z.B. in [BauGiePla 89] zusammengefaßt ist. Kennzeichnend für die *PROSIT-Methodik* ist neben der oben erwähnten vollständigen effektiven Analysierbarkeit der Modelle die systematische Vermeidung von Modellen verteilter Systeme, die gar nicht (vgl. die reiche Literatur über Unmöglichkeitsbeweise) oder nur durch Hinzunahme weiterer, im Modell nicht dargestellter, Protokollmechanismen implementierbar sind.

Nahezu das ganze Spektrum des Protocol Engineering ist den Netzmethoden zugänglich [Bau 90, Bil 87].

3. OSI-Testmethodik und TTCN

Voraussetzung für die allgemeine Akzeptanz von Testergebnissen sind eine allgemein verstandene und akzeptierte Testmethodik sowie Testsuites hoher diagnostischer Qualität. Lücken, Mehrdeutigkeiten und Widersprüche in den Protokollnormen machen sich beim Entwurf abstrakter Testsuites verschärft bemerkbar. Deren Schreiber werden daher zu wichtigen Lieferanten von Mängelberichten zu den Protokollnormen.

Im Normungsprojekt JTC 1.21.23 wurde in den letzten Jahren die *OSI-Testmethodik* entwickelt und in der mehrteiligen Norm [ISO 9646, vgl. BauGie 92] niedergelegt. Dieses derzeit noch in Erweiterung begriffene Regelwerk schließt die Testnotation TTCN (Tree and Tabular Combined Notation) als Teil 3 ein. Die anderen Teile sind den Testmethoden (sprich Testarchitekturen, -konfigurationen), der Erstellung von Testsuites (ATS, hierarchisch in Testgruppen gegliederte Mengen von Testfällen), sowie dem Prozedere der Testdurchführung gewidmet.

Formale Aspekte der Testmethodik werden in [FMCT] behandelt.

3.1. PICS und PIXIT

Protokollnormen spezifizieren in der Regel *parametrisierte Protokolle*, deren konkretes Verhalten erst durch eine Bindung der Parameter an feste Werte festgelegt wird. Die konkreten Parameterwerte einer Implementation werden in den Begleitdokumenten PICS (Protocol Implementation Conformance Statement) und PIXIT (Protocol Implementation Extra Information for Testing) festgehalten. PICS und PIXIT werden bei der Auswahl und Parameterwertebindung der Testfälle einer PICS-PIXIT-unabhängigen genormten Testsuite berücksichtigt. (In der Norm wird für die Parameterwertebindung der Begriff Parametrisierung verwendet, wovon wir hier absehen, damit nicht sowohl die Einführung von Parametervariablen in Spezifikationen als auch die Beseitigung von Parametervariablen durch Einsetzung konkreter Werte mit dem gleichen Wort bezeichnet werden.)

Die sogenannten Profile beschränken von vornherein die Wahlmöglichkeiten bei der Realisierung eines oder mehrerer aufeinander aufsetzender Protokolle.

3.2. Testmethoden

Die *Testmethode* mit den unmittelbarsten Beobachtungsmöglichkeiten einer zu testenden Implementation (IUT, impl. under test), nämlich dem unmittelbaren Zugriff auf die untere und obere Dienstschnittstelle der IUT (gelegentlich als konzeptuelle oder wirklich lokale TM bezeichnet), wurde (leider) nicht in die Norm aufgenommen. Sie ist bei erworbenen kommerziellen Implementationen i.a. nicht anwendbar, wohl aber bei Inhouse-Tests der Entwickler. Die genormten Testmethoden (lokale, verteilte, koordinierte und Fern-TM) beschreiben Testkonfigurationen einschließlich der zugehörigen *Steuer- und Beobachtungspunkte* (points of control and observation, PCO). Sie umfassen alle einen "unteren Tester", der eine entfernte Partnerinstanz der IUT einschließt, und teilweise einen oberen Tester, der mehr oder weniger direkt auf die obere IUT-Dienstschnittstelle zugreift.

Eine Testsuite bezieht sich gewöhnlich auf eine vorgegebene Testmethode, was sich entsprechend bei der Testgenerierung niederschlägt. Die in [ISO 9646],Teil 3, beschriebene Struktur der PCOs ist bereits bei der formalen Beschreibung zu erfassen.

3.3. Testziele

Nach der OSI-Testmethodik ist eine TSS&TP-Norm (test suite structure and test purposes) Voraussetzung für die Normung einer Testsuite für ein Protokoll. Ein *Testziel* (test purpose, TP) schält einen zu überprüfenden typischen oder kritischen Punkt des Protokollverhaltens heraus. Die TPs werden gemäß einer protokollspezifischen Struktur hierarchisch geordnet. Allerdings hat man bei dieser Gliederung Spielraum für Designentscheidungen, so z.B. die Testziele nach Funktionsgruppen, PDU-Typen, Protokollzuständen oder Verbindungsphasen usw. zu gruppieren. Diese Strukturierung taucht in den Testsuites in Form der "Testgruppen" wieder auf.

Automatische Verfahren zur Testzielgenerierung scheinen derzeit noch nicht das Expertenwissen über die "typischen Fehlerquellen" bei Protokollimplementationen ersetzen zu können.

3.4. Die Testnotation TTCN

TTCN (Tree and Tabular Combined Notation) ist gegenwärtig die einzige genormte Notation für Testfolgen. Wir skizzieren hier die Grundlagen von TTCN, und zwar in der graphisch orientierten Form TTCN.GR. Eine maschinenorientierte Form, TTCN-MP, in der eine Testsuite lediglich als Zeichenkette dargestellt wird, dient als Medium für die maschinelle Übertragung und Verarbeitung von Testfolgen.

In den *Deklarations- und Constraintsteilen* einer Testsuite werden testrelevante Daten und Datenmuster definiert. Im *Dynamischen Teil* werden Testfälle beschrieben, deren jeder testzielorientiert eine Menge von Testereignisfolgen mit den zugehörigen Testurteilen festlegt.

Testereignisse sind die aus Protokoll- und Dienstsicht atomaren Aktivitäten wie der Austausch einer einzelnen PDU oder eines ASP am PCO. Durch die Zusammenlegung gemeinsamer Anfangsstücke wird aus der Menge aller bei einem bestimmten Testfall möglichen Testereignisfolgen ein *Verhaltensbaum*, in dem die Vater-Sohn-Relation die zeitliche Nachfolge bedeutet und Geschwisterknoten Alternativen sind. Der Ereignisfolgenbaum kann mit einem Alternativensatz anfangen. Die Baumstruktur wird in TTCN.GR durch Einrückung dargestellt: Jedes Ereignis schreibt man in Form einer *Verhaltenszeile*, Alternativen bündig untereinander, nachfolgende Ereignisse eingerückt, z.B.:

```
! CONNECTrequest
    ? CONNECTconfirm
        usw.
    ? DISCONNECTindication
```

Send-Ereignisse aus Testersicht sind mit "!" gekennzeichnet, Receive-Ereignisse mit "?".

Viele Verhaltensbäume sind eigentlich unendlich verzweigt und/oder unendlich tief oder wenn endlich, dann zumindest unhandlich groß. TTCN umfaßt eine programmiersprachenähnliche Steuerung des Kontrollflusses durch sogenannte Konstrukte, die es erlauben, diese Bäume in endlicher Form, modular und möglichst klein darzustellen.

Verhaltenszeilen können außer Testereignissen auch die sogenannten Pseudo-Ereignisse für interne Rechnungen und Zeitmessungen des Testers sowie Konstrukte wiedergeben. Sie enthalten bei Bedarf auch Zeilennummern, Etiketten (als Sprungziel oder zu Testaufzeichnungszwecken), Verweise auf Daten(muster)definitionen im Constraintsteil, Testurteile und Kommentare.

Das Endziel der Abarbeitung eines Testfalles ist die Zuweisung eines *Urteils*, nämlich PASS, FAIL oder INCONCLUSIVE. Vom kompletten Verhaltensbaum her steht bei jedem Blatt, d.h. am Ende jeder Testereignisfolge, fest, ob die IUT in diesem Testfall kein Fehlverhalten zeigte und das Testziel erreichte (PASS), oder einen Protokollfehler beging (FAIL) oder ob die IUT bei protokollgemäßem Verhalten das Testziel nicht erreichte (vgl. Abschnitt 6.2.1) bzw. ob – z.B. auf Grund von Übertragungsproblemen tieferer Ebene – keine Unterscheidung möglich ist (INCONCLUSIVE). Es scheint nicht genau geklärt zu sein, mit wie vielen INCONCLUSIVEs im Test sich eine Implementation als normkonform bezeichnen darf. Mehr dazu in 6.2.1.

3.5. Die Testdurchführung

Drei wichtige Schritte vor dem eigentlichen Testen einer Protokollimplementation, d.h. der konkreten Ausführung des Dynamischen Teils der Testsuite, sind

- die statische Konformitätsprüfung von PICS und PIXIT auf die konsistente und normkonforme Angabe von implementierten Funktionen und Parameterwerten; ferner, anhand von PICS und PIXIT,
- die Selektion von durchzuführenden Testfällen, wobei auch testmethodenspezifisch einige Testfälle entfallen können, und
- die sog. Parametrisierung, die Wertebindung der Testsuiteparameter.

Endergebnis ist die PATS, die (selected and) parameterized abstract test suite.

Als vierter Schritt kann eine Vor-Übersetzung der abstrakten, in TTCN geschriebenen Testsuite in eine testwerkzeugorientierte sog. ausführbare Testsuite hinzukommen. Von der (zugelassenen) Möglichkeit einer anderen Reihenfolge dieser Schritte wird im Rahmen unserer Testgenerierung kein Gebrauch gemacht.

Bei der Abarbeitung des Verhaltensbaumes "sitzt" das Testsystem jeweils gleichsam vor einer Reihe von Alternativen, unter denen es sich die nächste auszuführende aussuchen muß. Dazu hält es den Momentanzustand der eingetroffenen Empfangsereignisse und Timerzustände fest (Schnappschußsemantik) und durchsucht die Alternativen gemäß der angegeben Reihenfolge nach einer passenden Verhaltenszeile. Scheitern alle Alternativen, wiederholt sich die Suche mit einem neuen Schnappschuß. Nach dem Gelingen einer Alternative bilden deren Kinderknoten die nächste Alternativenreihe.

4. Das PROSIT-Konzept der Testfallgenerierung

Der in PROSIT verfolgte Ansatz der Testgenerierung ist auf [ISO 9646] ausgerichtet. Ausgangspunkt für die Testfälle sind die Testziele aus der TSS&TP-Norm. Diese werden in Anfangsmarkierungen und ggf. Modifikationen eines auf die genormte Testmethode abgestimmten globalen Protokollmodells umgesetzt. Insbesondere wird weder "von den Daten abstrahiert", auf die sich ein Großteil der Testziele bezieht, noch werden eigene (normfremden) Testziele produziert.

Ergebnis der Netzanalyse- und Testgenerierungsschritte sind (mit den unten angegebenen Einschränkungen) Testfälle in TTCN, also der Dynamische Teil einer normbaren Testsuite, sowie eine Vorform ihres Constraintsteiles. Aus dem deklarativen Netzvorspann kann schließlich der Deklarationsteil des Testsuite leicht gewonnen werden.

Der PROVE-Projektbericht [CATG] gibt einen Überblick über eine Vielzahl nicht netzorientierter Methoden der rechnergestützten Testgenerierung.

4.1. Voraussetzungen

Zur Generierung von Testsuites sind keine beliebigen Netzmodelle der Protokolle geeignet. In diesem Abschnitt werden die für unsere Methode notwendigen Modellierungsvoraussetzungen angegeben.

Bei der Netzspezifikation des Protokolls wird die in [BauGiePla 89] beschriebene Modellierungsphilosophie zugrundegelegt. Insbesondere liegen die Dienstschnittstellen an Modulgrenzen, und das erzeugte TTCN ist in ASP-Ereignissen formuliert. Die Dienstschnittstellen werden durch Mengen von Kommunikationsstellen und die mit ihnen verknüpften Modulrandtransitionen dargestellt.

Jede Transition, die ein Send/Receive-Ereignis (ein ASP) darstellt, hat nur eine *Kommunikationsstelle* (für eben dieses ASP) zum Nachbarn. Testrelevante ASP-Parameter sind als Markenkomponenten der Kommunikationsstellen dargestellt. Die Kante zwischen der Kommunikationsstelle und der Send/Receive-Transition ist mit einem einzelnen Markenmuster beschriftet, nicht mit einer formalen Summe von mehreren.

Die Netze unterscheiden sich für unterschiedliche Testmethoden voneinander, da im Tester durch geeignet markierte zusätzliche Eingangsstellen Auswahlen von Alternativen vorab getroffen werden, die die anderen Netzteile noch "live" entscheiden dürfen. Ferner wird der Tester durch einen Zusatzmodul befähigt, testmethodenspezifisch fehlerhaft zu agieren, vgl. Abschnitt 2.1.

Für die *Timermodellierung* wird ein Standardmuster angenommen: Timeouts sind Alternativen zu rechtzeitigen Receives. Ist sowohl der zugeordnete Timer T abgelaufen als auch das erwartete ASP eingetroffen, wird in dubio pro reo verfahren: das ASP könnte vor dem Timeout angefallen sein; es wird daher ausgeführt. Dies wird im Netz mit einer Verbotskante modelliert. Im TTCN-Baum muß diese Priorität des ASP durch die entsprechende Alternativenreihenfolge

    ?RECEIVE
    ?TIMEOUT T

gesichert werden.

Wir setzen ferner (vgl. Abschnitt 6) voraus, daß das Protokoll nicht parametrisiert ist bzw. alle Parameter bereits durch konkrete Werte ersetzt sind, daß es keinen übermäßigen Nichtdeterminismus erlaubt und daß es formalisierbare abstrakte Fehlerklassen angibt.

Der gesamte Netzeditiervorgang wird graphikorientiert an der Produktnetzmaschine durchgeführt, die das eingegebene Netz auf syntaktische und statisch semantische Korrektheit (derzeit aber noch nicht auf die Erfüllung unserer generierungsspezifischen Anforderungen) überprüft.

4.2. Anfangsmarkierung

Da nun alle Verhaltensalternativen der Testermodule durch Zusatzbedingungen steuerbar geworden sind, können die meisten Testziele (vorläufig manuell) in zielorientierte *Anfangsmarkierungen* eines (für die gewählte Testmethode) einheitlichen Netzes umgesetzt werden. Nur in komplexeren Fällen wird das Netz zielspezifisch geringfügig erweitert, wenn nämlich innerhalb eines Testfalls eine mehrfach auftretende Alternative von Mal zu Mal unterschiedlich zu entscheiden ist.

[BauGiePla 89/90] geben erste Konzepte zur Validierung der Anfangsmarkierungen bezüglich der Testziele an. Eine weitere Ausarbeitung und Implementierung steht aber noch aus.

4.3. Erreichbarkeitsanalyse der beobachtbaren Ereignisse

Mit dem testzielorientiert markierten Netz wird eine (ggf. reduzierte [Och 88/90]) Erreichbarkeitsanalyse durchgeführt, eine Grunddienstleistung der Produktnetzmaschine. Zusätzlich werden testspezifische Zusatzinformationen eingegeben, siehe Abschnitt 5.

Schließlich wird der *Erreichbarkeitsgraph* auf die Schaltungen der testerseitigen Randtransitionen (die die PCOs ausmachen) und der Timeout-Transitionen der Tester-Moduln *eingeschränkt*. Der entsprechende Algorithmus der Produktnetzmaschine liefert den Minimalautomaten, dessen Sprache das Bild der Schaltungssprache unter dem Homomorphismus des Ausblendens der Schaltungen der nicht berücksichtigten Transitionen ist.

Erreichbarkeitsanalysen sind in der Regel sehr komplex; sie verbrauchen viel Zeit und Speicherplatz. Eine geeignet reduzierte Erreichbarkeitsanalyse [Och 88] eliminiert allerdings einen Großteil der redundanten Interleavingmuster und rückt vieles wieder in den Bereich des Machbaren. Alternativen zu Erreichbarkeitsanalyseverfahren sind nur dann praktikabel, wenn sie weder Wesentliches (z.B. ablaufwirksame Daten) weglassen noch eine maschinelle Erreichbarkeitsanalyse dadurch vermeiden, daß sie (z.B. bei monolithischen Automatenspezifikationen) den Spezifizierer zwingen, diese Analyse vorab informell und im Kopf auszuführen.

4.4. Daten

Um die konsistente Eingabe der zum generierten Verhaltensteil passenden Constraints (Daten und Datenmuster) der Testsuite zu erleichtern, werden gezielt die entsprechenden Details im Dialog abgefragt. Der Generierungsvorgang liefert eine *Vorstufe des Constraintsteils*. Die vollkommen TTCN-gemäße Constraintsgenerierung wurde aus Gründen mangelnder Manpower vorläufig zugunsten der uns wichtigeren Verhaltensbeschreibung zurückgestellt.

Aus den gleichen Gründen wurde eine Generierung der Testurteile zurückgestellt, obwohl [BauGiePla 89/90] bereits entsprechende Konzepte enthalten.

4.5. Generierte TTCN-Features

Nicht alle Ausdrucksmittel von TTCN werden in den generierten Testfällen benutzt. Das bedeutet aufgrund der "Redundanz" von TTCN in vielen Fällen keinen Verlust, so bei der Beschränkung auf das Einhängen von Teilbäumen zur Variation des Kontrollflusses, ohne Benutzung von GoTo oder Repeat. Etwas mehr dürfte sich bemerkbar machen, daß die nur einmalige Beschreibung gemeinsamer Teilbäume als sog. Testschritte noch nicht implementiert ist. Gravierend ist die momentan noch fehlende Benutzung von Variablen, so daß bei realistischen (d.h. massiv parametrisierten) Protokollen derzeit zu jedem PICS/PIXIT (bzw. deren Entsprechung beim Profiltesten) eine eigene PATS erzeugt wird, vgl. 6.1.

4.6 Ein Beispiel

Der zur Verfügung stehende Platz verbietet die Wiedergabe eines Beispiels zur Veranschaulichung der Methode anhand eines methodengemäß modellierten realistischen Protokolls. Wir beschränken uns daher auf ein triviales Miniprotokoll, das zudem weder norm- noch methodengemäß modelliert ist, und hoffen, das Vorgehen zumindest in den Grundzügen veranschaulichen zu können.

Ein asymmetrisches Protokoll verfolge den einzigen Zweck, daß der Sender (hier auch der Prüfling) eine einzige Nachricht A dem Empfänger (der hier der Tester sei) übermittelt, wozu den beiden als Diensterbringer ein gerichteter Kanal der Kapazität 1 zur Verfügung stehe, der eine abgeschickte Nachricht (sagen wir A oder B) entweder korrekt oder aber unleserlich (als X) abliefert, sie aber nicht wortlos verschluckt. Zeit spiele keine Rolle. Der Sender schickt entweder korrekt A ab (A-req) oder er schickt fälschlicherweise B ab (B-req, als triviales Fehlermodell für den Prüfling). Der Kanal liefert entweder A bzw. B unverfälscht ab (A-ind/ B-ind), oder stattdessen X (X-ind). Zur Abrundung haben Sender und Empfänger je eine Start- und eine Stopstelle. Die "Testmethode" entspricht dem Ferntesten. Die beschrifteten Tafeln im Hintergrund kommentieren informell die Testarchitektur und stellen keine Netzelemente dar.

Jede Transition kann schalten, wenn auf ihren Eingangsstellen den Kantenanschriften entsprechende Marken liegen. Kantenanschriften für "die anonyme Marke •" (formal eine beliebige konstante Marke, z.B. das 0-Tupel <>) sind weggelassen. Zu Beginn kann nur A_req oder B_req schalten, woraufhin <A> bzw. <B> auf K-Msg liegt. Danach ist eine Schaltung von A_ind bzw. B_ind möglich, aber auch Verfälschung, gefolgt von X_ind.

Abbildung 4 zeigt oben den Erreichbarkeitsgraphen des Systems, wobei die Markierungen hier teils verkürzt, durch Aufzählung der markierten Stellen, wiedergegeben sind. Der Restriktionshomomorphismus auf die "am PCO sichtbaren" Transitionen (A-ind/ B-ind/ X-ind) ergibt den Automaten links unten, der sich schließlich TTCN-orientiert, wie rechts unten gezeigt, schreiben läßt. Kursiv sind die jeweils fälligen

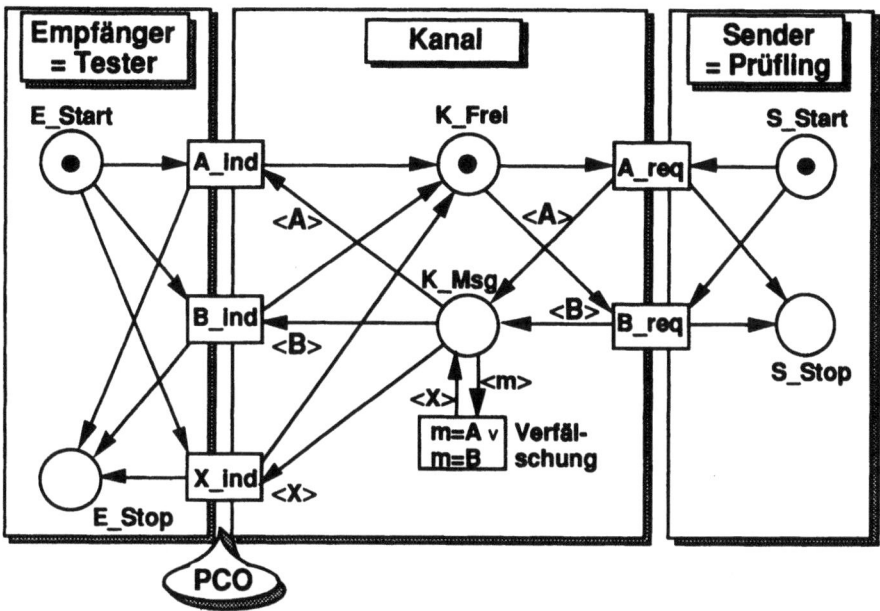

Abbildung 3: Ein globales Netzmodell für Tester, Dienstebringer und Prüfling

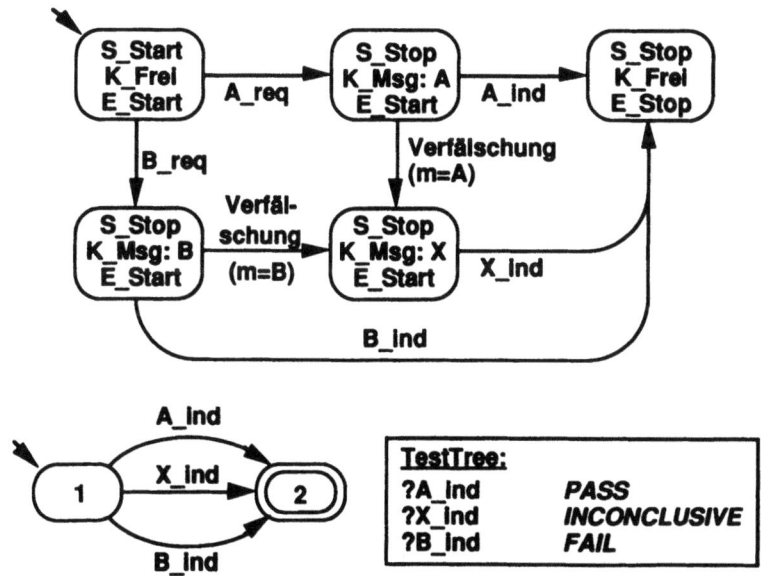

Abbildung 4: Die Testgenerierungsschritte zum vorstehenden System

Testurteile beigefügt. Wenn der Empfänger X empfängt, kann er nicht sagen ob es ein unleserliches A oder ein unleserliches B sein soll, d.h. ob sich der Sender protokollgemäß verhalten hat oder nicht.

5. Das Werkzeug zur Testfallgenerierung

Abbildung 5 gibt einen Überblick über die in Abschnitt 4 aufgeführten Generierungsschritte sowie über die dabei benötigten bzw. erzeugten Daten, die wir nun im einzelnen beschreiben.

Für die Einschränkung des Erreichbarkeitsgraphen auf die sichtbaren (PCO- bzw. Timer-) Transitionen müssen diese gemäß der gewählten Testmethode ausgezeichnet werden. Bei der Benennung der sichtbaren Transitionen der Netzmodule fehlen zudem gewisse für die Testfallgenerierung benötigte Informationen, z.B. der explizite Zusammenhang zwischen Kantenanschriften und den genormten ASP-Parametern bzw. PDU-Feldern. Diese Zusatzinformation wird vom System abgefragt und in einer dem Netz zugeordneten Tabelle *<Testsuitename>-TTCN-Info* gesammelt. Zusätzlich werden im Send- oder Receive-Falle constraintsorientierte Informationen abgefragt.

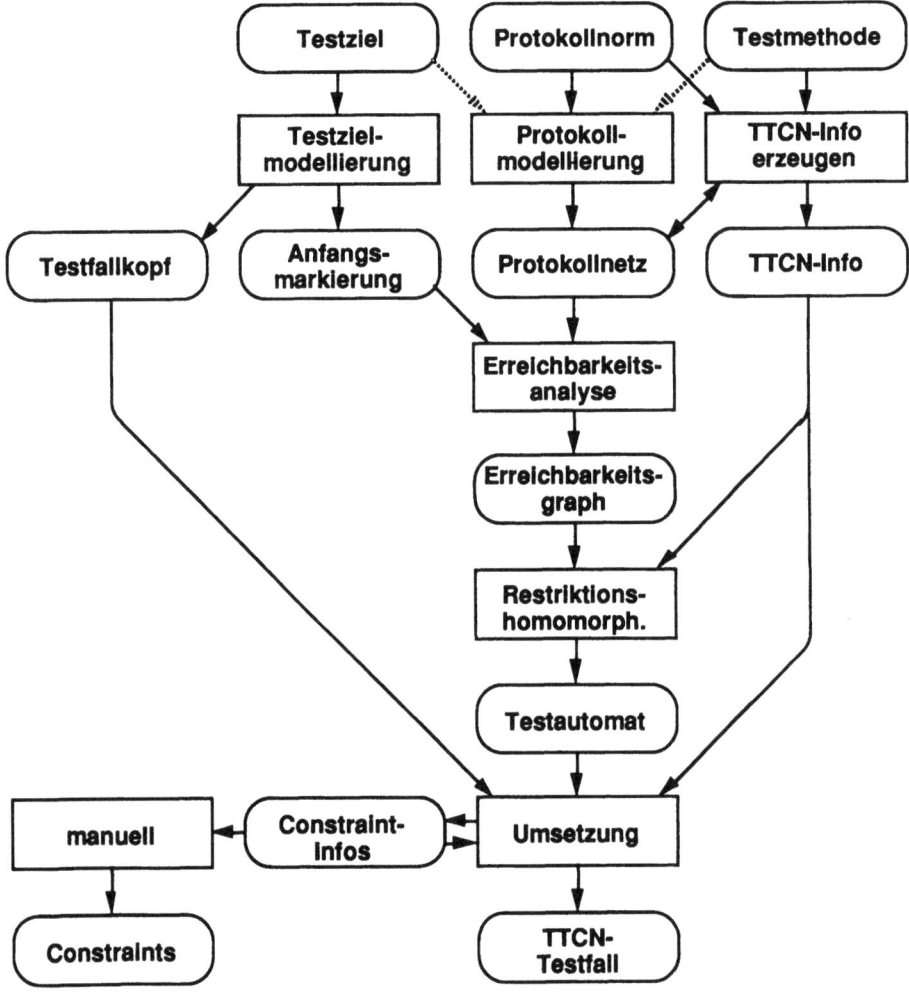

Abbildung 5: Generierungsschritte und Datenfluß

Der Generierungsvorgang erzeugt und füllt automatisch eine Tabelle *<Testsuitename>-Constraints-Info*. Diese besteht aus einer Liste von Constraint-Infos, die jeweils eine laufende Nummer, normgemäße ASP/PDU- bzw. Parameter/Felder-Namen sowie deren Wert bei der betreffenden Schaltung enthalten. Duplikate werden dabei vermieden.

Das die Behandlung eines Testfalles (nach der Erreichbarkeitsanalyse und deren Restriktion auf die testerseitig sichtbaren Transitionen) abschließende *Umsetzungsprogramm* transformiert den als Homomorphismusoutput erhaltenen Minimalautomaten in eine untereinander verknüpfte TTCN-Baumfamilie, d.h. in eine – derzeit noch bis auf die Urteile – komplette sog. Behaviour Description. Die Schaltungen werden dabei mittels der TTCN-Info in entsprechende TTCN-MP-Zeilen und Constraint-Infos transformiert, während die Markierungen anschließend nicht mehr explizit auftauchen.

Das Benutzerhandbuch [BauGiePau 91] beschreibt, wie die Testgenerierung in die Produktnetzmaschine eingebettet ist.

6. Probleme und Ausblicke

Wie bereits im Abschnitt 4.1 geschildert, ist unser Ansatz zur Testgenerierung momentan an eine Reihe mehr oder minder einschneidender Einschränkungen geknüpft. Die Beseitigung dieser Einschränkungen ist als Forschungs- und Entwicklungsaktivität geplant. Die qualitativen Probleme mit dem Nichtdeterminismus (6.2.1) scheinen allerdings bei [ISO-9646] bzw. den OSI-Protokollen zu liegen.

Im übrigen sind die geschilderten Probleme keineswegs "netztypisch", sondern müssen wohl generell bei der formalen Testgenerierung gelöst werden.

6.1. Parametrisierung

Viele Protokollbeschreibungen sind parametrisiert, und zwar einerseits durch die Auswahl von *Funktionsgruppen* und andererseits durch die Auswahl gewisser *Werte und Wertebereiche*.

Funktionsgruppen können vorhanden sein oder fehlen, sei es beliebig (optionale F.) oder unter gewissen gegenseitigen Einschränkungen (bedingte F.). Eine Protokollrealisierung benutzt daher quasi implizit eine Reihe Boolescher Parameter, die die Auswahl der nicht obligatorischen Funktionsgruppen beschreiben. Andere Parameter erlauben Auswahlen von Protokollwerten wie Fenstergrößen für unbestätigte Datenpakete, maximalen Stringlängen, Timerwerten, Adressen von Systemen und Instanzen.

Allen diesen Parametern werden durch PICS, PIXIT sowie ggf. durch Profile und deren zugeordnete analoge Dokumente konkrete Werte zugewiesen. Eine Testsuite für ein Protokoll wird in der Regel die gleichen Parameter wie das Protokoll selbst enthalten. Die Anfangsmarkierung und die Erreichbarkeitsanalyse des Protokollnetzes hingegen arbeiten derzeit nur mit konkreten Werten an jeder Position. Sie sind also nur für nicht parametrisierte bzw. für dedizierte parametrisierte Protokolle, d.h. mit bereits eingesetzten aktuellen Werten, möglich.

Nun wäre es aber unökonomisch, für jedes mögliche ausgefüllte PICS-PIXIT-Paar eine eigene Testsuite zu produzieren.Hier bietet sich der Ausweg an, die Parameterwerte symbolisch als Stringkonstanten durch die EA durchzuschleusen und erst im Testsuite-Deklarationsteil als Integerkonstante oder -variable zu deklarieren. Allerdings können diese Werte in komplexere Ausdrücke eingehen, so daß die EA zu diesem Zwecke dahingehend erweitert werden muß, daß sie Ausdrücke erlaubt, wo sie derzeit Werte erwartet.

Erschwerend kommt hinzu, daß symbolische Werte in Entscheidungen über Verhaltensalternativen benutzt werden könnten. Hier wäre abzuwägen, ob man eine (zwangsläufig umfangreichere) *parametrisierte EA* so definiert, daß sie diese Alternativen alle mit einschließt, oder ob man in jeder eingeschlagenen Alternative buchführt, für welchen Wertebereich von Parametern sie eigentlich möglich ist. Man kann nämlich im zweiten Fall Zweige von vornherein ausschließen, die mit keiner Parameterwertekombination möglich ist.

Wegen der Bedeutung parametrisierter Systembeschreibungen ist die parametrisierte Erreichbarkeitsanalyse von Netzen eine vorrangige Forschungs- und Implementierungsaufgabe.

6.2. Nichtdeterminismus

6.2.1. Qualitative Probleme

Die Rolle des *Nichtdeterminismus* ist in [ISO 9646] nicht ganz geklärt, was vielleicht auch verständlich ist, da es letztlich den Protokollnormen obliegt, das Protokoll eindeutig zu spezifizieren. Ein Hauptproblem liegt in der Frage, ob eine IUT, die laut Protokollnorm in einer gegebenen Situation auf verschiedene Arten reagieren darf, im Test "auf Befehl" jede einzelne dieser Arten korrekt demonstrieren können muß oder nicht.

Die Tendenz geht momentan dahin, zur Bescheinigung der Normkonformität eine beliebige, der IUT überlassene, Wahl zu gestatten, so daß evtl. ganze Verhaltensbereiche offiziell ungeprüft bleiben: Nehmen wir z.B. eine IUT, die nie gegen die Norm verstößt. Der Tester kann sie in eine Situation hineinsteuern, in der sie legalerweise A oder B antworten darf. Darf ein Testziel die Wahl von A (und nicht B) verlangen und ein anderes Testziel beim nächsten Mal die Wahl von B (und nicht A)? Kann der IUT-Vertreiber sich nicht vielmehr darauf berufen, es sei seine Privatsache, wann die IUT A sagt und wann B? Die Testmethodiknorm schreibt vor, zumindest nicht FAIL zu urteilen, wenn die IUT in der Wahl ihrer Alternativen "eigensinnig" ist, sondern höchstens INCONCLUSIVE. Aber wie liest der Testkunde oder die Zertifizierungsstelle einen Testbericht mit vielen INCONCLUSIVEs? Damit ist das Problem doch nur vertagt.

Es handelt sich indessen keineswegs um unlösbare Probleme; die Frage ist vielmehr, in welche Normen die notwendigen Festlegungen gehören, und wer sie wann vornimmt.

Zunächst ist bei diesen Alternativen eine gewisse Typisierung vonnöten:

Manche Alternativen sind für *Notfälle* gedacht, die nur selten vorkommen sollten, z.B. eine Ablehnung eines Verbindungsaufbauwunsches (oder ein Verbindungsabbruch) wegen Betriebsmittelengpasses. Bei ihrem Auftreten ist ein INCONCLUSIVE angebracht, was in einem gewissen Rahmen zur Testwiederholung führt. Als Testziele sind solche Alternativen meist nicht geeignet, da "externe" Tester die IUT i.d.R. schlecht in diese Notlage manövrieren können. Zudem sind Konformitätstests explizit nicht auf Hochlasttesten ausgerichtet. Die korrekte Handhabung dieses Ausnahmeverhaltens durch die IUT bleibt damit notwendigerweise (abgesehen von Inhousetests der Implementierer) ungeprüft.

Andere Alternativen gehen auf *unkooperatives Verhalten* besonders des Dienstbenutzers im IUT-System zurück, z.B. eine benutzerseitige Ablehnung eines Verbindungsaufbauwunsches. Dies kann derzeit nur im Ferntesten passieren, bei dem kein oberer Tester existiert. Dabei müßte laut derzeitiger Testmethodik im PIXIT angegeben sein, daß sich das IUT-seitige System ausdrücklich nicht zu positiveren Reaktionen (per sog. Implicit Send) verpflichten läßt. Damit würden nahezu alle Testfälle in der Testfallauswahl herausfallen, und im Testbericht würden zwar sehr wenige aber nur positive (PASS-) Urteile stehen! Dies wäre sicherlich aus praktischen Gründen unannehmbar.

Sinnvoller ist es,

- in den Protokollnormen die Notfälle und unkooperativen Fälle als solche zu kennzeichnen;
- die Notfälle ungetestet zu lassen;
- unkooperative Fälle mit FAIL zu belegen;
- bei den "wirklich gleichgültigen" Alternativen im PICS alle angeben zu lassen, die bei dieser IUT evtl. auftreten können, und alle angegebenen gezielt durchzutesten.

D.h. aber, daß die Norm in den beiden letzten Fällen verlangt, daß die nur scheinbar unbeeinflußbaren Alternativen zumindest während des Testens *im Testzielsinne zu steuern* sind (sie sind es aus praktischen Gründen wahrscheinlich ohnehin!). Dies wäre ein plausibler Ausweg aus der derzeit unbefriedigenden Situation.

6.2.2. Quantitative Probleme

Ein praktisches Problem für jede Art von Testgenerierung über eine vollständige Erreichbarkeitsanalyse ist der *hochgradige Nichtdeterminismus*. Im Falle nur weniger Alternativen der IUT ist es möglich, in der Erreichbarkeitsanalyse und im Verhaltensbaum alle Alternativen explizit einzeln aufzuzählen. Im Falle zahlreicher Alternativen - i.a. mehrfach hintereinander - wächst die Erreichbarkeitsanalyse unzumutbar schnell an. Das Verfahren wird impraktikabel.

Für hochgradigen Nichtdeterminismus gibt es sowohl protokollspezifische als auch modellierungstechnische Beispiele:

- In der Aufbauphase kann eine Protokollinstanz evtl. beliebige User Data mitsenden (z.B. 2^{256} verschiedene), die aber nicht ablaufrelevant sind und evtl. gar nicht mitmodelliert werden.
- Sie hat ferner evtl. einen Verhandlungsspielraum, z.B. eine vorgeschlagene Maximallänge auf irgendeinen Wert zwischen dem vorgeschlagenen und 128 zu kürzen. Praktisch könnte die IUT per PICS/PIXIT angehalten werden, sich vorher fest-

zulegen. Damit wäre das Problem des Nichtdeterminismus auf die Parametrisierung zurückgeführt.

- In der Datenphase darf der Sender evtl. beliebig segmentieren, der Empfänger unterschiedlich viele Pakete quittieren.
- Eine ausführliche Modellierung des Eintreffens eines an dieser Stelle inkorrekten ASPs würde eine Alternative für jede unerlaubte ASP-Parameterkombination erfordern. Hier ist zwangsläufig als Abstraktionsschritt ein Aufeinanderfalten gleichartiger Fehlerfälle vonnöten. Dabei besteht die Gefahr, daß viele unterschiedliche Ersatzdarstellungen entstehen, die mitsamt der anschließenden manuellen Nachbereitung den manuellen Aufwand und die Fehlerrate erhöhen. Zudem ergeben sich dabei neue Probleme, die im folgenden Abschnitt erläutert werden.

Durch die reichen Ausdrucksmittel von TTCN ist das Problem in der Testspezifikation selbst ohne weiteres zu lösen; ?OTHERWISE z.B. faßt alle verschiedenen empfangbaren ASPs zu einer Alternative zusammen und kann dann angewendet werden, wenn das Folgeverhalten des Testers in allen Fällen gleich ist. Datenmuster in Constraints können eine Unzahl im Detail unterschiedlicher Ereignisse zusammenfassen. Andere Alternativenmengen können mit den programmiersprachlichen Möglichkeiten von TTCN behandelt werden.

Alle diese Ausdrucksmittel stehen aber in der reinen Erreichbarkeitsanalyse nicht zur Verfügung, und es scheint nicht ganz einfach zu sein, sie unter Umgehung der Zustandsexplosion automatisch aus der formalen Beschreibung zu generieren.

6.3. Fehlerdatentypen

Wenn in einer Situation alle eintreffenden fehlerhaften Daten vom Tester gleich behandelt werden, kann leicht das OTHERWISE von TTCN bzw. ein ähnliches Konstrukt in der formalen Beschreibung benutzt werden. In den Fällen aber, in denen der Tester selbst gezielt fehlerhafte Daten produziert, und vor allem überall dort, wo das Protokoll differenzierte Reaktionen auf Fehlerdaten vorsieht, müssen fehlerhafte Daten explizit mitmodelliert werden.

Erfahrungsgemäß kann man die gültigen Datenobjekte und ihre Datentypen problemlos auf unterschiedlichen Abstraktionsebenen modellieren. Wie aber modellieren wir (allgemein oder für die gegebene Situation) inkorrekte Datenobjekte? Sie gehören i.a. einem anderen Typ als die korrekten an. In einer formalen Beschreibung müssen wir ja i.a. die entsprechenden Datentypen definieren, bei Produktnetzen z.B. im Vorspann des Protokollnetzes. Gesucht sind geeignete gemeinsame Obermengen, die sowohl die korrekten als auch die in der Norm beschriebenen fehlerhaften umfassen.

Zwei unterschiedliche Vorgehensweisen zeichnen sich ab: abstrakte Standarderweiterungen oder der Abstieg zur Bitebene.

6.3.1. Standard-Konstruktionsregeln für fehlerhafte Datenobjekte

Man könnte versuchen, (generell und protokollunabhängig) zu jedem Konstruktor (kart. Produkt, Folgenbildung, etc.) einen um abstrakte fehlerhafte Objekte erweiterten Konstruktor angeben, welcher die vielfältigen syntaktisch a priori ungültigen zusammengesetzten Werte zumindest symbolisch umfaßt. Dabei wäre zu berücksichtigen, daß es i.a. auch inhaltliche Bedingungen an Werte oder Wertekombinationen von Komponenten gibt, deren Nichterfüllung das zusammengesetzte Datenobjekt syntaktisch ungültig macht.

Wie weit aber soll nun die Phantasie im Erfinden fehlerhafter zusammengesetzter Objekte gehen? Besser wäre es, die zu behandelnden Fehlermöglichkeiten wären im Protokoll explizit beschrieben und müßten nicht protokollunabhängig erfunden werden.

Ein weiteres Problem liegt darin, daß bei der Methode der Konstruktionsregeln für fehlerhafte Datenobjekte ein und dasselbe Datenobjekt zu verschiedenen Fehlertypen passen kann. Dann wird es evtl. je nach Betrachtungsweise durch unterschiedliche Fehlerobjekte repräsentiert, beispielsweise einmal als inkorrekte PDU überhaupt und ein andermal als PDU mit fehlerhafter zweiter Komponente. Ist man in der Lage, formal unterschiedliche Datenobjekte im Modell als Repräsentanten ein und desselben konkreten Datenobjektes zu erkennen?

6.3.2. Ausmodellieren bis zur Bitebene

Häufig sind die in der Praxis auftretenden Fehlertypen eng mit den Codierungs- und Decodierungsmechanismen verknüpft und nicht so leicht auf abstrakterer Ebene erfaßbar.

Man kann sicherlich unter Berücksichtigung von Codierungs-Decodierungs-Regeln für jedes "atomare" Feld und für jede höher aggregierte Datenobjektebene konkret die Menge der *überhaupt möglichen* Werte genau angeben sowie ihre Partition in die Mengen der syntaktisch bzw. semantisch *gültigen* und *ungültigen* Werte. Nur ist es leider höchst unbefriedigend, wenn man in einer formalen Protokollspezifikation nicht in der Lage ist, von der Bitebene zu abstrahieren.

Der Weg zwischen Skylla und Charybdis hindurch muß erst noch gefunden werden.

Danksagung:

An dieser Stelle möchte ich Christa Paule, Heinz Jürgen Burkhardt und Alfred Giessler für zahlreiche Diskussionen und Kommentare sowie einem anonymen Gutachter für seine Verbesserungsvorschläge danken.

Literatur

[Bau 90]　　　　B. Baumgarten: *Petri-Netze, Grundlagen und Anwendungen*; BI Wissenschaftsverlag, 1990

[BauGie 92]　　B. Baumgarten, A. Giessler: *OSI-Testmethodik und TTCN*; voraussichtlich 1992

[BauGiePau 91]　B. Baumgarten, A. Giessler, C. Paule: *Testgenerierung mit Produktnetzen – Algorithmus, Implementierung, Benutzeranleitung*; Arbeitspapiere der GMD Nr. 605, 1991

[BauGiePla 89] B. Baumgarten, A. Giessler, R. Platten: *The Derivation of Test Cases from Net Models of OSI Protocols*; Arbeitspapiere der GMD Nr. 386, 1989

[BauGiePla 90] B. Baumgarten, A. Giessler, R. Platten: *Test Derivation from Net Models*; Protocol Test Systems; North-Holland, 1990

[Bil 87] J. Billington: *Protocol Engineering and Nets*; Proc. 8th Europ. Workshop on Application and Theory of Petri Nets, Univ. Zaragoza, 1987

[BurEckGie 86] H.J. Burkhardt, H. Eckert, A. Giessler: *Testing of Protocol Implementations – a systematic approach to the derivation of test sequences from global protocol specifications*; Protocol Specification, Testing and Verification V, North-Holland, 1986

[BurOchPri 89] H.J. Burkhardt, P. Ochsenschläger, R. Prinoth: *Product Nets – A Formal Description Technique for Cooperating Systems*; GMD-Studien Nr. 165, 1989

[CATG] RACE Project R1087 PROVE: *Computer Aided Test Generation; Methodology and Recommendation for Automation*; 1990

[CCITT X.200] CCITT recommendation X.200: *Reference model of open systems interconnection for CCITT applications*; 1988/89

[FMCT] I SO P.1.21.54, CCITT Q.10/X: *Formal Methods in Conformance Testing*; Den Haag, 1991

[ISO 7498] ISO 7498: *Information processing systems – Open Systems Interconnection – Basic Reference Model*; 1984

[Och 88] P. Ochsenschläger: *Projektionen und reduzierte Erreichbarkeitsgraphen*; Arbeitspapiere der GMD Nr. 349, 1988

[Och 90] P. Ochsenschläger: *Modulhomomorphismen*; Arbeitspap. der GMD Nr. 494, 1990

[Och 91] P. Ochsenschläger: *Die Produktnetzmaschine – Eine Übersicht*; Arbeitspapiere der GMD Nr. 505, 1991

[PROSIT 88] Fachgruppe 'Verfahren zur Konstruktion von Kommunikationssystemen': *PROSIT Schriftenverzeichnis 1982-1987* Arbeitspapiere der GMD Nr. 313, 1988

(demnächst erweitert als: *PROSIT Schriftenverzeichnis 1982 -1991*)

[Rei 86] W. Reisig: *Petrinetze – Eine Einführung*; Springer, 1986

[Sta 90] P.H. Starke: *Analyse von Petri-Netz-Modellen*; B.G. Teubner, 1990

[Stö 89] K.H. Stöttinger: *Das OSI-Referenzmodell*; Datacom, 1989

Die Offene Petrinetz-Methode zur Analyse und Darstellung des funktionalen Verhaltens verteilter Systeme

Heinz Dibold
Deutsche Bundespost Telekom
Forschungsinstitut beim FTZ
Postfach 10 00 03
D-6100 Darmstadt

Kurzfassung

Moderne Telekommunikationsnetze sind räumlich und steuerungsmäßig verteilte, rechnergesteuerte Systeme, die ihre Aufgabe in Echtzeit mit hohen Zuverlässigkeits-Anforderungen zu erfüllen haben. Die funktionale Spezifikation ihrer interaktiven und i.allg. nebenläufig agierenden Komponenten zum Zwecke der Standardisierung und Implementierung gestaltet sich als besonders schwierig, da sie einerseits möglichst formal, d.h. eindeutig und verifizierbar, andererseits aber auch anschaulich erfolgen muß.

Der vorliegende Beitrag stellt die Offene Petrinetz-Methode vor, die am Forschungsinstitut der Deutschen Bundespost Telekom zur implementierungsneutralen Spezifikation des funktionalen Verhaltens von Telekommunikationssystemen entwickelt wurde. Die *Analyse* der funktionalen Systemanforderungen wird durch Richtlinien unterstützt, die auf bewährten Methoden aufbauen, ohne jedoch die durch die Aufgabenstellung vorgegebene maximale Nebenläufigkeit von Systemaktivitäten zu vernachlässigen. Die *Beschreibung* des erarbeiteten funktionalen Systemverhaltens wird aus den Teilen Ereignisdefinition und Systemmodell gebildet.

Durch die Verwendung hierarchischer farbiger Petrinetze finden die Ergebnisse der Analysephase ihren unmittelbaren Niederschlag in den Elementen eines formalen und dennoch anschaulichen Systemmodells. Auch die maximal zulässige Nebenläufigkeit von Systemaktivitäten, deren Kenntnis eine wichtige Voraussetzung für eine wirtschaftliche Implementierung verteilter Systeme ist, kommt durch das Systemmodell zum Ausdruck. Die Ausführbarkeit des hierarchisch strukturierten Systemmodells erlaubt eine schritthaltende Prüfung des erarbeiteten Systemverhaltens und verbindet somit die ersten Phasen des Systemlebenszyklus im Sinne des "Rapid Prototyping".

Die der Offenen Petrinetz-Methode zugrunde liegenden Richtlinien, ihr Bezug zu bekannten Methoden, die Aufgabe der Ereignisdefinition und die Eigenschaften des Systemmodells werden vorgestellt. Anhand des allgemein bekannten Fernsprechsystems werden die Besonderheiten einer Systemspezifikation gemäß der Offenen Petrinetz-Methode erläutert, die Praxisrelevanz ihrer Eigenschaften wird aufgezeigt.

1. Einführung

Moderne Telekommunikationssysteme werden aufgrund ihrer Komplexität, der geforderten Flexibilität sowie aus Gründen der Wirtschaftlichkeit überwiegend mittels Rechnersteuerung realisiert. Sie stellen i.allg. räumlich ausgedehnte und steuerungsmäßig verteilte Systeme dar.

Die Spezialisierung in der Industriegesellschaft stellt den Netzbetreiber vor das Problem, ein von ihm gewünschtes Telekommunikationssystem in allen für die Erfüllung der Aufgabenstellung benötigten Eigenschaften eindeutig zu beschreiben. Grundsätzliche Überlegungen, Standardisierungsaspekte und die in Gang gekommene Öffnung der Fernmeldemärkte erfordern darüberhinaus, daß diese Systembeschreibung möglichst unabhängig von spezifischen Implementierungstechnologien gestaltet wird.

Für die Belange der Deutschen Bundespost Telekom wird am Forschungsinstitut beim FTZ der Frage nachgegangen, wie das funktionale Verhalten von Telekommunikationssystemen, d.h. von Vermittlungsstellen, Diensten, Dienstebausteinen und Protokollen, analysiert und möglichst technologieneutral beschrieben werden kann. Unter dem **funktionalen Systemverhalten** sei hierbei die Gesamtheit der Forderungen an die Interaktionsfähigkeit eines Systems mit seiner Umwelt verstanden, die zur Erfüllung der Aufgabenstellung des Systems unabdingbar sind. Ein erster Überblick über die Aufgabenstellung und die in deren Rahmen entwickelte "Offene Petrinetz-Methode" (**OPM**) wurde in [Dibo88] gegeben.

Ziel des vorliegenden Beitrags ist es, den im Rahmen der OPM erarbeiteten Ansatz zur *Analyse und Beschreibung* des funktionalen Verhaltens rechnergesteuerter Telekommunikationssysteme aufzuzeigen[1].

Die OPM ist speziell auf die ersten Phasen des Systemlebenszyklus [Balz85] abgestimmt, in denen die funktionalen Anforderungen zunächst analysiert und dann möglichst eindeutig beschrieben werden müssen. Im Gegensatz zu späteren Phasen existiert i.allg. noch keine formale Referenz, gegen die die erarbeitete Spezifikation verifiziert werden kann. Die Kommunikation mit dem Auftraggeber und/oder späteren Nutzer des Systems erfordert es, daß die Systemspezifikation möglichst anschaulich und aufgabennah gestaltet wird. Beiden Aspekten wird die OPM dadurch gerecht, daß ein ausführbares und für Techniker intuitiv verständliches Systemmodell verwendet wird. In Kapitel 2 wird zunächst in die Modellbildung eingeführt, die der OPM zugrunde liegt.

Um die Eigenarten der gewählten Systemspezifikation besser verstehen zu können, werden in Kapitel 3 die für eine methodische Unterstützung der Anforderungsdefinition vorgeschlagenen Prinzipien sowie ihr Bezug zu bekannten Methoden dargelegt. Die in der Phase der Anforderungsdefinition gewonnene Kenntnis des funktionalen Systemverhaltens wird mittels Ereignisdefinition und Systemmodell beschrieben, die in den Kapiteln 4.1 und 4.2 vorgestellt werden.

Das auf hierarchischen farbigen Petrinetzen basierende Systemmodell (Kapitel 4.2.3) kann die Ergebnisse der Analysephase *ohne* semantische Lücke eins-zu-eins wiedergeben und die durch die Aufgabenstellung für das System vorgegebene maximale Nebenläufigkeit zum Ausdruck bringen. Die Ausführbarkeit des Systemmodells erlaubt eine schritthaltende Prüfung der entstehenden Spezifikation, bei neuen Systemen gegen die Ideen vom

---

[1] Dieser Beitrag beruht auf der überarbeiteten und erweiterten Fassung eines Vortrags, den der Autor im November 1989 in Darmstadt auf der "Professorenkonferenz im FTZ der DBP TELEKOM" gehalten hat.

gewünschten Systemverhalten und bei bereits existierenden Systemen gegen das beobachtbare Verhalten des zu analysierenden Systems.

Zur Veranschaulichung der OPM und zur praxisnahen Demonstration der Vorteile einer Beschreibung verteilter Systeme auf der Basis hierarchisch strukturierbarer höherer Petrinetze wird in Kapitel 4.3 ein Ausschnitt aus dem OPM-Systemmodell des herkömmlichen Fernsprechnetzes vorgestellt.

2. Modellbildung

Jede Systembeschreibung stellt ein modellhaftes Abbild des in der Realität existierenden bzw. zu erstellenden Systems dar. Für die Wahl einer geeigneten Modellbildung ist die Frage nach den dominierenden Eigenschaften der zu beschreibenden Systeme zu stellen.

Die für Telekommunikationsnetze relevanten und folglich hier betrachteten rechnergesteuerten Systeme fallen unter die Kategorie **"planned response system"** [MP84]. Darunter wird eine Baueinheit oder eine Zusammenstellung von Baueinheiten verstanden, die - ausgelöst durch Ereignisse außerhalb des Systems - vorherbestimmte Handlungen ausführt.

Wesentlich für die betrachteten Systeme ist folglich ihr *Anreiz-Antwort-Verhalten*. Während sich für die Beschreibung nachrichtentechnischer Übertragungsfunktionen die klassische Systemtheorie anbietet, ist für die Beschreibung rechnergesteuerter Systeme i.allg. ein ereignisgesteuertes Modell von Vorteil.

2.1. Begriffe der ereignisgesteuerten Modellierung

Für die in der OPM unterstellte ereignisgesteuerte Modellierung wird, in Anlehnung an [MP84] und andere, die reale Welt mit ihren Objekten in die disjunkten Instanzen **System** und **Umwelt** unterteilt, die über eine **Schnittstelle** kommunizieren (**Bild 1**).

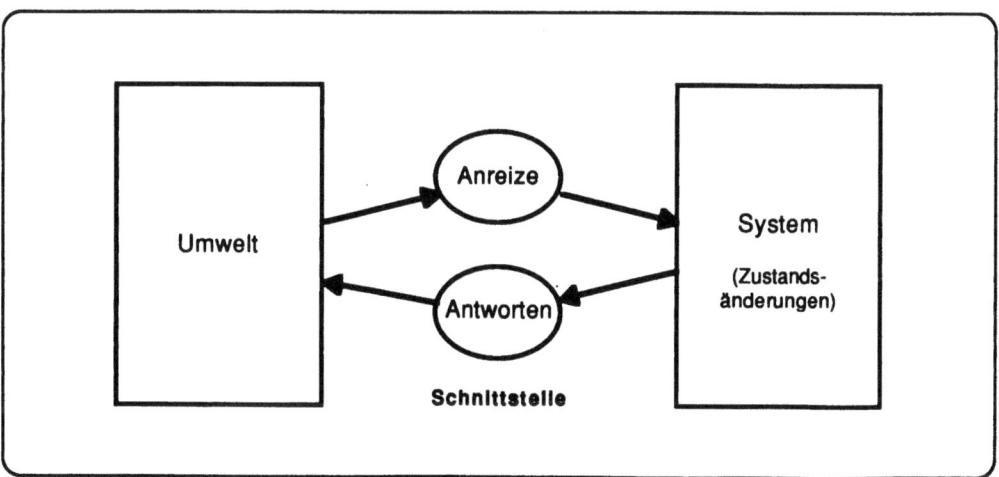

Bild 1: Ereignisgesteuerte Modellierung

Ein Ereignis, dessen Eintritt von der Umwelt veranlaßt wurde und auf das das System zu reagieren hat, wird mit **Systemanreiz** ("system stimuli") bezeichnet. Die Gesamtheit *aller* infolge eines Anreizes in vorherbestimmter Weise im System ablaufenden Einzelaktionen bildet eine **Systemaktvität** ("system activity"). Unter einer **Systemantwort** ("system response") wird ein für die Umwelt bestimmtes Ereignis verstanden, dessen Eintritt als Teil einer Systemaktivität vom System bewirkt wurde. Der Einfachheit halber werden im folgenden die verkürzten Begriffe "Anreiz", "Aktivität" und "Antwort" verwendet.

Ein Merkmal gedächtnisbehafteter Systeme ist es, daß die Erzeugung von Antworten nicht nur von dem auslösenden Anreiz, sondern auch von vorangegangenen Anreizen abhängen kann. Der Teil einer Aktivität, der die Erzeugung künftiger Antworten beeinflußt, sei als **Zustandsänderung** ("state change") bezeichnet.

Diese Betrachtungsweise beinhaltet bereits wesentliche Abstraktionen. Zum einen wird von der konkreten Erscheinungsform der Ereignisse Anreiz und Antwort in der realen Welt und den zur Kommunikation über ihren Eintritt erforderlichen Interaktionen zwischen System und Umwelt abstrahiert. Zum anderen wird die Betrachtung des Systems auf die Eigenschaft reduziert, die für die Umwelt vornehmlich von Bedeutung ist: Das **Anreiz-Antwort-Verhalten** des Systems, wie es sich nach dem Anstoß durch Anreize mit der Ausgabe von Antworten manifestiert.

2.2. Darstellung des Anreiz-Antwort-Verhaltens

Stellt man das Anreiz-Antwort-Verhalten des betrachteten Systems als strikt funktionalen Zusammenhang zwischen Anreiz und Antwort dar, so scheint die Forderung nach einer Systembeschreibung, die für die Systemimplementierung den größtmöglichen Freiheitsgrad zuläßt, in idealer Weise erfüllt zu sein ("Black Box"-Beschreibung in **Bild 2**). Da wir es hier nicht mit zeitkontinuierlichen Ein- oder Ausgangsgrößen zu tun haben, kann eine Zuordnung von Anreizen zu Antworten z.B. mittels Wertetabellen erfolgen.

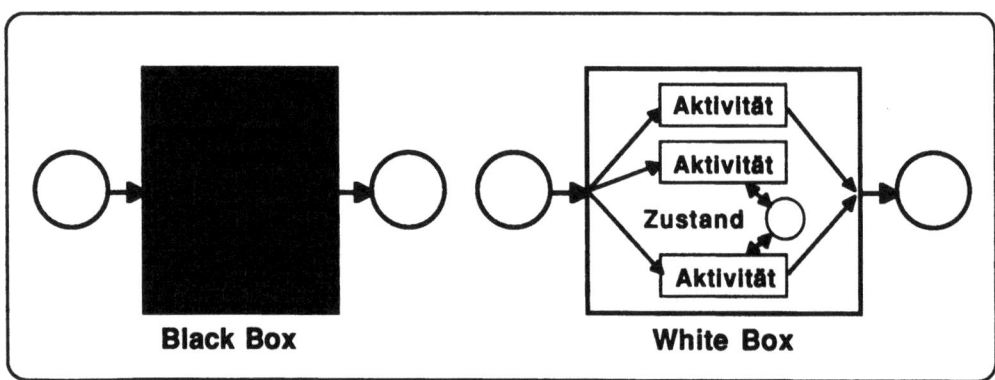

Bild 2: Zwei Möglichkeiten zur Beschreibung des Anreiz-Antwort-Verhaltens

In der Praxis stößt eine rein funktionale Darstellung nichttrivialer Systeme bei entsprechender Gedächtnistiefe und Anzahl unterschiedlicher Anreize jedoch schnell an die Grenzen der Darstellbarkeit und der Übersichtlichkeit (mehrdimensionale Tabellen, Abhängigkeit der Antworten von Anreiz-Folgen).

In Übereinstimmung mit [MP84], [DeMa78] und [HP87] wird in der OPM ein **operatives Modell** für die Beschreibung des Anreiz-Antwort-Verhaltens des Systems verwendet. Es zeichnet sich aus durch die explizite Sichtbarkeit der das Systemgedächtnis bildenden Zustandsspeicher und ihrer Wechselwirkung mit den Aktivitäten, die Antworten erzeugen und Zustände ändern können ("White Box"-Beschreibung in **Bild 2**). Der Forderung nach größtmöglicher Implementierungsfreiheit wird dadurch genüge getan, daß das zu implementierende System lediglich in seinem Schnittstellenverhalten (Anreiz-Antwort-Verhalten) und *nicht notwendigerweise* auch in seiner inneren Struktur mit dem beschreibenden Systemmodell übereinstimmen muß.

3. Analyse der funktionalen Systemanforderungen

Unter der Prämisse einer ereignisgesteuerten Modellierung ist zunächst das Problem zu lösen, diejenigen Forderungen an das Anreiz-Antwort-Verhalten des zu analysierenden Systems zu erkennen, die unabhängig von der Art der Implementierung zu stellen sind.

3.1. Die Methode "Essential Systems Analysis"

Ein bemerkenswerter Ansatz zur Lösung dieses Problems ist die Methode der "Essential Systems Analysis" (**ESA**) [MP84]. Ihr Anliegen ist die technologieneutrale Beschreibung der wesentlichen Eigenschaften von Anreiz-Antwort-Systemen ("interactive response systems"), unter besonderer Berücksichtigung der Richtlinien der "Structured Analysis" [DeMa78]. Das Systemmodell baut in [MP84] auf einem Datenflußgraphen auf, der mittels der fiktiven Eigenschaften einer perfekten Systemtechnologie zu interpretieren ist.

3.1.1. Vorteile der ESA

Kernstück der ESA ist die **Idee der perfekten Systemtechnologie**. Für alle Komponenten des zu betrachtenden Systems werden idealisierte Eigenschaften unterstellt: Sie sind fehlerfrei, ausfallsicher und kostenlos. Prozessoren können beliebig viele und beliebig komplexe Aktionen ohne erkennbaren Zeitbedarf erbringen. Speicher können beliebig viele und beliebig geartete Daten aufnehmen, auf die mühelos zugegriffen werden kann.

Die Anwendung dieser Idee bei der Erarbeitung eines neuen als auch bei der Analyse eines bestehenden Systems ermöglicht das Erkennen aller für das System **essentiellen** (d.h. wesentlichen) Eigenschaften. Die ausschließlich von bestimmten Technologien abhängigen und somit lediglich für bestimmte Implementierungsarten relevanten Systemeigenschaften sind "unwesentlich" und folglich nicht in die Systembeschreibung aufzunehmen.

Von Interesse ist die Frage wie die ESA hilft, bei bereits existierenden Systemen nachträglich die essentiellen Eigenschaften herauszuarbeiten. Hierbei bereitet die Zuordnung von Vorgängen, die "gleichzeitig" im System beobachtbar sind, zu dem sie jeweils auslösenden Anreiz, d.h. ihre Gruppierung zu Aktivitäten, erfahrungsgemäß Probleme. Ausgehend von einem Zustand, in dem das System inaktiv ("idle") ist, d.h. alle Aktivitäten vorausgegangener Anreize bereits ausgeführt sind, wird *ein einziger* Anreiz dem System angeboten. *Alle* in Folge dessen beobachtbaren Einzelaktionen - bis das System wieder inaktiv wird - bilden *zusammen* die zu diesem Anreiz gehörige Aktivität.

Die ESA-Prinzipien zum Erkennen der essentiellen Systemaktivitäten und der essentiellen Systemspeicher wurden in die OPM übernommen.

3.1.2. Nachteile der ESA

Als Nachteil der ESA sehen wir die Darstellung der erkannten essentiellen Systemeigenschaften durch ein Systemmodell an, in dem alle Aktivitäten sofort und ohne Zeitbedarf ablaufen:

Alle Anreize werden durch die reale Umwelt des Systems erzeugt; sie folgen somit mit einem zeitlichen Abstand größer Null aufeinander. (Abgesehen von den Erkenntnissen aus Sicht der Relativitätstheorie, die den Begriff der "Gleichzeitigkeit" als problematisch erscheinen lassen, läßt sich durch sukzessive Verfeinerung der Zeitskala stets eine Reihenfolge der Ereignisse fixieren). Alle durch einen Anreiz ausgelösten Aktivitäten laufen durch die Interpretation des Systemmodells mittels (hierbei falsch verstandener) perfekter Technologie sofort und ohne Zeitbedarf, d.h. in unmeßbar kurzer Zeit ab. Somit ist im ESA-Systemmodell sichergestellt, daß alle Aktionen einer Aktivität bereits abgeschlossen sind, bevor der nächste Anreiz Aktivitäten im System auslösen kann. Die Reihenfolge des zeitlichen Wirksamwerdens der Anreize wird folglich *unverändert* auf die zugehörigen Antworten und Zustandsänderungen übertragen.

Hinter dieser Eigenschaft des ESA-Systemmodells steckt die Absicht, daß Konflikte zwischen den einzelnen Aktivitäten, z.B. um den Zugriff auf ein und dasselbe Datum, erst gar nicht entstehen können, ergo eine Konfliktbehandlung auf dieser Abstraktionsebene (d.h. im Systemmodell) auch nicht modelliert werden muß. Damit handelt sich ESA jedoch den Nachteil ein, daß ein Systemverhalten, das auf der Nebenläufigkeit von Aktivitäten aufeinanderfolgender Anreize basiert, mit diesem Systemmodell *nicht* dargestellt werden kann.

Aktivitäten werden als **nebenläufig** bezeichnet, wenn sie in ihren Vorbedingungen (auslösender Anreiz und ggf. Systemzustand) und in ihren Nachbedingungen (Antwort und/oder Zustandsänderung) kausal derart unabhängig sind, daß das System sie in einer beliebigen Reihenfolge ausführen kann, ohne ihre Wirkung insgesamt zu verändern. Die Reihenfolge des Wirksamwerdens von Anreizen muß folglich *nicht* auf die Bearbeitungsfolge nebenläufiger Aktivitäten übertragen werden. Dies erlaubt eine unkoordinierte, insbesondere auch zeitlich überlappende Ausführung derselben.

Eine korrekte Implementation muß gewährleisten, daß das Schnittstellenverhalten des realen Systems mit dem Anreiz-Antwort-Verhalten des Systemmodells übereinstimmt. Im Falle des ESA-Systemmodells muß - in Abhängigkeit von den realen Umweltbedingungen und der gewählten Systemtechnologie - durch geeignete Synchronisierverfahren für *alle* Anreize sichergestellt sein, daß das reale System die zugehörige Aktivität abgeschlossen hat, bevor die Aktivität eines nachfolgenden Anreizes Auswirkungen zeigen darf.

Gerade in verteilten Systemen ist letztgenannte Forderung auf die Mehrzahl der Aktivitäten nicht übertragbar, da die Aufgabenstellung eine Nebenläufigkeit von Aktivitäten häufig zuläßt. Mehr noch - die Ausnutzung der der Aufgabenstellung innewohnenden Nebenläufigkeit ermöglicht zumeist erst die wirtschaftliche Realisierbarkeit des Systems. Diese Aussage trifft auch auf das Fernsprechsystem zu, worauf wir in Kapitel 4.3.1 zurückkommen werden.

Da der Implementierer sich an das vorgegebene Systemmodell halten muß, das ESA-Systemmodell die der Aufgabenstellung innewohnende Nebenläufigkeit jedoch nicht darstellen kann, verstößt ESA gegen seine eigene Zielsetzung und kann durch *Überspezifikation* zu technisch unnötig aufwendig implementierten Systemen führen.

Zur Vermeidung der geschilderten Nachteile wurden die ESA-Richtlinien für die OPM ergänzt: In einem zusätzlichen Analyseschritt ist die maximal mögliche Nebenläufigkeit der erkannten essentiellen Systemaktivitäten zu ermitteln.

Um diese Nebenläufigkeit auch beschreiben zu können, basiert das in der OPM verwendete Systemmodell auf höheren Petrinetzen [Baum90, Jens90, Pete81, Reis82].

3.2. Die Methode "Strategies for Real-Time System Specification"

Das durch die bisherige Modellbildung unter Berücksichtigung der ESA-Prinzipien gewonnene Systemmodell ist zu abstrakt, um das Funktionieren eines hiernach implementierten Systems in der real vorgegebenen Umwelt zu gewährleisten.

Diese Problematik wird in [HP87] ausführlich behandelt und ist in den dort definierten "Strategies for Real-Time System Specification" (**SRSS**) berücksichtigt.

Die Kernaussage der SRSS ist, daß die Analyse des betrachteten Systems zu erweitern ist: Die konkrete Ausbildung in der realen Welt der Anreize und Antworten bildenden abstrakten Ereignisse muß präzisiert werden (in SRSS als "in/out-processing" bezeichnet) und die von der Umwelt vorgegebenen bzw. für die Umwelt bedeutsamen zeitlichen Rahmenbedingungen (Wiederholrate der Anreize und zulässige Antwortzeiten) müssen fixiert werden. Beide Forderungen wurden in die OPM übernommen.

Eine weitere Aussage der SRSS ist der Vorschlag, bei gedächtnisbehafteten Systemen das Systemmodell in einen Daten- und einen Kontrollflußgraphen aufzuspalten. Während die Daten selbst wie bei ESA beschrieben ("minispecs", "data dictionary") und die Datenoperationen ebenfalls durch einen Datenflußgraphen festgelegt werden, wird das Verhalten des Kontrollteils durch endliche Zustandsautomaten ("finite-state machines") modelliert.

Bei der in der OPM verwendeten Modellierung mittels höherer Petrinetze ist weder die getrennte Darstellung von Daten- und Kontrollfluß noch die Verwendung unterschiedlicher Beschreibungsmittel für Datenoperationen und Steuerungsteil erforderlich. Daten- und Kontrollfluß können in einem einzigen Systemmodell konsistent und einheitlich beschrieben werden, eine Trennung der Darstellung derselben wird von uns auch nicht empfohlen.

4. Beschreibung des funktionalen Systemverhaltens

Nachdem die OPM-Prinzipien zur Unterstützung der Systemanalyse und ihr Bezug zu bekannten Methoden vorgestellt wurden, stellt sich die Frage, wie die erarbeiteten Systemanforderungen im Sinne einer Systemspezifikation zu dokumentieren sind.

Der Spezifikation des funktionalen Systemverhaltens fallen in der OPM folgende Aufgaben zu:
- Festlegen der für das System relevanten Anreize der Umwelt und ihrer Erzeugung
- Festlegen der für die Umwelt relevanten Antworten des Systems und ihrer Erzeugung
- Für alle Anreize: Festlegen der vom System infolge des Anreizes erwarteten Aktivität
- Aufzeigen der durch die Aufgabenstellung vorgegebenen maximalen Nebenläufigkeit von Aktivitäten
- Festlegen der durch die Umwelt für das System vorgegebenen zeitlichen Rahmenbedingungen

- Vermeiden von Anforderungen, die ihre Ursache nicht ausschließlich in der vom System zu erfüllenden Aufgabenstellung haben, da sonst von vornherein mögliche Implementierungsvarianten ausgeschlossen werden
- Beschreiben des Systems durch ein strukturierbares, intuitiv verständliches und selbstdokumentierendes Systemmodell, das auf symbolischer Ebene ausführbar sein soll.

Um diese Vorgaben zu erfüllen, wird die Spezifikation des funktionalen Systemverhaltens in der OPM aus den Teilen "Ereignisdefinition" und "Systemmodell" gebildet.

Die Eigenschaften der OPM-Spezifikation werden im folgenden sowohl allgemein als auch anhand eines konkreten Telekommunikationssystems erläutert.

Als Beispiel wird das Fernsprechsystem (Telefonnetz) gewählt, wie es sich in Form des gesamten Fernsprechnetzes den Fernsprechteilnehmern gegenüber bzgl. des herkömmlichen Fernsprechens verhält. Es umfaßt hierbei die Funktionalität des Zusammenwirkens *sämtlicher* technischen Einrichtungen, insbesondere die der Fernsprechapparate, Vermittlungsstellen und Übertragungswege (**Bild 3**).

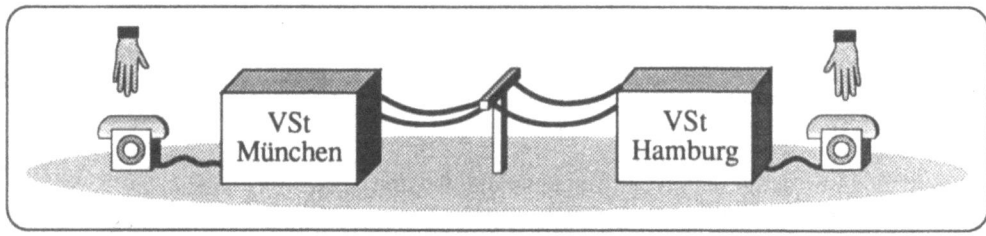

Bild 3: Das Fernsprechsystem als OPM-Anwendungsbeispiel

Die Funktionalität des Fernsprechsystems besteht nicht nur aus ereignisgesteuerten Abläufen, sondern umfaßt auch zeitkontinuierliche Funktionen. Hierzu gehört die Erzeugung von Tönen, z.B. Wählton, Besetztton, Rufton und Klingelzeichen, vor allem aber die zeitkontinuierliche Übertragung von Informationen zwischen Teilnehmern bei stehender Verbindung. Diese zeitkontinuierlichen Funktionen sind im Rahmen der ereignisgesteuerten Modellierung wie folgt zu behandeln:

Nicht die zeitkontinuierlichen Funktionen, sondern ihr *Beginn* und *Ende*, ggf. auch die Änderung wesentlicher Parameter, werden als Ereignisse modelliert. Die Beschreibung der zeitkontinuierlichen Funktionen selbst liegt außerhalb des Ansatzes der ereignisgesteuerten Modellierung und damit auch der OPM. Zu ihrer Beschreibung eignen sich die Methoden der klassischen Systemtheorie und zunehmend auch der digitalen Signalverarbeitung.

4.1. Ereignisdefinition

In der realen Welt gibt es vielfältige Wechselwirkungen zwischen einem implementierten bzw. zu implementierenden System und seiner Umwelt.

Die Ereignisdefinition legt fest, *welche* dieser Interaktionen bzgl. der Aufgabenstellung für das System relevant sind und damit an der abstrakten Schnittstelle des Systemmodells als Ereignisse (Anreize und Antworten) zu modellieren sind.

System und Umwelt sind zwei *getrennte* Instanzen. Ein Austausch von Informationen bzw. Objekten kann nur stattfinden, wenn beide an der erforderlichen Kommunikation teilnehmen und ein hierzu *vereinbartes Protokoll* einhalten. Die Ereignisdefinition legt für jeden Anreiz und jede Antwort das geforderte, für Umwelt und System gleichermaßen bindende konkrete Protokoll zur Erzeugung des abstrakten Ereignisses fest und gibt an, an welcher (Schnitt-) Stelle des abstrakten Systemmodells das jeweilige Ereignis wirksam werden kann.

Die Ereignisdefinition stellt somit den Bezug zwischen den von der Systemimplementierung und ihrer Umwelt geforderten *konkreten* Interaktionen einerseits und den *abstrakten* Aussagen des Systemmodells andererseits her.

Zur Erläuterung sei ein Auszug aus der Ereignisdefinition des Beispiels "Fernsprechsystem" wiedergegeben:

Anreiz "off_hook(A)" der Stelle "stimuli":

> Die Gabel des Fernsprechapparates an der Teilnehmeranschlußleitung "A"[2] ist durch Abnehmen des Hörers zu entlasten (Aufgabe der Umwelt), so daß eine im Fernsprechapparat zu erzeugende Gegenkraft die Gabel in ihre obere Stellung bewegen kann, was vom System zu erkennen ist (Aufgabe des Systems).

Das o.a. Beispiel deutet bereits an, daß einer rein formalen Systemspezifikation außerhalb des Systemmodells Grenzen gesetzt sind. Auch ist die Freiheit der Implementierung im Falle der Ausgestaltung der Schnittstelle(n) dadurch begrenzt, daß mit der Umwelt die konkrete Bedeutung und Bedienung derselben vereinbart sein muß.

Eine weitere Aufgabe der Ereignisdefinition ist es, alle *zeitbehafteten Rahmenbedingungen* aufzuführen, die sich aus den Eigenschaften der realen Umwelt für das System ergeben. Hierzu zählt die Angabe der minimalen Zeitabstände zwischen gleich- und ggf. auch verschiedenartigen Anreizen sowie der maximal zulässigen Systemantwortzeiten.

Für das Beispiel "Fernsprechsystem" könnte eine derartige Festlegung lauten:

> Auf die Erzeugung des Anreizes "off_hook(A)" der Stelle "stimuli" im Teilnehmerzustand "idle" (nichtgerufener Teilnehmer "A" hat abgehoben) hat das System längstens nach sieben Sekunden mit der Erzeugung der Antwort "dial_tone" der Stelle "responses" zu reagieren (Aufschalten des Wähltons).

Es ist zu erkennen, daß die exakte Formulierung dieser Art von Zeitbedingung auf informaler Basis problematisch ist. Die hier vorgeschlagene Verwendung eines operativen Systemmodells (Kap. 2) bietet den Vorteil, daß eine derartige Zeitbedingung der entsprechenden Aktivität im Systemmodell unmittelbar zugeordnet werden kann (siehe hierzu die Anmerkung in Kap. 4.3.3).

4.2. Systemmodell

Das Systemmodell beschreibt das geforderte Anreiz-Antwort-Verhalten des Systems auf der abstrakten Ebene der definierten Anreize und Antworten.

---

[2] "A" bezeichnet eine systeminterne Numerierung der Teilnehmeranschlußleitung, die nicht notwendigerweise mit der Rufnummer des Teilnehmers übereinstimmen muß.

Es wird durch ein hierarchisch strukturiertes höheres Petrinetz gebildet, dessen Eigenschaften durch den Netzgraphen, die Anfangsmarkierung und die Feuerungsregel eindeutig bestimmt sind. Es sind ausschließlich die im Systemmodell an der Schnittstelle zwischen System und Umwelt beobachtbaren Eigenschaften, die das geforderte Verhalten des zu beschreibenden Systems festlegen (operatives Systemmodell mit größtmöglicher Freiheit für die Implementierung).

4.2.1. Petrinetze

Spricht man von Petrinetzen, so sind i.allg. die klassischen Bedingungs-/Ereignisnetze gemeint. In der Praxis kommen, je nach Anwendungsgebiet und erforderlicher Modellierungsmächtigkeit, Vertreter recht unterschiedlicher Petrinetzklassen zum Einsatz. Ihnen allen ist gemeinsam, daß sie aufgrund ihrer graphischen, netzorientierten Darstellungsform sehr anschaulich und intuitiv verständlich sind. Petrinetze werden aus nur vier Grundelementen gebildet (bzgl. der verwendeten Fachterminologie muß auf [Baum90], [Jens90], [Pete81], [Reis82] u.a. verwiesen werden): Transitionen, Stellen, Kanten und Token. Damit lassen sich jeweils die essentiellen Aktivitäten, essentiellen Speicher, ihr Zusammenwirken sowie die essentiellen Ereignisse/Daten der Analysephase *unmittelbar* beschreiben.

Petrinetze sind formal und eindeutig. Der Netzgraph bestimmt die statische Systemstruktur. Eine einzige Feuerungsregel, die für alle Transitionen des Netzes gilt, beschreibt das dynamische Verhalten des Systemmodells. In Petrinetzen sind alle Transitionen und damit alle Aktivitäten von Haus aus nebenläufig. Dies vermeidet bei verteilten Systemen das Einbringen von "unwesentlichen" Systemeigenschaften durch das Beschreibungsmittel. Abhängigkeiten zwischen Transitionen, z.B. Sequenzen und Konflikte, sind explizit zu modellieren.

Mit Petrinetzen in ihrer grundlegenden Form lassen sich Systeme auf der Basis elementarer Bedingungen und Ereignisse modellieren. Für unsere Zwecke haben diese Netze eine zu geringe Modellierungskraft, zudem fehlen ihnen geeignete Strukturierungsmöglichkeiten ("flache" Netze). Ihre Verwendung als Systemmodell würde im Falle realer Telekommunikationssysteme zu übermäßig großen und inhaltlich schwer nachvollziehbaren Netzen führen.

4.2.2. Hierarchische farbige Petrinetze

Aus diesem Grunde baut das Systemmodell der OPM auf *hierarchisch* strukturierbaren *höheren* Petrinetzen auf. In [Dibo88] wird noch über den Einsatz "Offener Petrinetze" (auf PROLOG-Basis) berichtet, für die rechnergestützte Werkzeuge als Prototypen entwickelt wurden. Aufgrund der kommerziellen Verfügbarkeit eines kombinierten Netz-Editors/Simulators [Design/CPN] wird für das Systemmodell heute ein hierarchisches farbiges Petrinetz (**HCPN**, "hierarchical coloured petri net") verwendet.

Bzgl. der ausführlichen Syntax und Semantik von HCPNs darf auf [Jens90] verwiesen werden, wo auch Anwendungsbeispiele zur Systembeschreibung zu finden sind. Im folgenden werden nur die wichtigsten Eigenschaften des auf HCPNs basierenden Systemmodells vorgestellt, soweit diese zum tieferen Verständnis des gewählten Beispiels "Fernsprechsystem" (Bilder 4 bis 7) erforderlich sind. Der unmittelbare Bezug zwischen den Ergebnissen der Analysephase und den Elementen des Systemmodells wird aufgezeigt.

4.2.3. Das HCPN-Modell

Der hier vorgestellte Ansatz modelliert das Anreiz-Antwort-Verhalten des zu spezifizierenden Systems als ein HCPN, das unter Beachtung der Richtlinien der OPM strukturiert und gestaltet wird. Das so entstandene Systemmodell wird im folgenden als **HCPN-Modell** bezeichnet.

Das HCPN-Modell beschreibt ein System durch ein HCPN, das das System in aktive und passive Systemkomponenten unterteilt und ihr statisches wie dynamisches Zusammenwirken zu dem gewünschten Zweck aufzeigt. Darüberhinaus wird das System, eingebettet in seine Umwelt, derart modelliert, daß mittels HCPN das an der Schnittstelle zur Umwelt beobachtbare interaktive Systemverhalten festgelegt wird.

Die *statische Systemstruktur* ergibt sich aus dem Netzgraphen, der aus den folgenden Komponenten gebildet wird.

Passive Systemteile können Daten und Objekte speichern, jedoch nicht verändern. Sie werden durch **Stellen** (Ellipsen) dargestellt. Hierzu gehören die essentiellen Speicher sowie die Schnittstelle(n) zwischen System und Umwelt. Die Angabe eines Stellennamens ist optional (normale Schrift, i.allg. innerhalb der Stelle). In Übereinstimmung mit der für die Anforderungsdefinition unterstellten perfekten Systemtechnologie können Stellen prinzipiell beliebig viele und beliebig strukturierte Objekte aufnehmen. Je nach Anforderung kann ein zerstörender oder nichtzerstörender Datenzugriff modelliert werden.

Aus Gründen der Fehlervermeidung, Ausführbarkeit und Analysierbarkeit des Systemmodells sind HCPNs datentypisiert. Jeder Stelle ist eine **Farbe** zugeordnet (kursive Beschriftung), die das Ablegen von Daten und Objekten (Token) auf den (oder die) zugehörigen Datentyp(en) beschränkt. Die jeweilige Farbe wird in einem separaten Deklarationsteil festgelegt, der "global declaration node". Es stehen sowohl vor- als auch benutzerdefinierte Datentypen zur Verfügung (Beispiele hierfür sind der "global declaration node" des Fernsprechsystems im **Anhang** zu entnehmen).

Die **Stellenkapazität**, d.h. die maximale Anzahl der von einer Stelle aufnehmbaren Token, ist in der derzeitigen Version des verwendeten Tools generell unbegrenzt. Wo die Aufgabenstellung es erfordert, kann die Stellenkapazität durch entsprechende Hilfsstellen begrenzt werden. Dieser Aspekt ist vor allem für die Gestaltung der Schnittstelle zwischen System und Umwelt von Bedeutung (Kap. 4.3.1).

Aktive Systemteile können Daten und Objekte verändern, jedoch nicht speichern. Dies sind die essentiellen Aktivitäten, die durch **Transitionen** (dünn umrandete Rechtecke) dargestellt werden. Die Angabe eines Namens ist optional (normale Schrift, meistens innerhalb der Transition). Transitionen sind von Haus aus nebenläufig, soweit keine direkte Abhängigkeit explizit modelliert ist (z.B. Sequenz oder Konflikt).

Im Sinne der perfekten Systemtechnologie (Kap. 3.1.1) erfüllt jede Transition ihre Aufgabe in einer idealen, ausfallsicheren Art und Weise. Es ist wichtig darauf zu achten, daß Petrinetze *kausale* Abhängigkeiten modellieren und keinen impliziten Zeitbegriff besitzen. Die Dimension Zeit ist ggf. explizit zu modellieren. Das verwendete Tool hält für HCPNs eine entsprechende Unterstützung bereit (Datentypen und Simulationssemantik).

Stellen und Transitionen sind über **Kanten** (Pfeile) verbunden, die den Daten- und/oder Objektfluß zwischen aktiven und passiven Systemteilen beschreiben. Hierbei können nur Stellen mit Transitionen und umgekehrt verbunden werden. Kanten der untersten Abstraktionsebene des Systemmodells tragen datentypisierte **Kantenanschriften**, die die über diese Kanten "fließenden" Token nach Art und Anzahl festlegen. Kantenanschriften können Konstante (symbolische Namen, per Konvention mit einem Kleinbuchstaben beginnend), Variable (Platzhalter, mit einem Großbuchstaben beginnend) und Funktionen enthalten. Es können **Konsumations-** (Pfeil von der Stelle zur Transition), **Produktions-** (Pfeil von der Transition zur Stelle) und **Lesekanten** (Doppelpfeil, entsprechend der Überlagerung je einer Konsumations- und Produktionskante der gleichen Kantenanschrift) verwendet werden.

Die *Dynamik des Systemverhaltens* wird durch das Markenspiel ("token game") beschrieben, für das die Feuerungsregel des HCPN, die für alle Transitionen gleich ist, maßgebend ist.

Zur Beschreibung und Untersuchung des (dynamischen) Systemverhaltens werden die Anreize und Antworten sowie die essentiellen Systemdaten und -objekte (incl. der Initialisierung des Systems) als datentypisierte **Token** dargestellt, mit denen Stellen "markiert" werden. Die Farbe eines Tokens muß mit der Farbe der zugehörigen Stelle übereinstimmen. Die im Rechnerwerkzeug [Design/CPN] vordeklarierten und auf der Sprache SML (Standard Meta Language [HMT87, Wiks87]) basierenden Datentypen erlauben es, einzelne Daten entsprechend ihrer Beziehung zueinander zu einem Token zusammenzufassen (z.B. durch "product"-, "record"-, "list" und "union"-Konstruktoren).[3] Token können im Rahmen der vor- und benutzerdefinierten Datentypen beliebig verschachtelt sein. Der Anfangszustand des Systems (nach dem Einschalten oder nach Rücksetzoperationen) ergibt sich aus den Initialisierungsanweisungen der Stellen des Systems (unterstrichene Stellenbeschriftungen).

Eine Transition ist **aktiviert**, wenn alle ihre Vorbedingungen erfüllt sind, d.h. alle vorgelagerten Stellen mindestens die gemäß Kantenanschrift erforderliche Art und Anzahl von Token aufweist, und alle Nachbedingungen erfüllbar sind, d.h. die Aufnahmefähigkeit der nachgelagerten Stellen gewährleistet ist.

Eine Transition *kann* **feuern**, wenn sie aktiviert ist (der Zeitpunkt des Feuerns ist in nichtzeitbehafteten Petrinetzen zunächst ohne Bedeutung). Feuert eine Transition, so werden in einer *ununterbrechbaren* Handlung alle ihre Vorbedingungen "konsumiert", d.h. die entsprechenden Token den vorgelagerten Stellen entnommen, und alle Nachbedingungen "produziert", d.h. Token entsprechend den zugehörigen Kanten auf den nachgelagerten Stellen der Transition abgelegt. Die Prüfung auf Übereinstimmung der auf einer Stelle vorhandenen Token mit dem Tokenmuster einer Kante erfolgt durch Mustervergleich. *Wichtig* ist hierbei, daß je Aktivierung die Bindung der Variablen einer Transition für alle Variablen gleichen Namens gleich sein muß.

Insoweit aktivierte Transitionen nicht in einem Konflikt um gemeinsame Token stehen, können sie auch "gleichzeitig" (d.h. in einem Simulationsschritt) feuern. Insbesondere kann eine mehrfach akitivierte Transition auch nebenläufig zu sich selbst feuern[4].

Neben der Verwendung von Konstanten und Variablen in den Anschriften der Produktionskanten können die zu produzierenden Token auch als beliebige Funktion der konsumierten

---

[3] Die Beschreibung der Daten auf der Basis einer funktionalorientierten Sprache unterstützt die Forderung nach einer möglichst implementierungsneutralen Spezifikation auch im Detail, ohne auf die symbolische Ausführbarkeit des Systemmodells verzichten zu müssen.

[4] Hierbei sind jeder Aktivierung der Transition die entsprechend zu konsumierenden und produzierenden Token eigens zugeordnet. Die Bindung von Variablen an Token muß innerhalb der Aktivierung konsistent sein, kann jedoch für unterschiedliche Aktivierungen unterschiedlich ausfallen.

Token beschrieben werden (vor- oder benutzerdefiniert, aber *ohne* Seiteneffekte, z.B. "X+Y-1"). Darüber hinaus können Funktionen auch im Rahmen eines optionalen *"code segment"* ("C"-Kästchen mit "input ..; output ..; action ..;") aufgerufen werden. Als Bestandteil des Feuerns der Transition werden diese nach der Konsumation aber vor der Produktion von Token ausgeführt. Somit können Token berechnet und an noch freie Variablen der Produktionskanten gebunden werden. Die Einbindung von Programmcode *mit* Seiteneffekten erlaubt u.a. die Beobachtung und Beeinflussung der Simulation der Systembeschreibung.

Für das Systemmodell ist von Bedeutung, daß mit Hilfe des "code segments" auch ein Transitions-Verhalten in kompakter Form beschrieben werden kann, das außerhalb einer praktikablen Darstellung mittels der verwendeten Petrinetzklasse ist (dieser "Öffnung" der verwendeten Petrinetzklasse verdankt die OPM u.a. ihren Namen).

Die Beschreibung der Komplexität der Funktionalität eines vorgegebenen Systems kann in gewissen Grenzen auf die Netz- und Tokenstruktur sowie die Kantenanschriften verteilt werden. Hierbei spielen die **"guards"** (kursive Schrift innerhalb der Transitionen) eine Rolle, die die mittels Kantenart und Kantenanschriften gegebenen Vorbedingungen einer Transition um Boole'sche Zusatzbedingungen "und"-verknüpft erweitern. Neben Standard-Vergleichsfunktionen sind hier auch benutzerdefinierte Boole'sche Funktionen zulässig, solange sie seiteneffektfrei sind und sich lediglich auf die von der zugehörigen Transition zu konsumierenden Token beziehen.

Die nachfolgenden HCPN-Elemente dienen der *Strukturierung* des Systemmodells.

Subnetze (dick umrandete Rechtecke mit "HS"-Kästchen) stellen das wichtigste hierarchische Strukturierungsmittel und somit eine wesentliche Komponente der HCPNs dar. Sie lagern die Beschreibung komplexerer Aktivitäten auf jeweils eine eigene Diagrammseite aus, wo die zugehörige Verfeinerung modelliert wird (auch rekursiv anwendbar). Durch definierte Schnittstellen (Subnetz-Ports, grau hinterlegte Stellen mit "B"-Kästchen) können Subnetze ein und desselben Typs auch mehrfach in einem HCPN eingesetzt werden.

Mit Hilfe von **"global fusion sets"** (Ellipsen mit dickem grauen Rand und "FG"-Kästchen) kann auf den Inhalt ein und derselben globalen Stelle mittels Stellen auf (unterschiedlichen) Diagrammseiten zugegriffen werden, ohne das Netz durch eine Vielzahl von Kanten und Subnetz-Ports zu überladen. Dieses Konstrukt ist wegen der hiermit verbundenen Seiteneffekte nur mit äußerster Vorsicht anzuwenden. Ein geeigneter Anwendungsfall wäre z.B. der Zugriff auf eine globale Stelle, die die aktuelle Systemzeit bereithält.

Im übrigen sind HCPN-Modelle weitgehend selbsterklärend, was auch auf die Zulässigkeit langer Namen für Bezeichner zurückzuführen ist.

Fassen wir zusammen, so ergeben sich die folgenden *Vorteile* aus der Verwendung eines HCPN für das Systemmodell:

- Das Anreiz-Antwort-Verhalten des Systems kann auf rein kausaler Ebene beschrieben werden. Im Gegensatz zum ESA-Systemmodell werden im HCPN-Modell keine Annahmen über das zeitliche Verhalten "interner" und somit implementierungsabhängiger Systemaktivitäten benötigt. Auch eine Aufspaltung des Systemmodells in getrennte Modelle für den Daten- und Kontrollfluß, wie sie in SRSS vorgeschlagen wird, ist nicht erforderlich.
- Hierarchische Strukturierungsmechanismen helfen bei der Bewältigung der Komplexität realer Kommunikationssysteme.

- Die Mächtigkeit der Datentypen und -operationen farbiger Petrinetze sowie die Zulässigkeit von Variablen und Funktionen in Kantenanschriften ermöglicht kompakte und implementierungsneutrale Systemmodelle.
- Bereits in den frühen Phasen des Systemlebenszyklus kann die (entstehende) Systembeschreibung gegen die (informellen) Vorstellungen vom gewünschten Systemverhalten geprüft werden: Ausgehend vom Anfangszustand des Systems können Anreize, Antworten und Zustandsänderungen durch das Wandern der Token im Netzgraphen symbolisch nachvollzogen und visualisiert werden. (Ein "Schnappschuß" dieses Markenspiels für das Fernsprechsystem ist den Bildern 4 bis 7 zu entnehmen).
- Obwohl HCPNs zu den höheren Petrinetzen gehören, stehen die wichtigsten Analysetechniken der klassischen Petrinetztheorie prinzipiell zur Verfügung (an der Verminderung des Berechnungsaufwands wird noch gearbeitet).

4.3. HCPN-Modell des Fernsprechsystems

Nach der allgemein gehaltenen Beschreibung des HCPN-Modells sollen dessen Eigenschaften anhand einer konkreten Anwendung präzisiert werden. Vorgestellt wird ein kleiner Ausschnitt (Bilder 4 bis 7) aus dem HCPN-Modell des herkömmlichen Fernsprechsystems (Bild 3), das als Beispiel einer Modellierung gemäß der OPM ausgewählt wurde. Zum tieferen Verständnis des Systemmodells sind die im **Anhang** wiedergegebenen Datentypdeklarationen des Beispiels zu beachten.

4.3.1. HCPN-Modell, Ebene 1

Bild 4 zeigt das Systemmodell des Beispiels auf der obersten Abstraktionsebene und gibt uns Gelegenheit, auf die Bedeutung der während einer Simulation des Systemmodells beobachtbaren Token einzugehen. Was auf den ersten Blick wie eine einfache Blockdarstellung aussieht, ist *eine von mehreren* Sichten auf das farbige Petrinetz, dessen hierarchische Ebenenstruktur durch den HCPN-Editor/Simulator verwaltet wird. Das Anreiz-Antwort-Verhalten des Systems wird an der Schnittstelle zu seiner Umwelt formal und dennoch anschaulich spezifiziert.

Die *Ebene 1* des HCPN-Modells hat die Aufgabe, die zur Modellierung des Systems und seiner Umwelt als relevant erachteten *aktiven Einheiten* und die Gestalt der sie verbindenden abstrakten *Schnittstelle* aufzuzeigen. Die Ausgestaltung der Schnittstelle legt zugleich die maximal zulässige Nebenläufigkeit der (System-)Aktivitäten und somit des beobachtbaren Anreiz-Antwort-Verhaltens fest.

Im Beispiel wird das zu spezifizierende System durch das Subnetz "telephone network" modelliert. Die Angabe "..#3" hinter dem Kästchen "HS" zeigt uns an, daß die detaillierte Funktionalität dieses Subnetzes auf der Diagrammseite Nr. 3 beschrieben ist. Die Verfeinerung von Aktivitäten auf jeweils einer eigenen Diagrammseite ist eine der HCPN-Mittel zur Bewältigung der Komplexität realer Systeme.

Die Umwelt des Systems wird durch das Subnetz "users" gebildet. Da es uns primär um die Spezifikation des Fernsprechsystems geht, könnte auf die ausführliche Beschreibung des Verhaltens der Umwelt verzichtet werden. Die Verfeinerung der Umwelt auf der Diagrammseite Nr. 6 bietet jedoch die Möglichkeit, entweder das potentielle Verhalten der einzelnen Fernsprechteilnehmer oder, in Hinblick auf die schritthaltende Systemsimulation, einen

Testfallgenerator für das System zu modellieren. Wir haben uns in diesem Beispiel für letzteres entschieden (Kap. 4.3.4).

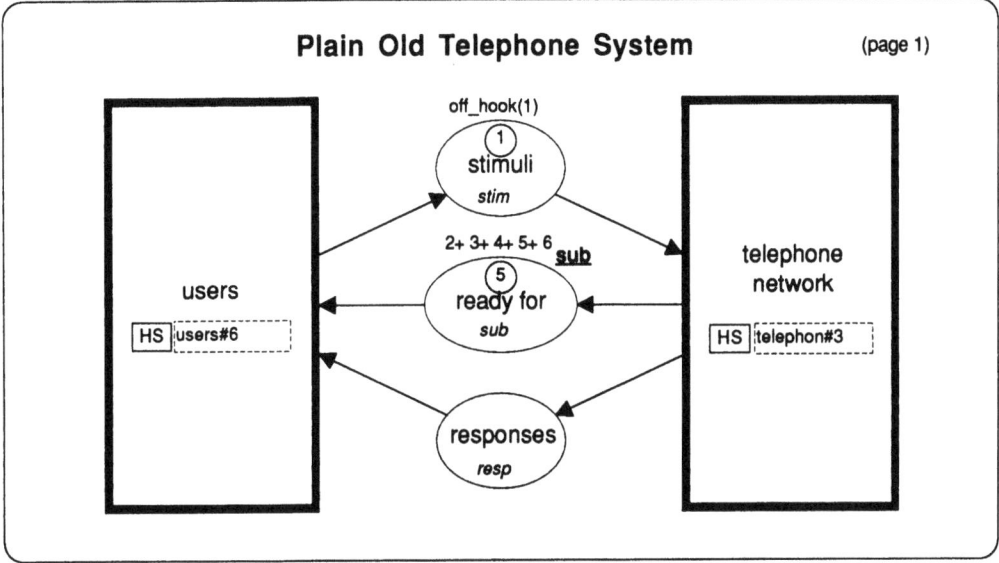

Bild 4: Das Fernsprechsystem und seine Umwelt

Die abstrakte und als passiv anzusehende Schnittstelle zwischen System und Umwelt wird durch die Gesamtheit der Stellen (Ellipsen) des HCPN der obersten Ebene modelliert. Das im Rahmen einer symbolischen Ausführung des HCPN-Modells auf diesen Stellen beobachtbare *Markenspiel* hat folgende Bedeutung:

Die Markierung einer dem System vorgelagerten Stelle mit einem Token "Stim" (z.B. das Token "off_hook(1)" liegt auf der Stelle "stimuli", siehe den in Bild 4 wiedergegebenen "Schnappschuß" aus einer Simulation der Systembeschreibung) bedeutet die Betrachtung des Systemmodells für den Fall, daß die Umwelt den durch "Stim" bezeichneten Anreiz (hier: das Abnehmen des Hörers von der Gabel des Fernsprechapparates des Teilnehmers Nr. 1) erzeugt hat. Die in der Realität hierzu erforderliche Interaktion zwischen System und Umwelt ist in der Ereignisdefinition festgelegt. Da das Auftreten eines Anreizes und damit das Vorhandensein eines Tokens auf der Stelle "stimuli" Vorbedingung für das Auslösen der zugehörigen Aktivität des Systems ist, wird das Token im Rahmen der von diesem Anreiz ausgelösten Aktivität vom Subnetz "telephone network" konsumiert und somit von der Stelle "stimuli" entfernt.

Das Fernsprechsystem muß sowohl die Anreize ihrer Art nach als auch gleichartige Anreize nach dem sie auslösenden Teilnehmer wohl unterscheiden können, da jeweils unterschiedlich zu reagieren ist. Die klassische Umsetzung dieser Vorgabe würde bedeuten, daß je Teilnehmer soviele Schnittstellen modelliert würden, wie Anreizarten zu unterscheiden sind. Das Markieren einer solchen Stelle mit einem (unstrukturierten) Token stellt dann den Eintritt des zugehörigen Anreizes des jeweiligen Teilnehmers dar. Diese Vorgehensweise ist i.allg. wenig praktikabel, vor allem wenn die Anzahl der Teilnehmer zu Simulationszwecken variabel gestaltet werden soll. Die Verwendung höherer Petrinetze für das Systemmodell der OPM erlaubt es, sowohl die gleichen Anreize unterschiedlicher Teilnehmer zu einer Anreizart unter Indizierung des auslösenden Teilnehmers zusammenzufassen (hier: "off_hook(A)" und "on_hook(A)") als auch zusätzlich gleichartige Anreize zu parametrisieren (hier: "digit(A,D)" für die Wahl einer Ziffer D durch den Teilnehmer A). Durch diese *Faltung* unterschiedlicher Stellen mit unstrukturierten Token in nur noch eine Stelle mit parametrisierten Token (hier: Stelle "stimuli" mit den oben erwähnten Token der Farbe "stim") wird das HCPN-Modell wesentlich kompakter, ohne an Verständlichkeit einzubüßen. Gleiches gilt auch für die Schnittstelle "responses" und ihre Token.

Die Markierung einer dem System nachgelagerten Stelle mit einem Token "Resp" (z.B. "responses" mit "dial_tone_on(1)") bedeutet, daß das System die durch "Resp" bezeichnete Antwort erzeugt hat (hier: Aufschalten des Wähltons für den Teilnehmer 1; die genaue Ausgestaltung der Antworten in der realen Welt wird ebenfalls in der Ereignisdefinition festgelegt). Im Rahmen der Umweltaktivitäten zur Erkennung/Verarbeitung dieser Antwort ist dieses Token zu konsumieren und somit von der Stelle "responses" zu entfernen.

Wie bereits erwähnt, fällt der Ausgestaltung der Schnittstelle auch die Aufgabe zu, die (in der Analysephase erkannte) maximal mögliche Nebenläufigkeit von Aktivitäten zu modellieren. In unserem Beispiel dient hierzu die Stelle "ready for". Als "teilnehmerspezifische Komplementärstelle" zu "stimuli" bewirkt diese Stelle, daß zu jedem Zeitpunkt auf "stimuli" nicht mehr als ein Anreiz-Token *je* Teilnehmer vorhanden sein kann, während die Anzahl von Anreizen *unterschiedlicher* Teilnehmer nicht eingeschränkt wird. Die in Bild 4 erkennbare Markierung "2+3+4+5+6" der Stelle "ready for" mit insgesamt fünf Token (Kreis mit der Zahl 5) ist nicht als arithmetische Funktion zu deuten, sondern sagt aus, daß auf dieser Stelle die Token "2" bis "6" ohne bestimmte Reihenfolge liegen (das "+" ist ein Operator des für Stellenmarkierungen in HCPNs verwendeten "multi set"-Datentyps). Durch die vorliegende Gestaltung der Schnittstelle enthält das Systemmodell folgende Aussagen:

- Die Reihenfolge des zeitlichen Wirksamwerdens aller Anreize *ein und desselben* Teilnehmers ist auf die Reihenfolge der zugehörigen Aktivitäten und somit auf das Wirksamwerden der entsprechenden Antworten und Zustandsänderungen zu übertragen.

    Für die Implementierung des Systems lassen sich zwei Konsequenzen ableiten:
    a) Die Umwelt kann im Rahmen der physikalischen Gegebenheiten Anreize "ungebremst" erzeugen. In diesem Fall muß das implementierte System in der Lage sein, auch bei dem zeitlich kürzesten Abstand zweier Anreize die Bearbeitung des jeweils vorhergehenden Anreizes abgeschlossen zu haben, bevor der nächste Anreiz desselben Teilnehmers Auswirkungen zeigen darf.
    b) Die Umwelt "kooperiert" und wartet mit der Erzeugung eines neuen Anreizes solange, bis das implementierte System den vorangegangenen Anreiz soweit bearbeitet hat, daß es den nächsten Anreiz "vertragen" kann. Dies setzt ein entsprechendes Signal des Systems an die Umwelt voraus.

Die Praxisrelevanz dieser für ereignisgesteuerte Systemmodelle bislang vernachlässigten Betrachtungsweise läßt sich eindrucksvoll am gewählten Beispiel des Fernsprechsystems belegen:
Das HCPN-Modell gibt vor, daß beim Auftreten der Anreizfolge "digit(A,"1"), digit(A,"2")" eines Teilnehmers A diese Reihenfolge auch auf das Wirksamwerden der zugehörigen Antworten und/oder Zustandsänderungen übertragen werden muß, d.h. das System im Endeffekt die Wahl der Rufnummer "12" und nicht "21" erkennen muß.

Die Implementierung des Systemmodells unter Verwendung eines Fernsprechapparates mit Mehrfrequenzwahl (Tastentelefon) folgt der o.a. Konsequenz a):

Betätigungen der Zifferntasten, die zeitlich schneller aufeinanderfolgen, als sie der Fernsprechapparat aufgrund der Mindesthalte- und Mindestpausezeiten für Mehrfrequenzwahltöne an die Vermittlungsstelle weitergeben kann, werden durch den Fernsprechapparat zwischengespeichert (solange die hierfür implementierte Speicherkapazität ausreicht) und in unveränderter Reihenfolge zeitlich verzögert weitergereicht.

Die Implementierung des Systemmodells unter Verwendung eines Fernsprechapparates mit Pulswahl folgt hingegen der Konsequenz b):

Die Feinmechanik der Wählscheibe des Fernsprechapparates stellt sicher, daß nach jeder Wahl einer Ziffer (Aufzug der Wählscheibe bis zur gewünschten Ziffer) für die Zeitdauer der Übertragung zur Vermittlungsstelle mittels Pulscode keine weitere Wahl erfolgen kann (die Wählscheibe muß erst zurücklaufen). Auch die zwischen zwei Ziffern benötigte Pausenzeit wird gewährleistet, indem die Wählscheibe weiter aufgezogen werden muß und somit länger zurückläuft, als es für die reine Pulsübertragung erforderlich wäre. Die Freigabe des Systems für die nächste Wahl wird der Umwelt durch das Zurückkehren der Wählscheibe in die Ausgangsposition "signalisiert".

- Das System kann auf Anreize *unterschiedlicher* Teilnehmer in einer willkürlichen Reihenfolge reagieren. Die Reihenfolge der Anreize muß nicht auf die Folge des Wirksamwerdens der zugehörigen Antworten und/oder Zustandsänderungen übertragen werden, d.h. das implementierte System kann die Anreize unterschiedlicher Teilnehmer in einer beliebigen Reihenfolge bearbeiten (soweit diese Freiheit in der Praxis nicht durch die Vorgabe einer maximalen Reaktionszeit für das System eingeschränkt wird).

Auch für diese Aussage bietet das Fernsprechsystem ein anschauliches Beispiel: Unterstellen wir im HCPN-Modell die Anreizfolge "off_hook(1),off_hook(2)", was in der Realität bedeuten kann, daß ein Teilnehmer in München den Hörer von der Gabel seines Fersprechapparates genommen hat und ca. zehn Sekunden später (d.h. eindeutig in dieser Reihenfolge beobachtbar) ein Teilnehmer in Hamburg desgleichen tat. Dem HCPN-Modell ist zu entnehmen (Bild 6, erläutert in Kap. 4.3.3), daß dem Anreiz "off_hook(A)" eines vorher inaktiven Teilnehmers A das Aufschalten des Wähltons und die Änderung des Teilnehmerzustandes in die Vorwahlphase zu folgen hat.

Das HCPN-Modell gibt weiter vor, daß die zu den beiden Anreizen gehörigen Aktivitäten nebenläufig sind, d.h. daß die Erzeugung der Antwort "dial_tone_on(1)" und die Zustandsänderung des Teilnehmers 1 von "idle(1)" auf "pre_dial_phase(1)" unabhängig von den entsprechenden Maßnahmen für den Teilnehmer 2 erfolgen kann.

Für das implementierte Fernsprechsystem bedeutet dies, daß das Aufschalten des Wähltons in einer beliebigen Reihenfolge erfolgen kann, d.h. daß der Teilnehmer aus München nicht notwendiger Weise den Wählton als erstes erhalten muß.

Es darf betont werden, daß ohne die Kenntnis und Nutzung derartiger Nebenläufigkeiten das Fernsprechsystem wirtschaftlich nicht zu realisieren wäre.

Das Systemmodell stellt mittels HCPN eine ausführbare Spezifikation dar. Aussagen der vorgenannten Art können sowohl deduktiv abgeleitet als auch durch symbolische Ausführung des HCPN-Modells rechnergestützt gewonnen werden.

4.3.2. HCPN-Modell, Ebene 2 des Systems

Bild 5 zeigt einen Ausschnitt aus der Abstraktionsebene 2 des HCPN-Modells. Wiedergegeben ist die Diagrammseite 3, die die konsistente Verfeinerung des Subnetzes "telephone network" darstellt. Dieses Subnetz stellt das eigentliche *operative Modell* für das Anreiz-Antwort-Verhalten des Systems dar. Verbindungen zu den Stellen der übergeordneten Netzebene, in diesem Fall zu den Stellen der Schnittstelle zwischen System und Umwelt, werden über *Ports* (grau hinterlegte Stellen mit "B"-Kästchen) geführt. Die als Transitionen modellierten oder in weiteren Subnetzen verborgenen essentiellen Aktivitäten sowie ihre Verbindungen zu essentiellen Systemspeichern sind explizit erkennbar.

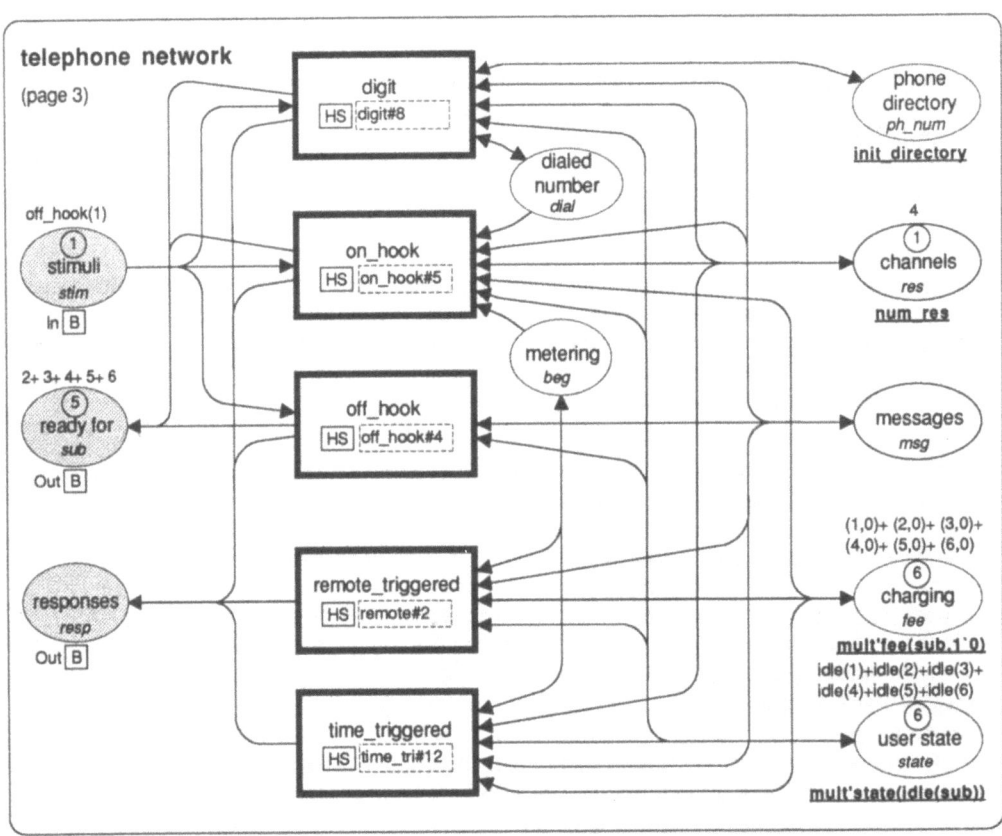

Bild 5: Das Subnetz "telephone network"

Gemäß den ESA-Richtlinien für die Unterstellung einer perfekten Systemtechnologie ist eine Beschränkung von Ressourcen, die lediglich von der gewählten Implementierungstechnologie abhängig ist, eine unwesentliche Eigenschaft. Das HCPN-Modell trägt dem Rechnung, indem *essentielle Speicher* durch Stellen mit unbegrenzter Stellenkapazität modelliert werden. Beispiele hierfür sind die Stellen "phone directory", "dialed number" und "metering". Die *essentiellen Daten* des Systems, d.h. die Daten, die sich das System unabhängig von der Art der Implementierung merken muß, werden durch strukturierbare Token des HCPN modelliert.

Im Beispiel sind die Zustände der einzelnen Teilnehmerverbindungen durch Token der Art "idle(A)" auf der Stelle "user state" abgelegt.

In Bild 5 ist auch zu sehen, wie - trotz unterstellter perfekter Technologie - eine *aufgabenbedingte* Beschränkung von Ressourcen modelliert werden kann: Die für jede Art der Implementierung aus wirtschaftlichen Gründen eingeschränkte Anzahl der gerichteten Kommunikationskanäle zwischen Teilnehmern ergibt sich aus der Zahl, mit der die Stelle "channels" aktuell markiert ist.

Der nach Inbetriebnahme oder nach einer Rücksetz-Aktion erwartete (verteilte) *Anfangszustand* des zu implementierenden Systems (wie auch des zu simulierenden Systemmodells) wird durch die Gesamtheit der Anfangsmarkierungen aller Systemstellen festgelegt. Die Angabe der *Anfangsmarkierung* einer Stelle (fette, unterstrichene Stellenanschrift) ist optional. So bedeutet im Beispiel die Konstante "num_res" (im Deklarationsteil definiert, siehe Anhang), daß die Stelle "channels" nach der Initialisierung über vier gerichtete Kanäle verfügt. Auch eine größere Anzahl von Token läßt sich durch mächtige SML-Funktionen leicht erzeugen: Die Anweisung "mult'state(idle(sub))" initialisiert die Stelle "user state" mit je einem Token "idle(A)" pro Teilnehmer A der Teilnehmermenge "sub".

Die durch Anreize, d.h. Token der Stelle "stimuli" auszulösenden *Aktivitäten* sind in weiteren Subnetzen verborgen. Es fehlen noch jegliche Kantenanschriften, da für das Verständnis des Systemmodells auf dieser Ebene nur die weitere Strukturierung der Aktivitäten sowie die prinzipielle Art der Tokenzugriffe (konsumierend, produzierend, lesend/verändernd) bekannt sein muß, nicht jedoch der genaue Daten- und Kontrollfluß.

Die *Strukturierung des HCPN-Modells* lehnt sich an die ESA-Richtlinien für den "event-partitioned data flow graph" an: Alle Aktivitäten, die durch dieselbe Anreizart ausgelöst werden (und sich damit nur noch in den weiteren Vorbedingungen bzgl. des Systemzustandes unterscheiden), werden zu je einem Subnetz zusammengefaßt. Hierdurch wird sichergestellt, daß sich die Struktur des Systemmodells in erster Linie aus der Aufgabenstellung ergibt, und erst in zweiter Linie von den Gewohnheiten des Systemanalytikers bzw. -modellierers abhängt.

In Bild 5 sind folgende Besonderheiten zu erkennen:
- Das Subnetz "remote_triggered" beherbergt Transitionen, die *nicht unmittelbar* durch einen Anreiz ausgelöst werden. Aufgrund einer verteilten Bearbeitung der zugehörigen Aktivität werden sie - als Folge eines Anreizes - durch ein "internes" Signal auf der Stelle "messages" angestoßen.
- Zeitgetriggerte Systemaktivitäten (Subnetz "time_triggered") werden *nicht* durch Anreize der Umwelt ausgelöst. Das Systemmodell spezifiziert für diese Aktivitäten, was im Falle des Eintritts der zugehörigen zeitbehafteten Bedingung (Eintritt eines bestimmten Zeitpunkts oder Ablauf eines Zeitintervalls) vom System zu tun ist. Der zur Auslösung dieser Aktivitäten erforderliche Zugriff des Systems auf die aktuelle Systemzeit wird als gegeben vorausgesetzt und daher nicht als explizite Vorbedingung der zugehörigen Transitionen modelliert.
Bei einer zeitbehafteten Simulation des HCPN-Modells, die durch das verwendete Rechnerwerkzeug unterstützt wird, wären die entsprechenden zeitbehafteten Vorbedingungen explizit zu modellieren.

Betrachten wir nun das Subnetz "off_hook" in Bild 5. Zu erkennen ist, daß darin befindliche Aktivitäten durch Anreize der Stelle "stimuli" ausgelöst werden. Ferner hat das Subnetz lesenden und schreibenden Zugriff auf die Stellen "user state" (Zustand aller Teilnehmerverbindungen, initialisiert auf den Zustand "idle") und "messages" (zur verteilten Bearbeitung von Aktivitäten). Alle weiteren Informationen sind der Verfeinerung des zugehörigen Subnetzes (Diagrammseite 4) vorbehalten.

4.3.3. HCPN-Modell, Ebene 3 des Systems

Mit **Bild 6** sind wir auf der untersten Abstraktionsebene des dreistufigen HCPN-Modells für das Fernsprechsystem angelangt.

Die Verfeinerung des Subnetzes "off_hook" legt im Detail und bzgl. des auslösenden Teilnehmers parametrisiert den *kausalen* Zusammenhang zwischen dem einzelnen Anreiz und der vom System erwarteten Aktivität (sich äußernd in Antwort und/oder Zustandsänderung) für alle Anreize des Typs "off_hook(A)" fest.

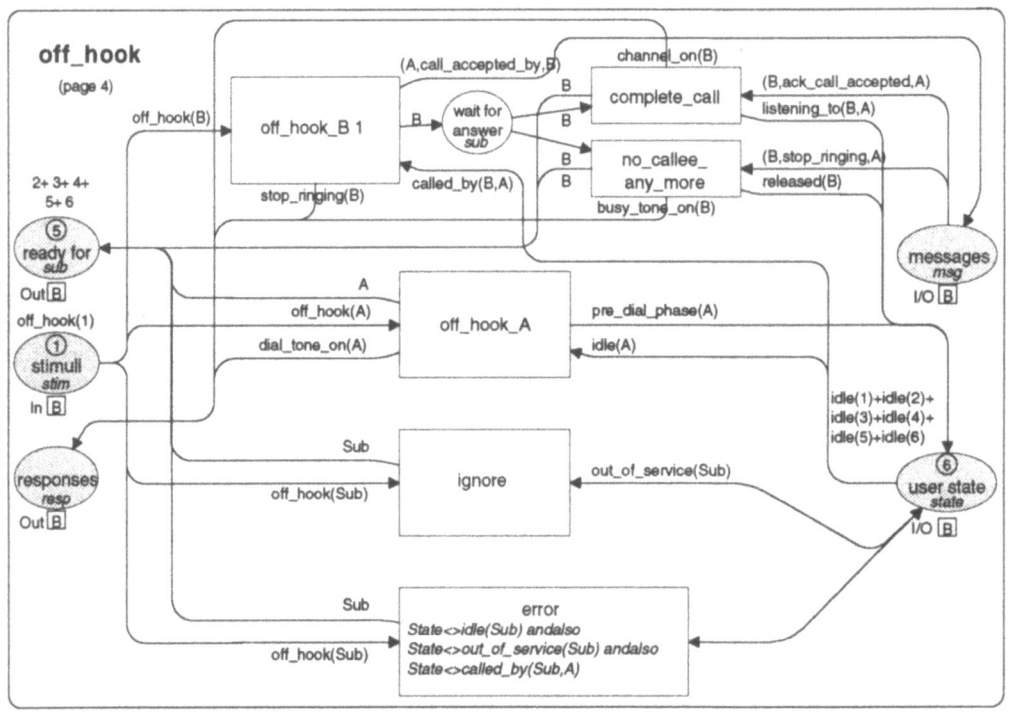

Bild 6: Das Subnetz "off_hook"

Die *essentiellen Aktivitäten* werden durch Transitionen modelliert (dünn umrandete Rechtecke). Art, Umfang und Randbedingungen der von ihnen zu erbringenden Daten- und/oder Kontrollflußmanipulation werden durch die Transitionskanten, die Kantenanschriften sowie die Transitionsinschriften im Zusammenhang mit der Feuerungsregel *formal* festgelegt.

Das Subnetz "off_hook" umfaßt alle Aktivitäten, die durch Anreize "off_hook(A)" der Stelle "stimuli" ausgelöst werden können. Welche Aktivität konkret ausgelöst wird, d.h. welche Transition feuert, hängt zusätzlich von der aktuellen Markierung der Stelle "user state" ab.

In der mittels Bild 6 wiedergegebenen aktuellen Markierung des Systemzustandes (Schnappschuß einer Simulation des HCPN-Modells) ist als einzige die Transition "off_hook_A" aktiviert. Feuert sie, so konsumiert sie zunächst ihre beiden Vorbedingungen, die Token "off_hook(1)" der Stelle "stimuli" und "idle(1)" der Stelle "user state" (Anreiz und aktueller Zustand des Teilnehmers 1). *Ohne Unterbrechung* erzeugt sie dann ihre drei Nachbedingungen, die Token "dial_tone_on(1)" (Einschalten der zeitkontinuierlichen Funktion zur Erzeugung des Wähltons für den Fernsprechapparat des Teilnehmers 1), "pre_dial_phase(1)" (Übergang des Teilnehmerzustandes 1 in die Vorwahlphase) sowie "1" (Freigabe der Schnittstelle für den nächsten Anreiz des Teilnehmers 1), und legt diese auf den entsprechenden Stellen ab.

Am Beispiel der Transition "ignore" ist auch die Bedeutung einer Lesekante zu erkennen: Das Feuern der Transition verändert den Teilnehmerzustand nicht, da zwar zunächst die Token "off_hook(Sub)" und "out_of_service(Sub)" konsumiert, dann aber die Token "out_of_service(Sub)" und "Sub" produziert werden.

Die Eigenschaft der Petrinetze, die für das Feuern einer Transition benötigten Token *physisch* zu konsumieren und das Feuern einer Transition *ununterbrechbar* durchzuführen, macht eine zusätzliche Konfliktbehandlung für den Zugriff mehrerer Aktivitäten auf ein und dasselbe Datum (z.B. den Zustand eines Teilnehmers) im Systemmodell überflüssig.

Im Gegensatz zum ESA-Systemmodell werden im OPN-Modell *keine* Angaben über die Zeitdauer "interner" und somit implementierungsabhängiger Aktivitäten gemacht. An die Stelle der ESA-Forderung nach "Null-Zeitbedarf" für die Auslösung und Durchführung der Aktivität tritt die der Wirklichkeit *angemessenere* Forderung der Ununterbrechbarkeit der die Aktivität bildenden Einzelaktionen, soweit dies durch die Aufgabenstellung für das System gefordert ist. Hierzu zwei Beispiele:

- Alle zu der vollständigen Bearbeitung des Anreizes "off_hook(A)" im Teilnehmerzustand "idle(A)" erforderlichen Einzelaktionen (s.o.) werden in der Transition "off_hook_A" *ununterbrechbar* zusammengefaßt.

- Insoweit eine Aktivität nicht durch eine einzige Transition modelliert werden soll oder kann, z.B. wegen der aufgabenbedingten Notwendigkeit einer (räumlich) verteilten Bearbeitung, ist die Ununterbrechbarkeit der Ablauffolge der einzelnen Transitionen - und damit die *Unveränderbarkeit des Resultats* der zugehörigen Aktivität - durch eine geeignete (Schnittstellen-)Modellierung sicherzustellen.

  Im Falle der Transitionen "off_hook_B_1" bis "complete call" bzw. "no callee any more" wird dies erreicht, indem das auslösende Token "off_hook(B)", gemeinsam mit dem Zustand des Teilnehmers B, als Vorbedingung von "off_hook_B_1" zwar durch diese Transition bereits konsumiert wird, jedoch erst durch die letzte Transition, entweder "complete call" oder "no callee any more", das Token "B" auf "ready for" und der sich neu ergebende Teilnehmerzustand auf "user state" abgelegt wird. Dadurch wird sichergestellt, daß einerseits die Schnittstelle "stimuli" nicht während der noch laufenden Bearbeitung des anstehenden Anreizes erneut mit einem Anreiz für den Teilnehmer B markiert wird und andererseits der noch in Änderung befindliche Zustand des Teilnehmers B durch Aktivitäten anderer Teilnehmer nicht unzutreffend gelesen oder gar inkonsistent verändert wird.

Noch eine Anmerkung zu der in Kapitel 4.1 vorgeschlagenen Beschreibung der zeitlichen Rahmenbedingungen für das System mittels HCPN-Modell:

Als Beispiel wird die Forderung zitiert, daß das System innerhalb einer vorgegebenen Höchstzeit auf das Abheben eines Teilnehmers im Zustand "idle" mit dem Aufschalten des Wähltons zu reagieren hat. Bei einer zeitbehafteten Simulation - die das Rechnerwerkzeug für HCPNs unterstützt - läßt sich diese Forderung unmittelbar der Transition "off_hook_A" in Bild 6 zuordnen, indem eine entsprechende obere Zeitgrenze für das Schalten der Transition nach deren Aktivierung festgelegt wird. Problematischer wird die Angelegenheit, wenn sich die Zeitvorgabe auf eine Aktivität bezieht, die durch eine Folge von Transitionen modelliert wird (z.B. die mit der Transition "off_hook_B_1" beginnende Folge aus insgesamt drei Transitionen).

An Syntax und Semantik der unmittelbaren Darstellung von zeitbehafteten Forderungen im HCPN-Modell wird noch gearbeitet. Aus diesem Grunde sind derartige Forderungen in den Bildern 4 bis 7 nicht enthalten.

4.3.4. HCPN-Modell, Ebene 2 der Umwelt

Wie bereits erwähnt, ist eine Verfeinerung des Verhaltens der Umwelt zur Beschreibung des Systems nicht zwingend erforderlich, zumal der Umwelt i.allg. kein bestimmtes Verhalten aufgezwungen werden kann.

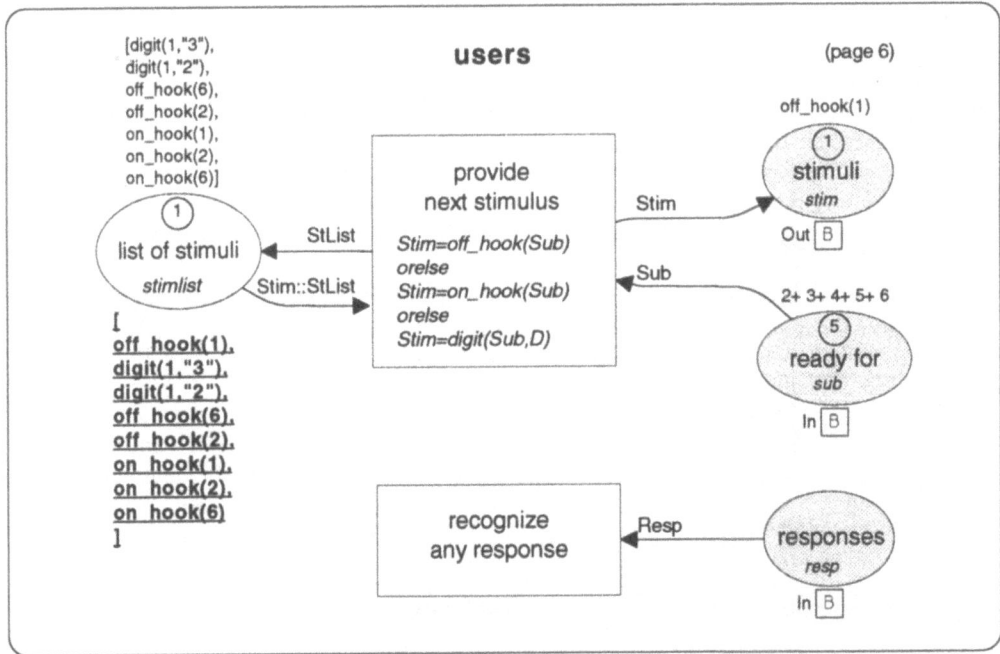

Bild 7: Subnetz "users"

Die Beschreibung des erwarteten (potentiellen) Verhaltens der Umwelt kann jedoch eine gewisse Hilfe für das Verständnis des Systems geben. Unter dem Aspekt eines ausführbaren Systemmodells bietet sich auch die Beschreibung der Umwelt in Form eines typischen Testszenarios für das System an.

Bild 7 zeigt die Verfeinerung des Subnetzes "users" unseres Beispiels "Fernsprechsystem", das einen derartigen Testgenerator bildet. Die Erzeugung von Anreizen wird durch die Transition "provide next stimulus" übernommen. Die Kombination aus der Verwendung einer Listenfunktion als Anschrift einer Konsumationskante ("::"-Operator, der aus einem Datum und einer Liste eine neue Liste mit dem Datum als erstes Element der neuen Liste bildet) und den Regeln zur konsistenten Bindung von Variablen zur Aktivierung einer Transition ermöglicht es, mit *einer* Transition das folgende Verhalten zu beschreiben:
- Der auf der Stelle "list of stimuli" per Anfangsmarkierung liegenden Liste mit Anreizen wird so lange ein Token nach dem anderen in der durch die Liste vorgegebenen Reihenfolge entnommen, bis die Liste leer ist, wobei
- die Aufnahmefähigkeit des Systems für den Anreiz eines bestimmten Teilnehmers beachtet wird ("hand-shake"-Verfahren über die Stelle "ready for").

Durch mehrfaches Feuern der Transition "provide next stimulus" kann die Stelle "stimuli" - soweit vom System aus zulässig - auch mit mehreren Token markiert und somit die nebenläufige Ausführung von Aktivitäten studiert werden.

In unserem Beispiel bildet die Stelle "recognize any response" die Wahrnehmung der Antworten durch die Umwelt nach, indem die entsprechenden Token von der Stelle "responses" ohne weitere Vorbedingungen abgeräumt werden.

Die hier vorgestellten Bilder 3 bis 7 sind nur ein Teilausschnitt aus dem HCPN-Modell des Beispiels Fernsprechsystem, das ohne die Verfeinerung des Teilnehmerverhaltens derzeit acht Bilder der gezeigten Art umfaßt. Das HCPN-Modell in seiner Gesamtheit wird rechnergestützt erstellt, simuliert und gewartet. Informationen über das verwendete Rechnerwerkzeug sind in [Design/CPN] zu finden.

Zusammenfassung und Schlußfolgerung

Vorgestellt wurde die *Offene Petrinetz-Methode* zur Analyse und Darstellung des funktionalen Verhaltens von Telekommunikationssystemen. Ihre Eigenschaften sind speziell auf die Erfordernisse eines Netzbetreibers in den ersten Phasen des Systemlebenszyklus abgestimmt. *Ziel* der OPM ist eine implementierungsneutrale, formale und dennoch anschauliche Beschreibung verteilter Systeme, die auch die zulässige Nebenläufigkeit von Systemaktivitäten darstellen kann.

Die Anwendung der OPM bietet eine methodische Unterstützung der schwierigen, da bislang nicht automatisierbaren Phase der Systemanalyse. Die zugrunde liegenden Richtlinien und ihr Bezug zu bekannten Methoden wurden dargelegt.

Fassen wir zusammen, so besteht die OPM im Kern aus den folgenden Schritten:
1. Definieren der Aufgabenstellung sowie der relevanten Umwelt des Systems
2. Ermitteln der zwischen System und Umwelt relevanten Interaktionen (Systemanreize und -antworten) und Festschreiben ihrer konkreten Ausbildung in der realen Welt (Ereignisdefinition)

3. Ermitteln der essentiellen (d.h. unabhängig von der Implementierungsart benötigten) Aktivitäten und der essentiellen Speicher des Systems in Hinblick auf die geforderten Systemantworten und Systemschnittstellen
4. Ermitteln der essentiellen Aktivitäten zur Pflege der Daten in den essentiellen Systemspeichern (Anfangsinitialisierung und zeitgesteuerte Aktivitäten)
5. Ermitteln der durch die Aufgabenstellung für das System vorgegebenen maximalen Nebenläufigkeit von essentiellen Aktivitäten
6. Beschreiben des Zusammenhangs zwischen essentiellen Aktivitäten (Systemantworten und Zustandsänderungen) und den sie auslösenden externen bzw. temporalen Ereignissen (Systemanreize) in Form eines operativen Systemmodells auf der Basis hierarchisch strukturierbarer höherer Petrinetze, derart daß
   - alle essentiellen Aktivitäten dem Ereignis zugeordnet werden, das sie unmittelbar auslöst,
   - alle essentiellen Daten/Objekte entsprechend ihrer Ursache in der Umwelt des Systems in essentiellen Speichern zusammengefaßt werden,
   - die maximal mögliche Nebenläufigkeit der essentiellen Aktivitäten zum Ausdruck gebracht wird
7. Schritthaltende Prüfung der Systembeschreibung gegen die informellen Ideen über das gewünschte Systemverhalten mittels symbolischer Ausführung des Systemmodells.

Die erarbeiteten funktionalen Systemanforderungen werden in der OPM folglich mittels Ereignisdefinition und ausführbarem Systemmodell beschrieben:

Die *Ereignisdefinition* fixiert den Bezug zwischen den abstrakten Anreizen und Antworten des Systemmodells und den zu ihrer Erzeugung in der realen Welt konkret erforderlichen Interaktionen zwischen System und Umwelt.

Dem *Systemmodell* fällt die Aufgabe zu, das Anreiz-Antwort-Verhalten des Systems an der Schnittstelle zu seiner Umwelt operativ zu beschreiben. Das gewählte Systemmodell erlaubt es, die erarbeiteten essentiellen Systemeigenschaften *unmittelbar* durch entsprechende Elemente des Systemmodells wiederzugeben, d.h. ohne semantische Lücke. Es wird durch ein hierarchisches farbiges Petrinetz gebildet, dessen Netzgraph gemäß den Richtlinien der OPM strukturiert und ausgestaltet wird.

Die Vorteile der Verwendung höherer Petrinetze für die Beschreibung verteilter Systeme wurden erläutert. Es sei betont, daß die OPM nicht auf die Verwendung von HCPNs fixiert ist. Als "offene" Methode können auch andere höhere Petrinetzklassen für das Systemmodell verwendet werden, soweit sie eine hierarchische Verfeinerung von Transitionen zulassen und symbolisch ausführbar sind. Beispiele hierfür sind die auf der Sprache PROLOG definierten "Offenen Petrinetze" [Dibo88] und die auf der Sprache Smalltalk basierenden Netze in [DGGK88].

Die wesentlichen Eigenschaften des Systemmodells wurden anhand von Ausschnitten aus dem HCPN-Modell des Fernsprechsystems vertieft, das als konkretes OPM-Anwendungsbeispiel ausgewählt wurde.

Derzeit untersuchen wir, inwieweit die OPM die Beschreibung von Diensten in einem Intelligenten Netz, die sich an der CCITT-Empfehlung I.130 orientiert, wirksam unterstützen kann. Die ersten Ergebnisse sind vielversprechend und belegen die Allgemeingültigkeit des OPM-Ansatzes für verteilte, ereignisgesteuerte Systeme [Dibo92].

Eine *Weiterentwicklung* der OPM ist in folgenden Punkten erstrebenswert:
- Verfeinern der OPM-Richtlinien zur Analyse der durch die Aufgabenstellung für das System vorgegebenen maximalen Nebenläufigkeit von Aktivitäten
- Aufnahme der zeitbehafteten Rahmenbedingungen für das System in das HCPN-Modell (Syntax und Semantik von zeitbehafteten Attributen derjenigen Transitionen bzw. Subnetze, die essentielle Systemaktivitäten modellieren)
- Nutzung neuer Forschungsergebnisse aus der Theorie hierarchischer farbiger Petrinetze zur analytischen Untersuchung von HCPN-Modellen (Ableitung von Systemeigenschaften aus dem Systemmodell)
- Nutzung von Parallelrechnern (z.B. vernetzte Transputersysteme) zur Interpretation von HCPN-Modellen, um die Gewinnung statistisch relevanter Aussagen auf simulativer Basis zu vereinfachen
- Erarbeiten von Richtlinien für ein methodisches Ableiten von Performance-Modellen unmittelbar aus dem funktionalorientierten HCPN-Modell (z.B. Übergang zur zeitbehafteten Simulation des HCPN-Modells zur Bewertung von Implementierungsalternativen, *ohne* das Beschreibungsmittel wechseln zu müssen)
- Untersuchungen zur Anwendbarkeit der OPM auf objektorientierte Systeme (Subnetze des HCPN-Modells und Objekte objektorientierter Sprachen weisen einige Gemeinsamkeiten auf: Sie "kommunizieren" über zu konsumierende Token/Nachrichten, sie agieren nebenläufig, sie sind in ihrer Wirkung auf das Gesamtsystem frei von Seiteneffekten und der Zugriff auf ihre internen Stellen/Daten erfolgt ausschließlich über ihre Schnittstellen/-Methoden).

Literaturverzeichnis

[Baum90]   Baumgarten, B.: Petri-Netze: Grundlagen und Anwendungen. BI Wissenschaftsverlag, Mannheim/Wien/Zürich, 1990. ISBN 3-411-14291-X

[Balz85]   Balzert, H.: Phasenspezifische Prinzipien des Software Engineering; In: Angewandte Informatik (1985), 3, p. 101-110.

[Design/CPN] Design/CPN Manual. Meta Software Corporation. Cambridge, USA, 1991.

[DeMa78]   DeMarco, T.: Structured Analysis and System Specification; Yourdon Press, New York 1978.

[Dibo88]   Dibold, H.: A Method for the Support of Specifying the Requirements of Telecommunication Systems; Tagungsband des Internationalen Zürich Seminars, Zürich, 1988, S. 115-122.

[Dibo92]   Dibold, H.: Hierarchical Coloured Petri Nets for the Description of Services in an Intelligent Network; Tagungsband des Internationalen Zürich Seminars, Zürich, wird März 1992 erscheinen.

[Jens90]   Jensen, K.: Coloured Petri Nets: A High Level Language for System Design and Analysis. Aarhus university, computer science department, report DAIMI PB - 338, November 1990. Paper also published in: G. Rozenberg (Ed.): Advances in Petri Nets 1990, Lecture Notes in Computer Science, Springer-Verlag.

[HMT87]   Harper, R.; Milner, R.; Tofte, M.: The Semantics of Standard ML, Version 1. Technical Report ECS-LFCS-87-36, University of Edinburgh, LFCS, Department of Computer Science, University of Edinburgh, The King's Buildings, Edinburgh EH9 3JZ, August 1987.

[HP87]   Hatley, D.J.; Pirbhai, I.A.: Strategies for Real-Time System Specification; Dorset House Publishing, New York, 1987.

[MP84]   McMemamin, S.M.; Palmer, J.F.: Essential Systems Analysis; Yourdon Press, New York, 1984.

[DGGK88]   Dähler, J.; Gerber, P.; Gisiger, H.-P., Kündig, A.: A Graphical Tool for the Design and Prototyping of Distributed Systems. Tagungsband des Internationalen Zürich Seminars, Zürich, 1988, S. 123-129.

[Pete81]   Peterson, J.L.: Petri Net Theory and the Modeling of Systems; Prentice Hall Inc., Engelwood Cliffs, N.J. 07632, 1981.

[Reis82]   Reisig, W.: Petrinetze - Eine Einführung; Springer-Verlag, Berlin Heidelberg, 1982.

[Wiks87]   Wikström, Å.: Functional Programming using Standard ML. Prentice-Hall International Series in Computer Science, 1987. ISBN 0-13-331968-7, ISBN 0-13-331661-0 Pbk.

Anhang

Im folgenden wird die *"global declaration node"* des OPM-Beispiels "Fernsprechsystem" wiedergegeben. Sie zeigt - in reduzierter Form - die für ein tieferes Verständnis des HCPN-Modells der Bilder 4 bis 7 erforderlichen Datentypdeklarationen auf.

```
(* declaration of constants *)                    (* global declaration node *)
val    num_sub    = 6; (* number of subscribers *)
val    num_res    = 4; (* number of unidirectional paths for channels
                           between subscribers *)
val    init_directory = ...;

(* declaration of data types *)
color intc      = int;              var Count,Fee, Start,NewStart  : intc;
color stringc   = string;           var Ctry, Area, Phone          : stringc;

color sub       = int with 1..num_sub declare sub;  var A,B,Sub: sub;
                  (* identification for subscriber line *)
color subXsub   = product sub*sub; (* two ordered subscriber ids *)
color d         = string with "0".."9" and 1..1;    var D          : d;
                  (* one digit to dial a phone number *)
color subXd     = product sub*d;    (* tuple (subscriber id,digit) *)
color stim      = union off_hook:sub + on_hook:sub + digit:subXd
                    declare mult;                   var Stim       : stim;
                  (* system stimuli with one or two parameters *)
color resp      = union dial_tone:sub + busy_tone:sub + ringback_tone:sub
                    + ringing_signal:sub + special_announcement:sub;
                                                    var Resp       : resp;
                  (* system responses with one parameter *)
color state     = union idle:sub + pre_dial_phase:sub + dial_phase:sub
                    + calling:subXsub + talking_to:subXsub
                    + called_by:subXsub + listening_to:subXsub
                    + announcement:sub + released:sub + out_of_service:sub
                    declare mult;                   var State      : state;
                  (* user state with one or two parameters *)
color msg_kind  = with call_attempt|ringing|busy|unavailable
                    |stop_ringing|ack_stop_ringing|call_accepted_by
                    |ack_call_accepted|transfer_info|ack_transfer
                    |end_listening_of|ack_end_listening|end_talking_of
                    |ack_end_talking;               var MsgK       : msg_kind;
color msg       = product sub*msg_kind*sub;
                  (* internal signal: (destination,message,source) *)
color res       = int with 0..num_sub;              var Res        : res;
                  (* available channel-resources, two per connection *)
color ph_num    = ...;
color dial      = product sub*stringc;              var Dialed     : stringc;
                  (* number which was dialed by subscriber until now *)
color stimlist  = list stim;                        var StList     : stimlist;
                  (* list for ordered application of stimuli *)
color fee       = ...;
color beg       = ...;

(* declaration of user-defined functions *)
fun    ...      = ...;
```

Differences between Estelle and LOTOS Descriptions of a Protocol

Jochen Grobholz and Richard L. Tenney

Mathematics and Computer Science
University of Massachusetts
Boston, MA 02125-3393 USA

Abstract: In *Guidelines for the Application of Estelle, LOTOS, and SDL*, several protocols and services were described in these three formal techniques. Although it was intended that in each case the three specifications should describe the same protocol, this was not verified after the specifications were written. In this paper we consider the Estelle and LOTOS descriptions of the Abracadabra protocol and note that they differ in both trivial and significant ways.

1 Introduction

Estelle [ISO, IS9074] and LOTOS [ISO, IS8807] are two standardized formal description techniques developed within ISO TC97/SC21 to describe computer communication protocols and services. These techniques are based on different theoretical models:

- Estelle is based on extended finite automata, and
- LOTOS is based on communicating processes.

To aid those interested in using Estelle and LOTOS as well as a third standardized technique, SDL, which was developed within CCITT, a group of representatives from ISO and CCITT worked together to write *Guidelines for the Application of Estelle, LOTOS, and SDL* [ISO, TR10167] in which several protocols and services are described in all three techniques. In each case, specifiers fluent with the technique being used were chosen and asked to produce the clearest specification possible. It was specifically intended that, through coordination among the writers, the resultant descriptions would be as close to each other as possible, given the differences in the techniques. However, this desideratum was not investigated after the formal descriptions were produced.

The present study begins to address the interesting question of whether or not it is feasible to show that formal descriptions written in different techniques describe the same protocol. As the first step of this larger project, we investigated the descriptions in Estelle and LOTOS of some of the protocols in *Guidelines*.

The techniques we have used for this report all rely on the existence of simulation and analysis tools, and the demands they place on these tools result in our being able to articulate additional features that we believe such tools should have. The simple protocols we have studied help make our points concrete; however they are but a few of those available and come from a limited domain. Additional examples should be studied with improved tools to enhance the theoretical basis for such analysis.

Both static and dynamic analyses were performed on selected descriptions, and the differences were noted. The dynamic analysis consisted of generating "interesting" paths for each description and then, using simulators for Estelle and for LOTOS, analyzing the behavior of the descriptions along these paths.

2 Tools

2.1 Estelle Tools

For Estelle we made use of *fd* and *fsm*, a pair of tools developed by Tom Blumer [1986] of Phoenix Technologies, Inc. The program *fsm* is a tool that can start from an Estelle protocol description and generate the corresponding finite automaton and then all possible paths through that automaton, subject to constraints set by the user.

It is possible to follow these different paths by hand, but the program *fd* can generate executable modules that can process input from a user or a file. These modules can display their internal data structures (*e.g.* queues, states, and variables), and thus one may view the functioning of the automaton as it follows a path.

2.2 LOTOS Tools

For LOTOS we used a simulation tool known as *hippo* [van Eijk, 1988] that works interactively, at each step offering the user a menu of choices to choose as the next step. These choices depend on the state the protocol simulation is in. If certain inputs are unexpected in a state, the simulator does not include these inputs among the choices it offers the user. The simulator can also print the path used to get to a state as well as the state itself. In addition, we checked some of our results using *ISLA*, a LOTOS tool set from the University of Ottawa [Logrippo *et al.*, 1988].

3 Differences

3.1 Overview

Despite any claims that both description techniques describe the same protocol, observable differences were found. These differences can be classified into three categories: illegal inputs, ambiguities, and data representation. We discuss these below.

Although we considered additional formal descriptions in *Guidelines*, we report here only on our results with the Abracadabra protocol. This is an alternating bit protocol augmented with connection and disconnection phases that are close to usual protocols. Although all aspects of this protocol are slightly simpler than what one would expect in an actual protocol, it is still complex enough to make the study of its observable behavior challenging.

In *Guidelines* there are four descriptions of the protocol available: the original English description as well as the formal specifications in Estelle, in LOTOS, and in SDL. We discuss differences between the Estelle and the LOTOS descriptions.

It is possible to generate sequences from either description using various test generation techniques and then to see if the two descriptions can produce the same observable behavior for each sequence.

3.2 Illegal Inputs

Estelle accepts any syntactically correct input at any time, while LOTOS offers only a limited number of inputs, depending on the current state of the protocol. Syntactically correct inputs that are not expected in this state are essentially refused as illegal.

For example, in the Abracadabra protocol specification, a connection is initiated by a connection request from the user using **ConReq**. Normally the user issues only one of these, and the English specification is silent on what should happen if the user becomes impatient and issues a second one before receiving an indication of what happened to the first. The Estelle and LOTOS specifications differ in their handling of this situation.

In the Estelle specification the sequence **ConReq ConReq** is legal. The second **ConReq** is consumed but causes no other changes. But in the LOTOS description, no process is prepared to accept a **ConReq** at gate a after the consumption of the first **ConReq**. It is in this sense that one might say that the input is refused.

The analysis in *Guidelines* recognizes this difference between the Estelle and LOTOS descriptions as a deficiency. It notes that this is simply an instance of a larger problem. Rather than resolving this, *Guidelines* says "the actual FDT being used affects how Service User misbehaviour should be most naturally described."

The way the Estelle description deals with this problem is to have transitions that handle inputs that are unexpected. In most cases, this is handled uniformly using an Estelle StateSet to create a series of transitions that consume the unexpected input without changing the control state or the values of any of the variables; thus effectively the input is ignored. The StateSet **xxxIgnore** is defined to comprise all those states in which the input **xxx** is to be ignored. The transition schema

```
from xxxIgnore to same
    when IP.xxx
        begin
        end;
```

would cause the input **xxx** at interaction point IP to be ignored. If such a transition were not available, the input **xxx** would remain at the head of its queue, preventing all other interactions in that queue from being processed until the state of the module changed to one in which **xxx** could be processed.

LOTOS on the other hand is based on processes. From the point of view of a protocol specifier, input is handled in such a way that that it is either consumed by a process or refused as illegal. The user is unable to initiate a second **ConReq** because there is no process willing to accept it.

3.3 Ambiguities

Another difference found in the two descriptions is the way that they process disconnect requests. We discuss in detail the situation with a **DisReq** initiated by the user, but similar comments apply to a DR PDU received through the Communications Medium.

In Estelle the handling of a DisReq is straight forward. After the user makes a DisReq the protocol machine sends a DR and ends in the state DRsent and eventually in the state Closed.

In the LOTOS description, on the other hand, an ambiguity arises: under certain circumstances, a DisReq can be consumed by different processes. This occurs as follows.

In figure 1 we show those portions of the complete LOTOS specification from clause 10 of *Guidelines* that we need for our purposes (all arguments to the processes are omitted).[1] As can be seen by following that skeleton of the LOTOS specification, opening a connection at one endpoint begins by invoking the process Connect, which can accept a ConReq from the user. This causes the process TryConnect to start. This TryConnect can be interrupted by either a UnitInd(DR) or by a DisReq. We discuss only the second possibility; the first leads to similar problems. TryConnect itself starts up the process StandBy, a process that loops, throwing away unwanted PDUs, until it is interrupted. One possible way it can be interrupted is by the completion of the process Disconnect. As can be seen, this Disconnect will exit (*i.e.* complete) if it accepts a DisReq.

Thus there are two distinct processes prepared to accept a DisReq. LOTOS semantics do not indicate which should do so, and both simulators referenced in section 2.2 offer both choices.

Each of these is necessary to fulfill the requirement of the protocol that it can be terminated at any time by sending a DisReq by either station. This leads to a situation where a hierarchy of processes is available to accept a DisReq. In the first case, where the DisReq is consumed by the a ? s : Address ! DisReq inside the process Connect, the processTryConnect is terminated and Connect invokes the process GiveUp, which terminates the connection. However, in the second case, where the DisReq is consumed by the process Disconnect inside of TryConnect, the connection is closed, but the process Connect remains active, waiting for another DisReq. In either case, up to N UnitReq(DR) PDUs may be sent, where N defines the maximum number of attempts to transmit a PDU without receiving an acknowledgement. In the second case, though, the user may issue another DisReq, in which case the specification behaves as though there were still a connection and may transmit up to N additional UnitReq(DR) PDUs. This explicitly violates the protocol specification.

The different observable paths are thus: (assume N, the maximum number of transmissions, equals 1)

ConReq	ConReq
UnitReq(CR)	UnitReq(CR)
DisReq (consumed inside Connect)	DisReq (consumed inside Disconnect within TryConnect)
UnitReq(DR)	UnitReq(DR)
UnitInd(DC)	DisReq (consumed inside Connect)
	UnitReq(DR) (protocol violation)

The problem we have identified is an error in the LOTOS description that can be fixed

---

[1] Other portions of the specification, such as those that deal with the possibility of opening a connection endpoint in response to a UnitInd(CR) transmitted through the communications medium are omitted from the figure.

```
process Connect
    (* this station starts *)
    a ? s : Address ! ConReq;
    (
        TryConnect
        [>
            (
                m ? u : Address ! UnitInd (DR);
                a ? s : Address ! DisInd;
                Connect
            []
                a ? s : Address ! DisReq;
                GiveUp
            )
    )
    ...
endproc (* Connect *)

process TryConnect
    ...
    StandBy
    ...
    [>
        Disconnect
        >>
        accept r : reason in
            GiveUp
    ...
endproc (* TryConnect *)
process Disconnect
    a ? s : Address ! DisReq;
    exit (UserDisc)
    ...
endproc (* Disconnect *)

process GiveUp
    ...
    [r = CMDisc] ->
        m ? u : Address ! UnitInd (DC);
        a ? s : Address ! DisInd;
        Connect [a, m] (N, P)
    ...
endproc (* GiveUp *)
```

Fig. 1. LOTOS Abracadabra Protocol Fragment

by rewriting the way `DisReq` and `UnitInd(DR)` are handled. But it points to a much more serious problem: there are no priority rules for processes at different levels in the execution hierarchy that can synchronize on the same event. One possible rule would be that the process highest in the hierarchy synchronizes on the event and terminates all subprocesses.

In the Abracadabra example it is also possible to accept an unbounded number of `DisReqs` at gate a for a single connection, causing an unbounded number of `UnitReq(DR)` PDUs to be sent. If the process `Disconnect` inside of `TryConnect` synchronizes on the `DisReq`, the connection is properly released and the process `Connect` is started again. But this has the effect that `Connect` is recursively started and all "outer" `Connect` processes act as if there were still an established connection.

The problems described above do not occur if the other station starts. In this case the sequence starts with a `UnitInd(CR)`, and only the process `Disconnect` is used to enter the disconnection phase.

3.4 Data Representation

Another difference in the two descriptions is the representation of the package sequences numbers. In Estelle the numbers are represented by 0 and 1. This conforms with the specification in English. In LOTOS the sequence numbers are represented by `True` and `False`. For an alternating bit protocol it is possible to choose such a representation, but it is hard to enhance the the description for a protocol that allows more sequence numbers.

More generally, LOTOS and Estelle differ greatly in their handling of data types. LOTOS uses abstract data types (ACT ONE) to represent data and possible operations on data. All data types must be fully specified. Estelle, on the other hand, offers only Pascal data typing facilities. However, the exact specification of data types may be deferred by use of the ... or any constructs. This makes it possible to have syntactically correct incomplete specifications.

4 Future Work

Future work lies in two areas, formulation of "interesting" tests and development of an integrated test environment. The discrepancies noted in this paper were discovered by experimentation and by human analysis: they were figured out "by hand". There is a literature regarding the generation of test sequences for conformance, but the constraints of that particular field differ from those of this study. In particular, in conformance, it is necessary to regard the Implementation Under Test as a "black box" into which the tester is not allowed to peer. However, this is not the case with specifications; the user of a specification is encouraged to peer inside to understand the requirements of the protocol. As a consequence, we believe that there is merit to augmenting the techniques used in conformance testing (see *e.g.* [Sidhu, 1990] for a recent survey) to consider the case where there are two automata available for analysis.

The available tools were important in our work, but it was impossible to coordinate the use of Estelle and LOTOS tools. It was not even possible to prepare a single set of data that could be passed through a filter and then be used by both sets of tools.

One reason for this is that the Estelle tools required inputs in a fairly rigid format, naming the interaction point, the interaction, and its parameters. One set of Estelle tools that we considered but could not use did not allow presentation of arbitrary inputs at an interaction point; although it was possible to make arbitrary modifications to an element in the queue. The LOTOS tools, on the other hand, required that all interactions be entered as menu choices from a dynamically changing menu. The LOTOS tools also required that all data be entered as ACT ONE entities, thus, for example, if three transmissions were to be allowed, the user must enter Succ(Succ(Succ(0))) when prompted for N.

For our purposes, it would be advantageous if all simulation tools admitted a clear translation between a textual representation of PDUs and accepted inputs, and if it were possible to observe—and even modify in arbitrary ways—features of the state of the simulated specification, such as the values of variables, the state of the automaton, *etc.* Note that we are not suggesting that all such tools directly accept these textual inputs, but it would be desirable if it were possible to write simple filters to convert a text file describing interactions to be presented at specified interaction points, as well as modifications to the state, into a file that could be presented to the tool as input. Nor do we suggest that this be the only interface to the tool; rather we merely ask that it be one of the possible interfaces.

With improved validation techniques and improved tools, it should be possible to increase our confidence in the equivalence of multiple descriptions of a protocol.

References

Blumer, Tom P. [1986]. *Estelle Development System, Users Manual.* Phoenix Technologies Ltd. Cambridge, Massachusetts.

ISO, IS8807. *Information processing systems — Open systems interconnection — LOTOS, an FDT based on the Temporal Ordering of Observation Behaviour.* Geneva: International Organization for Standardization, 1989.

ISO, IS9074. *Information processing systems — Open systems interconnection — Estelle — A formal description technique based on an extended state transition model.* Geneva: International Organization for Standardization, 1989.

ISO, TR10167. *Information processing systems — Open systems interconnection — Guidelines for the Application of Estelle, LOTOS, and SDL.* Geneva: International Organization for Standardization, 1991.

Logrippo, Luigi, Obaid, A., Braind, J. P., and Fehri, M. C. [1988]. An interpreter for LOTOS, a specification language for distributed systems. *Software — Practice and Experience,* 18 (4), 365–385.

Sidhu, Deepinder P. [1990]. Protocol testing: The first ten years, the next ten years. In *Protocol Specificiation, Testing, and Verification, X* (Logrippo, Luigi, Probert, Robert L., and Ural, Hasan, eds.). Amsterdam: Elsevier Science Publishers B.V. (North Holland).

van Eijk, Peter [1988]. *Software Tools for the Specification Language LOTOS.* Ph. D. thesis, Twente University of Technology, Enschede, The Netherlands.

This article was processed using the LaTeX macro package with LMAMULT style

Autorenverzeichnis

Baumgarten, B.	177
de Roever, W.P.	66
Dibold, H.	195
Effelsberg, W.	22
Födisch, R.	110
Freudenmann, J.	35
Gritzner, T.F.	144
Grobholz, J.	222
Heck, E.	1
Hofmann, B.	22
Hooman, J.J.M	66
König, H.	110
Müller-Clostermann, B.	1
Prinoth, R.	125
Tenney, R.L.	222
Vogel, A.	48

If you have any concerns about our products,
you can contact us on
ProductSafety@springernature.com

In case Publisher is established outside the EU,
the EU authorized representative is:
**Springer Nature Customer Service Center GmbH
Europaplatz 3, 69115 Heidelberg, Germany**

Printed by Libri Plureos GmbH
in Hamburg, Germany